Agricultural and Food Biotechnologies of *Olea europaea* and Stone Fruits

Editor

Innocenzo Muzzalupo

(Consiglio per la Ricerca e la sperimentazione in Agricoltura)
Agricultural Research Council
Olive Growing and Olive Oil Industry Research Centre
(CRA-OLI)
C.da Li Rocchi
87036 Rende (CS)
Italy

Co-Editor

Sabrina Micali

(Consiglio per la Ricerca e la sperimentazione in Agricoltura)
Agricultural Research Council
Fruit tree research centre (CRA-FRU)
00134 Rome
Italy

CONTENTS

FOREWORD

The rapid progress of agricultural and food chemistry, and the numerous researches conducted under the auspices of agri-food companies and private industries in both this and other countries, render it a difficult task to interpret facts which accumulate on a daily basis.

The e-book "Agricultural and Food Biotechnologies" edited by Dr. Innocenzo Muzzalupo and Dr. Sabrina Micali provides a number of papers that are excellent examples of advanced works applied to relevant problems. This book covers key areas in agricultural science, namely crop improvement, production, nutrients, pests, disease, genetic, genomic and the food industry.

The contributions by the authors include manipulation of variables and genetic resources of inheritance of quantitative genes, crop rotation, soil water, and the effect of temperature on production. Aspects such as protecting crops against pests and diseases whilst ensuring the protection of human health are also taken into account.

This book forms a valuable addition to the existing body of knowledge and is especially intended for university students and researchers in agriculture, biochemistry and biotechnology.

Francesco Paolo Montemurro
Deputy Director of
Agricultural Research Council
Study of Cropping Systems
Research Unit (CRA-SSC)
Metaponto (MT), Italy

PREFACE

This e-book presents an overview of the key aspects involved in the whole agro-industrial chain of olive (*Olea europaea* L.) and stone fruit crops, with the aim to provide to researchers and, in general, to all interested parties, a valuable tool for operational decision making. It emphasizes biological and practical aspects as well as latest genetic and bio-molecular knowledge and industrial transformation and processing applications. The information reported may be useful for policy and practical decisions to support national and international efforts aimed at the conservation and utilization of the world's plant resources.

The e-book is divided into the olive and the stone fruits sections and covers the latest aspects related to agronomy, biotechnology, plant nutrition, plant breeding, pomology, postharvest physiology, plant pathology of these two species.

Among cultivated plants, olive *(Olea europaea* L.) is the sixth most important oil crop in the world, presently spreading from the Mediterranean region of origin to new production areas, due to the beneficial nutritional properties of olive oil and to its high economic value. Over 750 million olive trees are cultivated worldwide, about 95% of them in the Mediterranean region. About 80% of the global olive oil production in 2011–2012 came from the European Union, with 77% of it concentrated in Spain, Italy and Greece. European Union with about 32% is also the world's leader in the production of table olives. Also in this case the European major producing countries are represented by Spain, Greece and Italy. Italy has about 600 olive cultivars and holds the world's record for the number of cultivated varieties, representing 25% of the known global olive germplasm.

Peaches, nectarines, plums, apricots, and cherries all belong to the *Prunus genus* and are often referred to as stone fruits because their seed is very large and hard. The global production in 2011 amounted to 35,555,562 metric tons for an estimated value of about 21 billion US dollars, being peaches and nectarines by far the most important commodities with a global world production of 20,152,690 metric tons and a value of about 11 billion US dollars. The edible part of these species plays an important role in the daily human diet as it contains compounds of high nutritional value, including vitamins that are not synthesized by the human body, sugars, aromas and flavour compounds, and raw material for food-processing industries.

Being perennial tree species, breeding of olive and stone fruits poses special problems compared to herbaceous crop species, mainly due to their long unproductive juvenile period which implies several years for a single breeding cycle and their large size affecting field management and harvesting costs. For these reasons, breeding strategies of these two species must be efficiently designed, with the aim to develop more competitive, sustainable and safe food production systems, where high nutritional and sensorial quality and traceability are ensured through the characterization of varieties, the adoption of new analytical techniques, and an overall improved understanding of ecosystem management, maintenance and enhancement of plant diversity and biological activity of soils. The safeguard and exploitation of genetic diversity being at the basis of any breeding program. In this e-book a section is devoted to the importance of studying the degree and distribution of plant genetic resources diversity, to their characterization and valorization as well as to their exploitation as sources of useful genes for breeding purposes.

We hope and trust that the information in this report will be used as the basis for policy and technical decisions to strengthen national efforts towards the conservation and utilization of the treasures incorporated in the world's plant genetic resources, in the view of addressing the urgent problems faced by agriculture today and tomorrow.

We would like to express our deepest gratitude to all the authors who contributed to this book by sharing their valuable works with us. Finally, thanks to the publishing house that provided us with great professionalism in the realization of the book.

We would also like to thank Bentham Science Publishers and Salma Sarfaraz for her clerical assistance, advice and encouragement during the development of this eBook.

Innocenzo Muzzalupo
Agricultural Research Council
Olive Growing and Olive Oil Industry Research Centre (CRA-OLI)
C.da Li Rocchi
87036 Rende (CS)
Italy

&

Sabrina Micali
Agricultural Research Council,
Fruit tree research centre (CRA-FRU)
00134 Rome
Italy

List of Contributors

Leonardo Bruno

Università della Calabria (University of Calabria – DiBEST), 87036 Arcavacata di Rende (CS), Italy

Adriana Chiappetta

Università della Calabria (University of Calabria – DiBEST), 87036 Arcavacata di Rende (CS), Italy

Guido Cipriani

Consiglio per la Ricerca e la sperimentazione in Agricoltura (Agricultural Research Council, Fruit Tree Research Centre, CRA-FRU), 00134 Rome, Italy and Università di Udine, (University of Udine DiSA, 33100 Udine, Italy

MariaTeresa Dettori

Consiglio per la Ricerca e la sperimentazione in Agricoltura (Agricultural Research Council, Fruit Tree Research Centre, CRA-FRU), 00134 Rome, Italy

Daniela Giovannini

Consiglio per la Ricerca e la sperimentazione in Agricoltura (Agricultural Research Council, Fruit Tree Research Unit, CRA-FRF), 47121 Forlì, Italy

Nino Iannotta

Consiglio per la Ricerca e la sperimentazione in Agricoltura (Agricultural Research Council, Olive Growing and Olive Oil Industry Research Centre, CRA-OLI), C.da Li Rocchi, 87036 Rende (CS), Italy

Alessandro Liverani

Consiglio per la Ricerca e la sperimentazione in Agricoltura (Agricultural Research Council, Fruit Tree Research Unit, CRA-FRF), 47121 Forlì, Italy

Sabrina Micali

Consiglio per la Ricerca e la sperimentazione in Agricoltura (Agricultural Research Council, Fruit Tree Research Centre, CRA-FRU), 00134 Rome, Italy

Innocenzo Muzzalupo

Consiglio per la Ricerca e la sperimentazione in Agricoltura (Agricultural Research Council, Olive Growing and Olive oil Industry Research Centre, CRA-OLI), C.da Li Rocchi, 87036 Rende (CS), Italy

Flora Valeria Romeo

Consiglio per la Ricerca e la sperimentazione in Agricoltura (Agricultural Research Council, Citriculture and Mediterranean Crops Research Center, CRA-ACM), Corso Savoia, 95024 Acireale (CT), Italy

Alisea Sartori

Consiglio per la Ricerca e la sperimentazione in Agricoltura (Agricultural Research Council, Fruit tree research centre, CRA-FRU), 00134 Rome, Italy

Stefano Scalercio

Consiglio per la Ricerca e la sperimentazione in Agricoltura (Agricultural Research Council, Fruit Tree Research Centre, CRA-FRU), 00134 Rome, Italy

Muhammad Siddiq

Department of Food Science, Michigan State University, East Lansing, MI 48824, USA

Pietro Toscano

Consiglio per la Ricerca e la sperimentazione in Agricoltura (Agricultural Research Council, Olive Growing and Olive oil Industry Research Centre, CRA-OLI), C.da Li Rocchi, 87036 Rende (CS), Italy

Elisa Vendramin

Consiglio per la Ricerca e la sperimentazione in Agricoltura (Agricultural Research Council, Fruit Tree Research Centre, CRA-FRU), 00134 Rome, Italy

Ignazio Verde

Consiglio per la Ricerca e la sperimentazione in Agricoltura (Agricultural Research Council, Fruit Tree Research Centre, CRA-FRU), 00134 Rome, Italy

CHAPTER 1

Botanical and Agricultural Aspects: Agronomic Techniques and Orchard Management

Pietro Toscano[1,*], Nino Iannotta[1] and Stefano Scalercio[1,2]

[1]Consiglio per la Ricerca e la sperimentazione in Agricoltura, Centro di Ricerca per l'Olivicoltura e l'Industria olearia - CRA-OLI (Agricultural Research Council, Olive Growing and Olive Oil Industry Research Centre), C.da Li Rocchi, 87036 Rende (CS), Italy and [2]Consiglio per la Ricerca e la sperimentazione in Agricoltura, Unità di Ricerca per la Selvicoltura in Ambiente Mediterraneo - CRA-SAM (Mediterranean Forestry Research Unit), C.da Li Rocchi, 87036 Rende (CS), Italy

Abstract: In this chapter the main botanical and agricultural aspects of the olive (*Olea europaea* L.) are summarized. In the section on botany, the functional parts and the biological cycle of the plant are described; while in the agricultural one, the environmental and physiological demands of the plant, and the most diffused training shapes are reported. In the agronomical section are described the main soil practices, with the aim of to preserve and improve ground conditions, such as setups, tillages, weeding, grassing; and trees practices, regarding plant growth and yield, such as irrigation, fertilization, and pruning. Then, a section regarding harvesting systems in different olive orchard typologies, and the control of olive pests and diseases in traditional, sustainable and organic farming, completes this chapter.

Keywords: Agricultural aspects, agronomic techniques, biological aspects, cultivation techniques, increase productivity, integrated farming, olive ecosystem, olive growing, olive production, orchard management, pests and diseases.

1. DIFFUSION

The olive tree (*Olea* spp.) is among the most ancient cultivated arboreal species. A native of the middle east (Iran, Syria, Lebanon) and domesticated around 3000 to 4000 B.C. [1], it progressively spread all around the Mediterranean basin with the expansion of commerce and domination of the Phoenicians, Carthaginians, Greeks and Romans. During the last 500 years, the olive has been introduced to

***Corresponding author Toscano Pietro:** Agricultural Research Council - Olive Growing and Olive Oil Industry Research Centre (CRA-OLI), C.da Li Rocchi, 87036 Rende (CS), Italy; Tel: +399844052204; Fax: +39984402099; E-mail: pietro.toscano@entecra.it

Innocenzo Muzzalupo & Sabrina Micali (Eds.)

the rest of the planet, nearly in all climatically compatible countries, between 30° to 45° north and south latitude, characterized by a temperate-warm climate, with long and dry summer seasons: in the Americas (United States, Mexico, Peru, Chile, Argentina), Oceania (South Australia), Southern Africa (South Africa), Asia (India) up to the extreme east (China, Japan) (Fig. **1**).

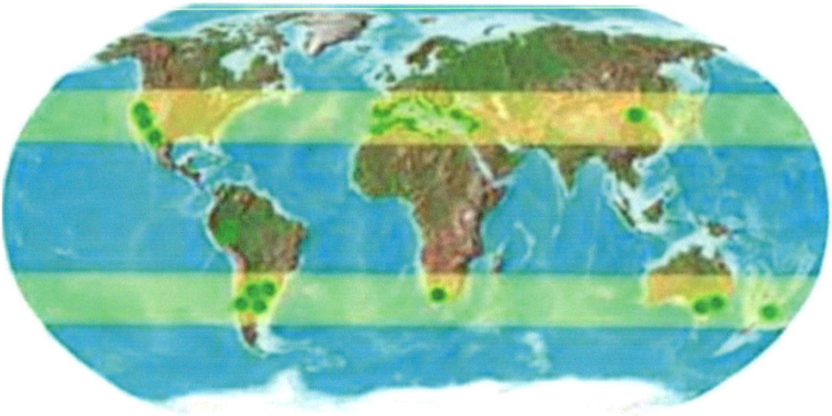

Figure 1: Olive diffusion in the world.

Areas occupied by olive orchards across the globe have been estimated to cover around 9.5 million hectares, distributed as reported in Table **1** [2].

Table 1: Olive harvested areas in the world

Country	Olive Harvested Areas (ha) 2010	
Afghanistan	2200	O
Albania	42700	F
Algeria	316300	F
Argentina	55700	F
Australia	30300	F
Azerbaijan	1856	O
Bosnia and Herzegovina	110	F
Chile	12874	O
China	280	F
Croatia	17096	O
Cyprus	10116	O
Egypt	128700	F

Table 1: contd...

El Salvador	4800	F
France	19453	O
Greece	834200	O
Iran	29700	F
Iraq	4700	F
Israel	24000	F
Italy	1190800	O
Jordan	60879	O
Kuwait	30	F
Lebanon	62500	F
Libya	205000	F
Malta	6	F
Mexico	6818	O
Montenegro	3000	F
Morocco	735400	O
Occupied Palestinian Territory	108100	F
Peru	11438	O
Portugal	345683	O
Slovenia	1018	O
Spain	2092800	F
Syrian Arab Republic	647500	O
Yugoslav Republic of Macedonia	5000	F
Tunisia	1645100	F
Turkey	826199	O
United States of America	14569	O
Uruguay	2900	F
Uzbekistan	100	F

O = Official data | F = FAO estimate
FAOSTAT | © FAO Statistics Division 2012

BOTANICAL AND AGRICULTURAL ASPECTS

2. BOTANICAL ASPECTS

The olive (*Olea europaea* L.) is a sub-tropical evergreen tree, with a multi-centennial biological cycle. In wild and spontaneous plants, it assumes a bushy shape, while in cultivated ones, it is usually constituted by a single trunk, with a cylinder-conic or globular canopy, and dimensions that vary according to the cultivar, the environment and the cultural conditions and range from 2 m in the specialized super-intensive olive orchards, up to 20 m and over, with trunks of over 2 m in diameter, such as the monumental 'Ottobratica' and 'Sinopolese' Italian olive cultivars in the Gioia Tauro plain (Italy) (Fig. **2**). Like all arboreal plants, the olive tree is endowed with a hypogeum apparatus, that contains the buried part of the stub and the roots, and an epigeous apparatus that includes the trunk, the branches and the canopy [3].

Figure 2: Typical olive-orchard of the Gioia Tauro plain (Italy).

2.1. Hypogeum Apparatus

The stub is made by the continuous production and overlap of neoplastic formations, called ovules or spheroblast, that make the basal part of the trunk more bulky which, in very old plants, can reach a diameter 3 to 4 times greater than the stem. The underground ovules generate the roots, while from the adventitious gems of the superficial ovules root-suckers develop that, in spontaneous or not cured plants constitute as many stems, thus forming the typical bushy aspect of the uncultivated plant [3].

At the beginning of development of the plantlet born by seed, the radical system is tap-root; but, already from the third to fourth year, the radical gems of the basal ovules originate new adventitious roots that gradually constitute a new more superficial radical system, replacing the primary derived from the embryonic root.

In agamically propagated plants, as well as those born by seed in the nursery, the radical system is a secondary type, because in the first ones the embryo does not exist, and in the second ones the embryonic rootlet is removed upon transplantation [4].

The roots are radially distributed from pedal, to distances up to three times higher to the radius of the canopy projection area, and to a depth varying on soil characteristics and water availability. In cultivated soils and in hilly zones, the greatest part of the roots is to be found in a layer at a depth of between 15 and 80 centimeters. Under conditions of extreme aridity and in sandy soils, the roots can reach a depth of up to 6 meters [5, 6]. Contrarily, in an irrigated crop, the radical apparatus is more superficial and, in localized watering systems, is gathered nearest to the irrigators [7-9].

The roots differ for age, diameter, corky degree and function and are distinguished as follows: principal, or conductors, transition, and absorbent. The first represents around 20% of the whole radical system, and serves functions of anchorage and vascular transport of substances absorbed by the others, that constitutes the remaining 80%. Other essential functions of the radical system are of hormonal synthesis, such as auxins, cytokines, gibberellins, ethylene and abscisic acid; as well as storage of reserve substances, such as starch, soluble carbohydrates, proteins and amino acids [10, 11].

2.2. Epigeous Apparatus

In cultivated plants, generally a single trunk develops from the stub that, in young plants, has a circular section, with a smooth green-greyish bark of different tonality, according to the cultivars. In the adult plant the trunk becomes twisted and gibbous, because of the uneven growth owed to the development of the principal branches that induce the formation of ribs and the generation of ovules.

With regards to the training shape, two or more main branches develop from the stem, that ramify into secondary branches of various orders, to constitute the leading structure of the plant (skeleton). On these branches constitute the canopy which is formed by shoots, buds, leaves and gems.

2.2.1. Branches

The branches are lignified elements, 1 to 3 years old, with leaves, and are distinguished for age, vigor and type of gems. The shoots represent the vegetative production of the year, which has not yet completely lignified, and is divided as follows: suckers, wood bearing, fruit-bearing and mixed shoots.

The suckers are straight, vigorous, and have small leaves, anticipated twigs and wood gems, and are generally borne by the ovules of the stub (root-suckers) and main branches. The wood-bearing shoots are endowed with gems that twigs originate from; while the fruit-bearing shoots are endowed with flower gems. In the mixed shoots, the gems will produce both twigs and inflorescences, with a ratio mainly depending on the tree's nutrient endowment and balance [3, 12].

2.2.2. Leaves

In the olive tree the leaves are thick, leathery, arranged opposite in pairs, opposite one another on the rachis. The two leaves are inserted on the twig, with a slight space in the attachment of two petioles: such space is more evident in very vigorous wood branches and in the suckers, thin to observe whorls apparently constituted by three leaves.

The olive tree leaf is simple, with whole border and flat edge, and can be elliptic, lanceolated, or mucronated. The color is intense green on the adaxial surface; opaque silvery-grey on the abaxial one due to the presence of peltate trichomes that protect the stomata and mesophyll from UV radiation, and restrict water loss consequent to wind [13-15]. Form and dimensions vary, also in the same plant, based on the age, the vigor of the branches, the nutritional state and on the season of development [16].

2.2.3. Inflorescences

Flower bud inflorescence generates at the petioles of the leaves, and is constituted by a central axle, with different orders of ramifications on whose apexes yellow-white flowers are inserted. The length of the inflorescence vary from 10 to 70 mm; the total number of flowers per inflorescence varies averagely between 10 and 40, also with accented polymorphism in the same plant, based on the shoot typology, and nutritional and vegetative-productive state of the plant.

The flower is formed by a short calyx with 4 sepals, a short-tubed corolla with 4 yellow-white petals, two opposite stamens and filaments supporting large pollen-bearing anthers on either side of the two-loculed ovary, with two ovules per carpel, that bear a short tick style and large and capitate stigma [16].

The pollen grains contained in the anthers have specific forms and dimensions, and can be used for cultivar recognition [17]. Pollination requires wind.

Flower anomalies are frequent in the olive tree, such as excess elements (corollas with 5 petals; or 3 stamen); or defective or aborted ovaries, and constitute the so-called staminate flowers (with aborted pistils and functional stamens), that nevertheless do not negatively affect plant productivity, as the percentage of physiological fruit set in the olive tree does not exceed 2%. Despite the presence of hermaphroditic flowers, authogamy in the olive tree is rare. Most cultivars are self-incompatible or self-sterile, and the ovules cannot fecundate with the pollen of its own flower, nor with that of the other flowers of the same plant or plants belonging to the same clone. Such physiological sterility, derives from the missed germination of the pollen on the stigma due to genetic incompatibility [3, 13, 16, 18]. The individualization of the self-fertile and self-sterile cultivars, and respective pollinators, is a fundamental factor in optimizing the choice of cultivars to associate in the realization of new olive-groves.

Among the most diffused Italian cultivars, resulting as self-sterile: ‘Ascolana’, ‘Bella di Spagna’, ‘Bosana’, ‘Caninese’, ‘Carboncella’, ‘Carolea’, ‘Cassanese’, ‘Castiglionese’, ‘Cellina’, ‘Cima di Mola’, ‘Coratina’, ‘Corniolo’, ‘Intosso’, ‘Itrana’, ‘Leccino’, ‘Leccio del Corno’, ‘Leucocarpa’, ‘Majatica di Ferrantina’, ‘Maurino’, ‘Minuta’, ‘Moraiolo’, ‘Moresca’, ‘Nocellara del Belice’, ‘Nocellara etnea', 'Olianedda', 'Ogliarola', 'Ortice', 'Ottobratica', 'Passalunara', 'Pendolino', 'Pidicuddara', 'Pisciottana', 'Pizzuta', 'Racioppa', 'Rosciola', 'Rotondella', 'Sinopolese', 'Santagatese', 'Tonda Iblea', 'Tonda di Cagliari', 'Tondina'.

Among the self-fertile cultivars are: 'Casaliva', 'Correggiolo', 'Dolce di Rossano', 'Frantoio', 'Grossa di Gerace', 'Ogliarola messinese', 'Rossellino cerretano', 'Razzola'.

Partially self-fertile are: 'Bella di Cerignola', 'Biancolilla', 'Dolce Agogia', 'Giarraffa', 'Nocellara Messinese', 'Santomauro', 'Taggiasca', 'Zaituna' [19].

2.2.4. Fruits

Olive fruits are drupes with form and dimensions which are typical for every cultivar, varying from spherical to oval, more or less lengthened. The color of the drupe is green, which is more intense in immature fruits, changing from whitish to reddish to purplish, up to black at complete ripening.

The drupe is constituted by 1.5 to 2.5% skin (pericarp), by 70 to 80% pulp (mesocarp), and by 15 to 20% stone (endocarp). The stone contains the seed, composed by the tegument, the endosperm and the embryo. The middle composition of the olive fruit, in percentage, is reported in Table **2** [3, 16].

Table 2: Avg. % composition of olive fruits (*Olea europaea* L.)

	Drupe	Pulp	Stone	Seed
Water	50	58	15	35
Oil	21	26	1	28
Protein (N * 6.25)	1.5	2.0	0.3	8.0
Non-azoted extracts	20	9	40	26
Raw fiber	6	4	38	2
Ashes	1.5	1	3	1

2.3. Biological Cycle

The olive natural biological cycle is multi centennial, with some over millennial monumental trees. In cultivated plants, particularly in profitable intensive orchards, the existence of plants is linked to the economic convenience of the crop. In such conditions, the classical development phases of youth, increasing productivity, maturity and senescence, are replaced with the two epochs of formation and production. In the productive stage, the phenological phases of the annual biological cycle, that interest the management of specialized olive orchards, are individualized as reported in Table **3**.

Table 3: Phenological phases of the annual biological cycle of the olive (*Olea europaea* L.)

Phenological phases	Period	Appearance
Dormancy	December-January	
Flower bud differentiation	February	
Vegetative resumption	End of February	Growth of new vegetation
Blossoming buds	Mid March	Inflorescences development
Flowering	Early May - Mid June	
Fruit set	End of May - End of June	Fading of flowers and appearing of fruit lets
First phase of fruits growth	Second half of June Mid July	
Pit hardening	End of July	Stop of fruit growth
Second phase of fruits growth	August - October	Appearance of lenticells - inolition
Veraison	October - November	Coloration of the fruit changes from green to purplish red
Ripening	November - December	Coloration of the fruit changes from purplish red to black

2.4. Inolition

The quantities of oil extractable from the drupes depend on the genetic characteristics of the cultivar and also on external factors that condition the inolition process, such as crop load, environmental conditions and management (nutrition, irrigation, pest defense). The olives' maturation is scalar and, in a Mediterranean climate, normally occurs in October-November; while the oil formation period is inclusive between the first of August and the veraison of drupes. Subsequently, the increase of oil is apparent, as it is due to the reduction of water content in drupes. The oil content in olives also varies from 20 to 40% on dry matter, according to the cultivars, the environment, the orchard management and the ripening stage, as well as the phytosanitary state of the drupes; while the industrial effective yield percentage can reach, on average, 12 to 22 kg of oil q^{-1} of olives, depending on the olive mill typology and the management of the extraction process.

3. AGRICULTURAL ASPECTS

The olive is a perennial, long-lived, evergreen and rustic tree, very resistant to drought, and capable of also living in marginal soils. Nevertheless, as for all arboreal fruit plants, the expression of the productive potentialities depends on the satisfaction of the physiological necessities in terms of environmental habitability, such as climate, soil, water and nutritional availability and phytosanitary conditions [3, 16, 20]. All of these factors must be considered and harmonized in modern olive growing management, to achieve the best cultural responses, in terms of both income and environmental sustainability.

3.1. Thermal Demands

High temperatures do not penalize the vitality of the plant, provided that there is no lack of water. The olive is instead more sensitive to frost: during winter dormancy, the tree can tolerate temperatures up to -5 °C while temperatures ranging between -5 and -10 °C may cause greater damage to shoots and young limbs, which may lead to their death. Temperatures below -10 °C kill large limbs and even the entire canopy of the tree and also the trunk. Higher frost, however, usually does not jeopardize the vitality of the stub, from which the plant can reconstitute itself with the suckers produced by the gems of the stub ovules. The thermal equivalents, in the phenological phases of the annual biological cycle, are reported in Table **4**.

Table 4: Thermal equivalents in the phenological phases of the olive biological cycle (*Olea europaea* L.)

from vegetative resumption to blossoming buds	10 °C
from blossoming buds to flowering	15 °C
from flowering to fruit set	18 °C
from fruit set to veraison	20 °C
from veraison to ripening	15 °C
from ripening to winter dormancy	5 °C
from winter dormancy to vegetative resumption	- 5 °C

3.2. Water Demands

The Olive is a species with a high degree of drought tolerance, capable of growing and producing a yield under prolonged summer water shortage, by means of physiological, biochemical and morpho-anatomical responses to reduce water loss and tolerate dehydration [21-23]. However its capacity to withstand severe and prolonged drought periods, causes reductions in photosynthetic performance and elaboration of assimilates, that negatively act on olive growth and productivity [24-27]. Olive tree water demands vary and depend upon factors such as soil type, climate, plant density, age of trees, cultural management (*i.e.* fertilizing, pruning) and watering system. The olive nevertheless has some critical periods during the annual cycle, in which the plant mostly needs water. The first one extends from bud differentiation up to flowering and therefore to the fruit set; in these phases a water deficit can create problems with regards to flower development with a smaller number of flowers for inflorescence, increasing ovary abortion, and a lower fruit set. Subsequently, the first phase of fruit growth is the most sensible to drought, while at the pit hardening period, the olive is most resistant to water deficit. Finally the olive needs water in the second phase of fruit growth and during the inolition [28-32].

3.3. Ground Demands

The soil is a non-renewable resource that supports and conditions the life of animal and vegetal species. It is composed by solid mineral particles of different sizes, and variable percentages of organic matter, bound together into structural aggregates to constitute different soil typologies.

Soil also acts as a storage of elements, and its structural arrangement is directly correlated to the availability of nutrients and water; and consequently to the plants' development and yield.

The textural class is the first parameter that defines soil properties, and is determined by the relative percentage of the three major soil mineral compounds: clay, silt and sand.

Clayey grounds are characterized by particles of a diameter of less than 0.002 mm; these soils have a low porosity and water permeability that can induce root asphyxia phenomena in wet conditions; while in the dry state they show a high tenacity, and tendency to form cracks.

Silty soils have elementary particles of diameter between 0.02 and 0.002 mm, of low structural stability and high bulk density, which involves formation of mud in wet conditions, and pulverization when in dry state.

Sandy soils have particles of a diameter between 2 and 0.02 mm, with high porosity, high water permeability and air circulation; consequently, these soils have low capacities of water-holding and fast mineralization of organic matter.

Soil textures with balanced proportions of clay, silt and sand, in presence of sufficient quantities of organic matter, constitute better structural aggregates, with optimal porosity, water and nutrients availability, that are the parameters directly correlated to the levels of “physical”, “chemical” and “biological” fertility of agrarian soils [32].

Even though the olive tree prefers deep loam textured and well drained soils, with adequate management of cultural techniques it can grow in various kind of soils, from alluvial plains, to terraces, to slopes, and shallow and marginal soils, in arid and semi-arid areas. According to the different environmental conditions and technological level of management, different olive-growing typologies correspond, from extensive olive orchards, with 100 or less trees ha^{-1} (Fig. **3**), up to the modern super intensive plantations, with 1200 to 1600 trees ha^{-1} (Fig. **4**) [93]. The most common Italian specialized olive-orchards are scaled between 6x4 (416 trees ha^{-1}) and 6x6 (277 trees ha^{-1}).

Figure 3: Traditional olive-orchard (photo: *Toscano*).

Figure 4: Super-intensive olive-orchard.

3.4. Nutritional Demands

In profitable olive growing, nutritional demands vary in relation to phenological phases, to climate, to orchard typology, to the trees' productive potentiality and the presence of other cultural techniques, such as soil grassing and irrigation. For these reasons, fertilization planning cannot be approached as a standard procedure. In every cultural situation, the purpose is to realize a correct balance between the vegetative and productive activity of trees. Vegetable organisms are constituted for 96% by carbon, hydrogen, oxygen and nitrogen; and in a smaller quantity, by potassium, phosphorus, calcium and magnesium; in the least part by iron, manganese, boron, copper, zinc and chlorine. These elements, distinct in macro-, meso- and microelements, are present in the soil in ionic or complex form in the circulating solution and are absorbed, moved and metabolized by the plants. In the olive, the level of global nutrition, expressed as percentages on the leaves dry matter for nitrogen, phosphorus and potassium, results as being 3.5% divided, respectively, in 2.1:0.35:1.05, with a physiological relationship of 6:1:3. The determination of the content of the principal elements in the leaves permits knowledge of the nutritional state of the plants and the variations in relation to their phenological phases [32].

3.5. Training Systems

The training system is, with the choice of the cultivar and tree spacing, the third essential parameter in creating a new olive orchard, and is based on cultural objectives and environmental limitations. In modern Mediterranean olive orchards, the most common shapes are [12, 33]:

Vase - with several variants as cone, cylinder, multiple cones, the vase is the most common shape in intensive specialized olive-growing. Usually the vase has a single trunk of about 80 to 100 cm height, branching into 2 to 4 primary branches, equally spaced and tilted about 45 to 50 degrees, from which the secondary branches develop, that generate shoots and twigs, to constitute the canopy. The advantages of this shape are: *i)* high surface-to-volume ratio per tree; *ii)* good light distribution in the canopy interior; *iii)* suitability for different growth habits, for table olives, for mechanized harvesting (combing), on pendulous cultivars, as well as for mechanical harvesting (shaking), if trees were trained to form a rigid and rising structure.

Vasebush - is a vase without a proper trunk, and with primary branches originating from the root-suckers of the stub, and secondary ones arranged similarly to the vase. The main advantage of this shape is that the trees are kept shorter, useful for the hand picking of table olives or mechanized harvesting.

Globe - is a shape with a single trunk and a globular canopy; it differs from the vase in the center of the canopy that is occupied by vegetation. This shape is usually adopted where sunlight is high, to protect the bark with high-density foliage. This shape is suitable for different growth habits and for mechanical harvesting by trunk shakers.

Single Trunk Free Canopy - all systems that require little or no pruning are included in this shape, combining the features of a single trunk with low cost and flexibility of minimum pruning. These shapes are suitable for different growth habits as well as for full mechanization.

Bush - is a free canopy system obtained with minimum pruning during the training phase, as well as on mature trees, allowing the canopy to grow as free as possible, so that the final shape is similar to naturally growing plants. This shape is suitable for intensive orchards, hand picking of table olives and mechanized harvest; and can easily be converted to other free vase forms.

Monocone - in this shape, the primary branches are inserted in a spiral on the trunk, that constitute the central axis of the tree, with length decreasing from the base to the top, and fruiting shoots uniformly distributed on the external edge of the canopy, resulting in a conical shaped tree, that is suitable for full mechanization in high-density orchards.

Hedgerow - is a training system in which trees grow freely, usually on a single-trunk vase bush or monocone-like, so that the canopy forms a productive wall along the row. This shape is the most suitable for full mechanization in common intensive orchards (6x4 to 6x6 scaled), using mechanical pruners and continuous wall-harvesters [32].

AGRONOMIC TECHNIQUES AND ORCHARD MANAGEMENT

In sustainable olive-grove management, the cultural techniques and their correct application are essential to exalt the productive potentialities of the plants, preserve the environment and maximize profit. They can be distinguished in: soil practices, directed to preserve and improve soil conditions (setups, tillages, amendments, weeding, grassing); and tree practices, directed to plant growth and yield (irrigation, fertilization, pruning, harvesting, pest defense).

4. SOIL PRACTICES

Soil management techniques must be planned and adjusted according to the specific pedoclimatic conditions in which the plants will be cultivated, beginning from the field setup for the plantation, and yearly to ensure the best habitability condition for the trees [34-36].

4.1. Plantation Preparation

The first action to carry out in the constitution of a new olive-grove is soil setup that usually consists of preliminary stone removal or grinding, when necessary, and ground leveling, followed by trenching. Trenching usually consists in ripping, to break up the soil, followed by a 90 to 100 cm deep ploughing for the overturning of layers and the incorporation of the basal dressing, to constitute a nourishing reserve. For basal dressing phospho-potassic and ammoniacal chemical fertilizers, and organic amendments (manure or compost) are used and are spread over the soil prior to tillage, in doses established according to the results of the soil analyses. Trenching must be effected some months before the plantation, to allow the ground to settle, and avoid movement of the plants after planting. On uncultivated or brought soils, trenching must be effected on the whole surface of the field; while in the case of fertile, or well-structured grounds, it can be limited to the orchard row areas.

4.2. Soil Tillage

Soil tillage has the purpose of burying fertilizers, improving airing, facilitating water regimation, limiting the loss of rainwater due to runoff or evaporation; and eliminating weeds. There are various kinds of machinery for soil tillage, hauled or PTO powered, that effect only an incision on the soil (chisels, subsoilers), or also overturn the worked layer (plows, disks harrows), up to shattering and remixing of the soil (horizontal and vertical rotary hoeings). However, in arboreal plants, soil tillage has some negative aspects such as: smaller transitability on bare damp ground, faster mineralization of organic matter, runoff erosion on slopes, formation of tillage pan, compaction and destructuration in the machine pathways. These negative effects can be attenuated by employing tools that operate without or with low shattering of worked layers.

4.3. Weed Control

In olive-groves, the uncontrolled development of weeds causes competition for water and nutrients, increases the risk of fire in the dry season, and hinders cultural operations and yard operativity [37]. Nevertheless, the presence of a vegetable coverage improves water regimation, organic matter balance and the physic-chemical characteristics of the soil; reducing nutrient leaching and erosion on slopes [35, 38, 39]. In sustainable cultivation, both in extensive and specialized groves, the temporary flora must be considered as a renewable resource in integrated management plans, and controlled using mechanical or chemical weed systems, as well as controlled grassing plans, according to the specific environmental conditions. Generally, the presence of temporary flora is useful in rainy seasons; while it is necessary to control flora development during the dry seasons, particularly in olive groves that are not irrigated. Mechanical weeding must be executed with superficial soil tillage (10 to 15 cm of depth), using over turner tools, such as disk harrows or multi ploughshares or disk plows.

When mechanical weeding is difficult or not useful, chemical herbicides can be used. Such products are distinguished as pre and post-emergency herbicides, according to their kind of action. The first ones act by radical absorption on germinate seeds, and are adopted when it is useful to maintain the soil clear, *i.e.* to facilitate olive harvesting with collector brooms, apart from environmental consequences, such as pollution or soil destructuration. Instead, post-emergency products are more useful and functional in the sustainable and integrated cultivation strategies. They are applied on the vegetation in action, and are distinguished in systemic or contact herbicides. The systemic herbicides can also be applied at low dosages, to reduce vigor of the weeds, maintaining the advantages of soil coverage; while the contact ones, not having systemic activity, can also be used to eliminate root-suckers. The use of herbicides in weeding, represents an effective option to reduce the crop overheads; since phytotoxic phenomena and environmental pollution risks occur in the case of improper or irrational use [35].

A further method of weed control is shredding of the temporary flora, burying the biomass as green manuring or not, at the end of the rainy season; or implementing the controlled grassing technique in the orchard management.

4.4. Controlled Soil Grassing

Controlled soil grassing in olive growing can be applied in different ways, according to the orchard and environmental and management typologies. Soil grassing can be natural or artificial; partially or wholly surface covering; in permanence or only for the rainy season. Grassy management consists in periodic shredding of the turf, to avoid the excessive development of biomass, and adjustments of the fertilization plans, in terms of doses, epochs and supply methods of the fertilizers, to avoid both water and nutritional competition toward the arboreal crop. Turf benefits in the olive grove are manifold, such as an increase of the organic fertility, improvement of the soil structure and carrying capacity, increase of rainwater infiltration and retention, reduction of runoff and erosion, and an improvement of edaphic biocenosis.

The simpler kind of soil grassing is "natural and permanent", which consists in the covering of the whole ground by spontaneous flora, that is periodically shredded and left on site to constitute a protective coverage and source of organic matter with its progressive degradation.

Temporary grassing is adopted in zones subject to prolonged summer drought, eliminating grass before the dry season with a superficial tillage or through herbicides, with or without burial of the biomass; and allowing coverage to naturally re-establish itself with the resumption of autumnal rains.

Artificial grassing involves the seeding of specific essences, chosen on the basis of specific demands, such as rapid emergency, carpeting, resistance to stamping and to coverage, of a small size to reduce mowing. The limits of such a technique are economic and managerial: besides the choice of the essences and the cost of seeding, normally after some years the lawn must be reconstituted, to avoid the proliferation of any unwanted flora.

The best machine for grassy management is the shredder that, endowed with suitable tools, is also used to grind pruning residues, and equipped with an inter-stock rotating mower, which allows grass cutting along the rows, avoiding damage to the trunks (Fig. **5**), [35].

Long-time experiences of controlled grassing in different non-irrigated olive orchard soils have confirmed the effectiveness of permanent grassing in improving soil properties, in rainwater regimation and soil erosion control, and on olive tree productivity [40-43]. While other studies on different applications of grassing methods, have shown that green manure in some environments in summer has better agronomic and productive results, avoiding competition for water by permanent grassing (Fig. **6**), [32, 44].

Figure 5: Shredder with inter-stock rotating mower (photo*: Toscano*).

Figure 6: CRA-OLI olive orchard on tilled and grassed soil (photo*: Toscano*).

5. TREES PRACTICES

As for the soil management techniques, tree practices are also highly important to maximize cultural results, satisfying the metabolic demands and ensuring a suitable sanitary state of trees, to optimize yield quality and economic and environmental sustainability of the orchard.

5.1. Irrigation

As seen above, the olive has a notable resistance to water stresses; nevertheless a suitable soil water availability, correlated with other agronomic techniques, is highly effective in enhancing shoot growth, reducing the unproductive period and the alternate bearing of the plants, increasing yield such as fruit size, pulp-to-pit ratio and oil content, and affecting the fruit maturation time [45-48].

The unitary water consumption of the olive tree, such as the water amount that must be transpired to synthesize a gram of dry matter, was estimated at around 1 lt m^{-2} of leaf, transpired daily in August. This index, with opportune calculations, is useful to compile the crop water balance, such as a comparison among the entity of the rain input and the losses of damp by evapo-transpiration, and determine the water deficit or excess in the different periods of the year, to establish watering volumes.

For the calculation of the water amounts to be used in an olive-orchard, formulas based on climatic data, such as the raininess and the potential evapo-transpiration (ETP) are used in adopting different cultural coefficients, varying according to the tree spacing, the age and shape of trees, and the season.

The criteria to be adopted in irrigation plans must be based on the watering needs of the crop, and on the knowledge of the critical phases of the vegetative cycle of the plants, to maximize the efficiency of the available amount of water, according to the economic principle of marginal productivity in the use of productive factors. The calculated seasonal watering volumes will therefore be reduced, considering the threshold of convenience, in relation to the efficiency of the irrigation system, the cost and availability of water [49, 50].

The most critical phenological phases in which water stress should be avoided are floraison, fruit set, fruit growth and inolition. In recent years many studies have tried to apply deficit irrigation strategies to olive trees, based on the asynchronous growth of olive fruits and shoots that reduces competition for assimilates at critical stages. The controlled irrigation deficit is a water management method that does not completely satisfy the tree's water requirements during the growing season. It causes a temporary and regulated water deficit in some specific phenological stage (*i.e.* pit-hardening), improving the water use efficiency (WUE), without affecting the olive oil yield [51-54]. To satisfy the needs of an intensive olive-grove, the results of different watering trials pointed out that for the olive it is sufficient to supply 30 to 50 % of the evaporated water, to increase the yield up to two times that of dry ones [55, 56]. In any case, table olives cannot be cultivated without irrigation [57]. Finally, it is necessary to note that the olive tolerates brackish water up to a salty residue of 4 g lt^{-1}, allowing irrigation of olive-groves with waters that are not usable for other crops [58, 59].

Irrigation can be realized in different ways and the choice of the optimal method should be made according to each single olive-grove typology and environment. The most diffused are constituted by sprinkling methods, using giant irrigators or wings, and by localized systems, with drippers or microjets. The first ones have the advantage of facility of moving and transfer, but generally with low water efficiency, also because the watering volumes must be calculated for the whole orchard surface. On the contrary, for the latter, the watered surface is 10% for drippers and 25% for microjets, both also allowing irrigation of orchards on slopes and supply of nutrients (see fertigation ahead) [32].

5.2. Fertilization

The olive is still considered to be a rustic plant, having little nutritional requirements and capable of surviving even in rough environments, with minimal care and management. As well as for irrigation, the olive could also live without fertilization but, as with all other cultivated fruit plants, it requires a suitable annual supply of nutrients to express its productive potentialities. The purpose of fertilization is therefore to ensure suitable quantities of nutrients to the plants, in a balanced relationship among them. In planning olive nutrition all factors that act on the plant metabolism, as soil fertility, water availability, vigor and productivity of the cultivar must be considered. The compilation of fertilization plans, in terms of choice of kind, doses, epochs and supply methods of the nourishing elements, must be based on a soil analysis and leave nutrient diagnostic, to assess the nutritional needs of the crop [60-62].

The fertilization is distinguished into organic and mineral. The first one has the purpose of improving the physical characteristics of the soil, such as structure, porosity, permeability, tackiness, consistency, water retention, and the pH, using green and animal manures, amendments, and composts. The second one is destined to nourish the plants, using chemical fertilizers spread on soil, or through watering systems, or foliage.

The principal nourishing elements, called macro-elements, are nitrogen, phosphorous and potassium.

Nitrogen is fundamental for plant growth: it participates in the synthesis of amino acids and proteins, in the formation of flowers, in the fruit set and in fruit development. Lack of nitrogen causes a reduction of growth, formation of defective flowers, low yield and alternate bearing.

Phosphorus is a growth regulator, essential in cellular division and in the development of the meristematic tissues, enhancing fruit set, fruit growth and maturation, and lignification of the shoots. The effects of phosphatic fertilization are shown nevertheless with extreme slowness because of both the relatively modest demands of the olive, and its immobilization in the ground. Lack of phosphorus, however rare, is manifested with a reddish or purplish leave coloration, and metabolic issues that reflect on growth and on fructification.

Potassium promotes the accumulation of carbohydrates such as starch, energetic reserve for the metabolic processes; it regulates the water balance of the plant increasing water retention of tissues and the regulation of transpiration; it is also an enzymatic activator, enhances inolition and increases resistance to extreme temperatures and some fungal diseases. Potassium is absorbed in a relatively elevated quantity by the olive, but the agrarian grounds are generally well endowed with it, above all clayey soils. Like phosphorus, is minimally soluble and fixed by the ground. A lack of potassium is very rare, eventually manifesting with decoloration and apical necrosis of the oldest leaves.

Other important nourishing elements for the olive are magnesium, calcium and boron.

Magnesium is an essential component of chlorophyll, but is generally not considered in fertilization plans, as it is sufficiently contained in many fertilizers.

Calcium is another element essential for growth, being a constituent of cell walls that contributes to the mechanical resistance of tissues, also acting as an activator of some enzymes. A lack of calcium can be due to soil acidity, and will be corrected with an adequate calcium supply, *i.e.* such as carbonate.

Boron acts in pollen growth, fruit set and plant productivity. A lack of boron is manifested with apical chlorosis of the leaves, followed by necrosis and leaf drop. Slight boron deficiency causes low fertility of the flowers, and an increase of ovary abortion. Boron deficiency is nevertheless easily resolvable through leaf treatments during the pre-flowering stage.

In intensive, specialized olive orchards, the supply of fertilizers is carried out in different ways, according to the state of the crop and the purpose of its application. Fertilizers are normally spread on the soil, because the nutrition of trees depends physiologically on the absorption of nourishing elements through the roots.

The supply of nitrogenous fertilizers must be carried out yearly and divided in at least two times: the first one at the vegetative resumption stage, at the end of winter, and the second in the first fruit growth phase, in late spring. For phosphorus and potassium, soil fertilization must be carried out every 4 to 5 years, in autumn, burying fertilizers with a tillage of 20 to 25 cm max deep, to limit damage to the roots. In irrigated orchards it is possible to use the watering system to distribute fertilizers during the watering season.

To calculate the amount of nutrients to supply, it is useful to adopt the restitution criteria, based on nutrients removed by yield: for 100 kg of drupes, the olive needs around 900 g of nitrogen, 200 g of phosphorus and 1000 g of potassium. In the fertilization plans such doses must be triplicated, considering losses for leaching, volatilization and soil fixation [32].

5.3. Fertigation

Fertigation consists in the supply of fertilizers to the trees through the watering system. Macroelements (N, P, K) are usually distributed by fertigation, while microelements (Mg, Fe, Bo) are supplied to the plants, when necessary, through leaves.

The advantages of such a practice consist in the easiness of application and in the efficiency of fertilizers, allowing a reduction in the requirement of fertilizers of up to 30% in comparison to soil distribution, and a sensitive reduction of the management costs in terms of purchase, transport and distribution of fertilizers, enhancing their efficacy to grant a suitable nutritional level to the trees, to maximize yield, oil production and profitability [63].

Mixtures of water-soluble or liquid nitrogen, phosphate and high strength potassium fertilizers, single or in various ratios, are employed in the practice and are also added to by secondary elements (Mg, Fe) and microelements (B, Zn). Such solutions, in strengths defined by fertilization plans, are conveyed in the watering system, using fertinjectors that work using the water flow to aspirate, by the Venturi principle, dilute and carry the nourishing solution (Fig. 7) [32].

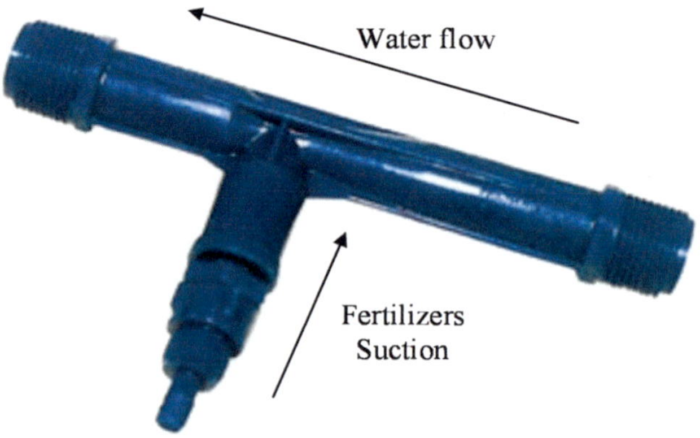

Figure 7: Fertinjector.

5.4. Foliar Fertilization

Besides that through the roots, olive tree can also absorb nutrients through foliage, which allow a quick and effective satisfaction of the plants demands both to support fructification in the on years, and to resolve lacks of microelements, or in extreme cases in which it is not possible to apply fertilizers to the ground.

The advantages of this technique are manifold: timely intervention, nutrients are supplied at the moment of greatest necessity and effective in a short time, integral use of elements that allows a reduction in the amount of fertilizers required by soil application. Even though foliar fertilization cannot entirely replace nutrition through the roots, results of many experiences carried out in different environments and olive orchards, using both single and variedly associated commercial fertilizers [64-68], and olive-specific Israeli commercial products [69-72], confirmed the effectiveness of this olive nutrition practice among sustainable cultivation strategies [32].

5.5. Pruning

Pruning is a practice adopted to modify the natural shape and structure of the trees, with the purpose of: *i)* of balancing vegetative and productive plant activity; *ii)* allowing a suitable illumination of the whole canopy to optimize photosynthesis and circulation of air, that also contribute to reduce the occurrence of pest and disease; *iii)* enhancing growth and disposition of fruiting shoots; *iv)* reducing alternate bearing; and *v)* facilitating harvesting. Academically, pruning of formation and pruning of production are distinct: formation pruning serves to give the selected form to the olive tree, according to the planned orchard typology and management; while production pruning is finalized to preserve the form and the size of the canopy, to eliminate inefficient or unproductive structures, and to facilitate the functional positioning of fruiting shoots. In practice both kinds of pruning coexist, because the olive bears fruits on the previous year's shoots so, in order to have fruit every year, a suitable vegetative growth must be achieved yearly in the optimal shape. The execution of pruning should maintain a high canopy-wood ratio, avoiding both an excess of primary branches and an excessive overlapping of secondary branches, leaving the greatest possible number of leaves. Extraordinary pruning is practiced, when necessary, to restructure the canopy to another shape, to restore health or to rejuvenate the plants.

Pruning should be performed between the end of winter and flowering. Cutting stimulates metabolism and growth and reduces the cold resistance of the plants favoring diseases from fungi or other parasites. However, pruning should not be delayed until after full bloom, since it will remove tissues towards which nutrients and carbon reserves have already been remobilized, resulting in a net loss for the plant. The time of pruning also influences the plant response. Removing shoots at bud break results in a much more vigorous growth of the remaining shoots than if the same operation were performed at the beginning of the summer. In late summer a second pruning must be performed, to eliminate suckers and water sprouts within the canopy [73, 74].

To achieve the best results, pruning must be rationally managed, and based on the harvesting system. According to the kind of mechanical harvest adopted, with shakers or mechanized tools, the shape of trees must be built in completely opposite form: for shakers, the canopy must have relatively short and erect and rigid shoots, to favor the transmission of vibrations to the drupes; while in case of manual tools, the pruning must bring up longer and pending fruiting shoots, to facilitate combing of the canopy [75, 76].

Although pruning is the most expensive practice in olive grove management, reaching up to 40% of total cultivation costs, it is nevertheless essential to optimize cultural responses. The current tendency is to prune olive trees as little as possible. Minimum pruning strategies should be applied in all possible cases to reduce costs and simplify pruning management, consisting in reduction of the frequency of pruning, and adoption of a free-canopy shape. Annual pruning is necessary in profitable olive trees in which a specific shape must be achieved, *i.e.* in table cultivars or when fruiting shoot growth must be optimized for harvesting system efficiency. A biennial pruning can easily be adopted in the majority of cultural conditions, but intervals that are longer than three or four years are unsuitable, because yields markedly decline, and pruning will have to be more intense, with consequences on the vegetative-reproductive balance of the tree.

Cultivars with an upright habit and those sensitive to foliage disease are, however, less suitable for infrequent pruning because the excessively thick canopy and upright growth will make harvesting and pest control more difficult and more expensive.

The intensity of pruning should also be adjusted according to the orchard condition such as the age of trees, cultivar, soil fertility, and water availability. As a general rule, the greater the intensity of cutting, the stronger the trees' vegetative response will be. The intensity of pruning should also consider the crop load, for their consequences on alternate bearing: in heavy cropping years, vegetative growth is reduced, so pruning must be limited to suckers. Conversely, more intense pruning must be carried out after low yield years, to eliminate the excess of vegetation [77].

In profitable olive orchards, mechanical pruning is essential to reduce costs and regain operative timeliness, even though it penalizes the productive efficiency of trees. To balance the economic needs of orchard management and ensure good production of fruiting branches, there are several ways to apply mechanical pruning, such as alternating topping and hedging, or implementing minimum pruning strategies, adopting annual hand pruning to eliminate suckers, and mechanical pruning every three or four years, to rejuvenate the canopy and restrain the trees in a suitable volume.

Some trials, performed in different olive orchard typologies, have studied the feasibility of mechanical pruning [76-81] also assessing the performances of toothed disk pruners in restructuration of long-time unpruned olive-orchards, in terms of working capacity and cutting quality [82, 83] (Fig. **8**). Other trials have been carried out to assess the performances in normal operative conditions of both toothed disk and scissor blade (Figs. **9** and **10**) pruning machines. Both resulted in greater efficacy and operative capacity; nevertheless, the first ones resulted as being more efficient on woody branches, up to 10 cm in diameter, while the second ones work better on twigs, even though they can cut branches up to 5 cm in diameter (unpublished data) [32].

Figure 8: Long-time unpruned olive-orchard restructuration with toothed disk pruning machine (photo: *Toscano*).

Figure 9: Toothed disk pruning machine (photo: *Toscano*).

Figure 10: Scissor blade pruning machine (photo: *Toscano*).

6. HARVESTING

In olive grove management, the harvest is the other most expensive practice, together with pruning. Therefore, harvesting systems must be rationalized, in order to reconcile operative costs with the necessity to harvest the maximum yield, of the best quality. In the past, these two demands were satisfied by a manual harvest and the availability of low cost manpower; with the increase in manpower costs, the manual harvest results as being too expensive, and can only be conducted on table olives, to avoid damage to the fruits and depreciation of their market value.

For the olives destined for oil extraction, the harvest must be effected at the veraison stage of fruits, when the pulp starts to become colored: this stage corresponds to the maximum yield in oil per hectare since, even if subsequently a light increase of oil is had, the losses for natural fruit drop reduce the advantage, and also the oil quality deteriorates. An anticipated harvest allows the obtainment of oil of the highest quality, rich in antioxidants and aromatic compounds, fruited and with high resistance to the oxidation. With maturation, the chemical composition of the fruit decreases in saturated fatty acids and phenolic compounds percentage, while it increases in polyunsaturated fatty acids, that penalize the stability and preservation of oil; also the volatile and aromatic compounds decrease, with negative consequences for the organoleptic characteristics of the oil [84].

Oil quality is strongly dependent on the integrity, health and quality of the olives, and therefore also on the way in which the harvest is carried out. In extensive and old orchards, the harvest of olives, which dropped naturally, from the soil by using nets of mechanical collecting brooms (Fig. **11**) is still practiced. This is the simplest and least expensive harvesting system, but it is also the most dangerous for the quality of oil and tree productivity. In smaller intensive olive orchards, the harvest can be mechanized using hand-held pneumatic or electric powered combs (Fig. **12**), operating from the soil or on platforms (Fig. **13**) to increase the productivity of the yard and grant oil quality, reaching up to 130 kg h^{-1} man^{-1}, operating on well structured trees [85].

Figure 11: Mechanical collecting broom on reversible tractor (photo: *Toscano*).

Figure 12: Pneumatic combs.

Figure 13: Platform with integrated net (photo*: Toscano).*

In the largest intensive olive orchards, harvesting is carried out using various kinds of shakers, from the simplest (Fig. **14**), to tools for articulated arms (Fig. **15**), up to self-moving tools, also equipped with a reverse umbrella interceptor (Figs. **16** and **17**), that allow considerable advantages in terms of harvesting timeliness, yard performance and cost reduction [86-89]; reaching a productivity of over 90% and up to 2.3 t h^{-1} of yield harvested in a single step on well-structured trees [90].

Figure 14: Simplest shaker (*Photo: Toscano*).

Figure 15: Shaker on articulated front-mounted arm (photo*: Toscano*).

Figure 16: Tractor rear-mounted shaker with reverse umbrella interceptor (photo*: Toscano).*

Figure 17: Shaker with reverse umbrella interceptor on self-moving multi-purpose telehandler.

In super intensive or hedgerow shaped olive orchards, the harvest is effected continuously, using harvesters derived from grape pickers (Fig. **18**), that work upon the row on the whole plant; or wall pickers (Fig. **19**), that work on a single wall of trees, with an operative capacity of 0.5 to 1 ha h^{-1}, and harvested yield percentages similar to shakers [32, 74, 90-93].

Figure 18: Self-moving continuous harvester, grape-pickers derived, in a super-intensive olive-orchard (photo: *Toscano*).

Figure 19: Self-moving continuous wall harvester with towed wagon for collected olives in a hedgerow olive-orchard (photo: *Pulcinelli*).

7. PESTS AND DISEASES

For several centuries, olive cultivation in the Mediterranean Basin has been based on the equilibrium between the environment and animal and plants living within it, taking under control the outbreak of pests and diseases.

In the past century agriculture oriented towards the increasing production strongly supported chemical control, causing strong environmental perturbations and increasing toxicological risks. In order to reduce these negative effects of chemical control on the environment [94] and on human health, farmers are increasingly adopting environmentally friendly control strategies, increasing biodiversity (plant and animals) in order to enhance the sustainability of agricultural ecosystems [95]. The natural control of pest outbreaks is the main goal for obtaining a rational control strategy. From this perspective, alternative control strategies should be employed involving agronomic management, biotechnological methods, use of repellents and natural biocides, integrating them by a chemical control, if strictly necessary.

7.1. Olive Ecosystem

The agro ecosystem could be considered as being formed by an abiotic environment and animals and plants living within it. Arthropods are the most abundant animals that lead to positive interaction (symbiosis, commensalism, *etc.*) and/or negative interaction (parasitism, competition, *etc.*) [96]. The agro ecosystem is a simplification of a natural ecosystem due to anthropic actions addressed towards increasing production, requiring continuous human input. Plants cultivated according to monocoltural practices are more sensitive to pest attacks. Furthermore, it is known that plant susceptibility to phytophagous insects and pathogens is influenced by fertilization, increasing more in the case of chemical fertilization because of the higher nitrogen content in plant tissues. Ecosystems with a high level of biodiversity are certainly more stable than those with low diversity rates, as they are more resistant to disturbance, and more resilient with greater recovery rates after a disturbance. Increasing biodiversity, and the related complexity, have positive effects on sustainable agricultural strategies [95]. In the near future there will be a challenge which will involve all researchers, namely to define ecological friendly control strategies, overcoming the limitations of monocultures. The goal should be to stabilize the agricultural system for a long time, not to reach the highest possible annual production.

7.2. Phytosanitary Control in Olive Growing

A modern phytosanitary control in olive growing should be based on the safeguarding of the environment. The concept of sustainable agriculture to date has been strongly supported promoting an agricultural development based on environmental respect. Common Agricultural Policy is the main supporter of sustainable agriculture and thanks to its recommendations, a strong increment of organic and integrated olive orchards has been registered in Europe. In conjunction with environmental safeguarding, these kind of management systems also have the goal of a qualitative increment of production, mainly decreasing toxicological risks [97] which are usually related to the conventional management of olive orchards characterized by the frequent use of chemical products.

Then, the phytosanitary control of the olive should be addressed to individuate and to utilize control strategies with high environmental sustainability and a high qualitative standard of production. For these reasons several control strategies have been developed as alternatives to chemical control.

Furthermore, in some areas, where environmental conditions are favorable to pest development, it is necessary to utilize chemical control strategies integrating all available strategies with the objective of containing damage, to safeguard ecosystems and reduce toxicological risks under the better economic conditions for the olive farmer.

Around 300 species of animal pests live within the olive ecosystem, but those causing significant economic damages are only some units (primary pests) and for these it is necessary to utilize specific control strategies. Secondary pests only occasionally cause damage and in very restricted olive areas.

Phytosanitary control is very important in order to obtain a high quality product [98]. In fact, several pest attacks causing quantitative damages, also cause damage to olive oil quality. In addition, they also cause risks for human health due to the indirect effects of chemical treatments. A rational application of control strategies must take into account the quality factor, evaluating damage thresholds for any parasites.

Table 5: Control methods of main insect pests

Damages	**Agronomic control**	**Organic control**		**Microbiological control**	**Biotechnique control**	**Control with products allowed in organic farming**	**Repellents**	**Chemical control**
		Conservative	**Inoculative**					
Olive fly								
Fruit	Early harvesting, low susceptible cultivar., increment of plant diversity	Ecosystem equilibrium	*Opius concolor*	*Bacillus thuringiensis*, *Beauveria bassiana*	Mass trapping	azadirachtin	Kaolin	Dimethoate, deltamethrin, spinosad
Olive moth								
Leaves, flowers, fruits	Low susceptible cultivars, increment of plant diversity	Ecosystem equilibrium		*Bacillus thuringiensis*	Sexual confusion	azadirachtin		Organophosph ates
Black olive scale								
Plant	Periodic pruning, increment of plant diversity	Ecosystem equilibrium	*Metaphycus bartletti*, *M. hevolvolus*, *M. swirskii*			White oils		Carbammates
Jasmine moth								
New shoots	Increment of plant diversity	Ecosystem equilibrium		*Bacillus thuringiensis*		azadirachtin		Organophosph ates
Olive weevil								
Leaves	synthetic band traps on trunks, increment of plant diversity	Ecosystem equilibrium			Bait trapping			Organophosph ates
Leopard moth								
Twigs	Pruning, increment of plant diversity	Ecosystem equilibrium		*Bacillus thuringiensis*, nemathodes, *Beauveria bassiana*	Sexual confusion, mass trapping	azadirachtin		Organophosph ates
Olive bark beetle								
Branches	Burning pruning residues, increment of plant diversity	Ecosystem equilibrium						

7.3. Control of Main Insect Pests

Insects are the main pests of the olive tree causing great qualitative and quantitative damages to production. The primary pest, the most dangerous for olives, is the olive fly (*Bactrocera oleae*). Other important insect pests are the olive moth (*Prays oleae*) and the black olive scale (*Saissetia oleae*). Phytophagous insects of secondary importance only occasionally causing economically appreciable damages could be considered to be the jasmine moth (*Palpita unionalis*), the olive weevil (*Otiorrhynchus cribricollis*), the olive bark beetle (*Phloeotribus scarabeoides*), the leopard moth (*Zeuzera pyrina*), and some others.

Against these species several control methods can be utilized, chosen by the farmer according to the kind of production (olive oil/table olive, integrated/organic management), damage threshold, *etc.* (Table **5**). The following paragraphs are devoted to illustrating damages and control methods of main insect pests and their biology.

7.3.1. Olive Fly - Bactrocera Oleae (Rossi, 1790) - Diptera Tephritidae

The olive fly is the key-pest for olive orchards, it is widespread in the Mediterranean Basin and in other olive areas through the World (Fig. **20**). The entity of damages is variable depending on the localization of olive groves (latitude and altitude) in areas with different climatic conditions.

Figure 20: Adult female of *Bactrocera oleae.*

Morphology: black thorax, hyaline wings 4 to 5 mm long with iridescent reflections. Males slightly smaller than females. Female with a sting ovipositor on the last segments of the abdomen. Egg elongated with rounded extremities, 0.7 mm long. Apod larvae, elongated and subconical, yellowish-white in color.

Biology: Widespread in nearly all the areas where the olive is cultivated, but with strong ethological and ecological variations, producing a complex biological mosaic. The number of generations per year varies according to climatic parameters and availability of olive drupes on plants attaining, in most favorable years (wet summers) and places, 5 to 6 generations per year. Adults feed on nitrogen and sugar substances. Females lay their eggs during stone hardening of drupes, usually the first half of July. A female can lay up to 1,000 eggs. The egg is laid under the drupe epidermis and hatches after a variable period, usually from 3 days to 3 weeks, depending on temperature. When the temperature is lower than 6°C or higher than 35°C larvae development is inhibited. Larvae have three mountings before pupating, feeding on the drupe pulp.

Population density and intensity of attacks depend on several parameters such as weather conditions, cultivar susceptibility, plant productivity, and natural enemies.

Damages: The olive fly can cause a strong decrease in olive production due to pulp removal and premature fruit drop, and a strong decreasing of olive oil quality. In fact, infestation causes some biochemical alterations within drupes altering oil acidity and peroxide numbers, chemical parameters of olive oil (phenolic and sterolic composition, spectrophotometric constants), and sensory characteristics of olive oil [99].

Several studies have confirmed a close relation between infestation and chemical alteration of olive oil that increases with the increasing of storage time before olive pressing. In order to limit the decreasing olive oil quality it is necessary to press the olives as soon as possible after harvesting, mainly in the case of a high percentage of infested drupes.

Damage thresholds, which were individuated and chosen when quantity of production was the main objective of farmers, have increased from 10 to 15% and greater. In table olive production the female oviposition sting is yet an aesthetic degradation of economic importance and could be the initial point of other olive degradations due to bacterial or fungal agents.

Sampling: Control strategies which are strictly linked to the presence of adults in the field and to the damages produced could be called “adulticide control”. The aim of these kind of strategies is to reduce the population density of adults, while the control of preimmaginal stages could be called “larvicide control”. In the first case, it is important to sample the adult population by using specific traps with attractants. Chromotropic traps (visual attraction), pheromone traps (sexual attraction), olfactory traps (ammonium salts), and alimentary traps (hydrolysed proteins), having differentiated attractiveness, costs and effectiveness are available on the market. Chromotropic traps are the most utilised for their effectiveness (male and female are captured equally and it is not influenced by temperature and humidity), their low cost and easy application in the field, but they are not selective towards non target insects. Control strategies should usually be applied when more than three females (or 8 adults of both sexes) trap/week are captured.

Sampling methods of pre-immaginal stages consist in a periodic sampling of drupes (7 or 10 days) in homogeneous environmental and agronomic conditions. Usually, it is sufficient to collect one drupe/plant randomly according to orientation and distance from the soil level, or to collect 10 drupes/plant on 10% of the olive trees. The threshold of 15 to 20% of infestation could be considered compatible with a good qualitative standard of olive oil production [100].

Control: <u>Agronomic control</u> (available for conventional, integrated and organic farming) – Genetic resistance – To research factors of genetic resistance within known olive cultivars is a focal point for pest control planting of low susceptible cultivars in new orchards [101]. Low susceptibility derives from both physical characteristics of drupes (hardness, dark color) and the amount of some molecules (oleuropein, cyanidine) within drupes (Table **6**). Oleuropein seems to inhibit the development of the first preimmaginal stages (eggs and young larvae) of the olive fly [102]. Some cultivars having early veraison, with high drupe content of cyanidine showed a low susceptibility, probably due to the difficulty of the female in recognizing drupes for laying eggs. - Harvesting – It is very important to know the optimal harvesting period of any cultivar in their optimal pedoclimatic conditions, resulting from inoliation trends and fruit drop. Anticipated harvesting often prevents damages caused by strong infestations and it is possible to obtain a high quality oil strongly reducing pesticides applications.

Biological control - (available for conventional, integrated and organic farming) – It is essentially based on the "conservative" biological control, that means to control pests maintaining ecosystem stability. In fact, the olive fly has several antagonists. "Classical" biological control based on the rearing and release of *Psyttalia concolor* (Hymenoptera Braconidae), has recently been re-utilized [103], but to date it is still difficult to rear this species and it is too expensive [104]. Recent studies on the use of *Bacillus thuringiensis* (Bt) enlarged the field of application of microbiological control, including the control of the olive fly with some specific strands. To date, some Bt strands are under experimentation in conjunction with attractants. Some encouraging tests are also being carried out on other micro-organisms (*Beauveria bassiana*).

Table 6: List of some cultivars of *Olea europaea* L. showing low and high susceptibility to the olive fly

Low susceptibility
Oleuropein > 30 mg/g
Bardhi i Tirana
Gentile di Chieti
Carboncella Pianacce
Nociara
Cima di Mola
Cellina di Nardò
Leccino
High susceptibility
Oleuropein < 30 mg/g
Giarraffa
Cucco
Picholine
Nocellara del Belice
Cassanese
Carolea
Maurino
Peranzana

Biotechnical control - (available for conventional, integrated and organic farming) – Mass trapping is the most used and efficient control technique [105, 106]. It employs attractive traps of different shape, color and materials, with contact insecticides. The objective of this kind of control strategy is to reduce the adult population utilizing a selective system (attract and kill) which is safe from a toxicological point of view. This technique seems to be particularly effective when utilized on large surfaces and against late infestations. The use of particularly effective traps with pheromones and with deltamethrin as the contact insecticide has recently been allowed in organic farming [107]. Today mass trapping devices with long efficacy (July-September) (Fig. **21**) without the need of maintenance are available on the market, and are utilized in a lower number than before (100 to 150 traps ha^{-1} instead of 400).

Figure 21: A new mass trapping device.

Control with natural substances - (available for conventional, integrated and organic farming) – Natural substances useful in control strategies are obtained from minerals or plants. They could be subdivided into biocides, killing insects (pyrethrum, rotenone, *etc.*), repellents, acting on adult behavior (sodium silicate, kaolin, soya lecithin), feeding disruptors, acting on the alimentary behaviour of insects (neem extracts), and bioregulators, acting on the development of preimmaginal stages. Natural pyrethrum showed a low efficacy, whilst rotenone (obtained from *Derris elliptica*) and azadiractin (neem extract) showed good efficacy also in areas favorable to the development of the olive fly such as Southern Italy. However, these active agents seems to have residues within olive oil and are harmful for the environment. Copper-based substances are also of natural origin and are allowed in organic farming. Recent studies proved the efficacy of copper-based products against the olive fly [108]. In fact, these substances reduce microbial flora on the olive canopy inhibiting the development of preimmaginal stages and adult reproduction.

Among repellents, kaolin showed very good efficacy. It is a mineral substance sprayed on the canopy and covering all leaves and drupes [109]. The white color of the canopy and fruits disorients females that do not recognize the fruits for laying eggs and the protective film acts as mechanical barrier against oviposition stings (Fig. **22**). Furthermore, olive oil coming from olives treated with kaolin are free from undesirable pesticide residues.

Figure 22: Olives sprayed with kaolin.

Antibacterial substances - (available for conventional, integrated and organic farming) – Recent studies carried out in Tuscany (Central Italy) [108] demonstrated that copper-based substances, usually employed against fungal diseases, are also effective against the olive fly because they greatly reduce the saprophytic flora of the leave surface eaten by the adult olive fly. Further studies carried out in Calabria (Southern Italy) have confirmed these results and suggest the simultaneous application of propolis (natural antibacterial substance) with copper greater efficacy.

Chemical control – (available for conventional and integrated farming) – It is based on the use of pesticides against adults, by spraying trees or associated with feeding attractants (hydrolysed proteins) in baits locally distributed on olive trees, or against preimmaginal stages by spraying trees.

More frequently utilized active agents are some organophosphates and a pyrethroid (deltamethrin). The most utilized active agent by farmers is the dimethoate against both adults and larvae. It can also be utilized at low concentrations (50g hl^{-1}) against larvae. New formulates, spinosad and the imidacloprid, have recently arrived on the market and can be used against the olive fly at low concentration rates.

The uncontrolled utilization of these active agents has serious and negative ecological and toxicological consequences, therefore they must be carefully utilized and included in integrated control programs, having a great bioecological knowledge of the olive fly, and after monitoring adult populations with pheromone or chromotropic traps, applying damage thresholds (>15% of larval infestation) and, if possible, applying forecasting infestation models [110, 111].

7.3.2. Olive Moth - Prays Oleae (Bern) - Lepidoptera Yponomeutidae

Biology: Gray micromoth with silvery reflections and black spots on the forewings. Larvae are polyphagous, feeding on flowers (anthophagous generation), drupes (carpophagous generation), and leaves (phyllophagous generation). Females lay up to 250 eggs. Larvae have 5 instars and their development is 4 to 6 weeks long, except for the hibernant generation.

Damages: Anthophagous and carpophagous generations are certainly the most dangerous reducing the quantity of olive production [112]. The carpophagous generation causes anticipated fruit drop because larvae feeding inside stones leave fruits with a hole 2mm large at the basis of the peduncles. Although often the field density of olive moths is high in pheromone traps, damages rarely have economic importance. This could be due to several factors, such as the high larval mortality and the high predation pressure of antagonists usually living in the olive orchards [113, 114].

Control: Agronomic control: Increasing the environmental complexity could be considered very useful and can be obtained by soil grassing, hedge planting, *etc.*

thus enhancing the suitability of olive orchards for antagonists. Some cultivars with small fruits are very resistant to olive moth infestation (cv. Semidana, Palma, Bosana) [115].

Biological control: An approach consists in increasing the presence of the several natural antagonists of the olive moth. Another strategy is microbiological, utilising *Bacillus thuringiensis* against larvae of the anthophagous generation, consequently also reducing the subsequent harmful carpophagous generation. Scarce results are obtained with the field release of *Trichogramma* spp. in some Mediterranean countries.

Control with natural substances - (available for conventional, integrated and organic farming) – On this subject very few tests have been carried out and trials with pyrethrum have shown insufficient results. Actually, rotenone and azadirachtin are under experimentation, but results are not yet exhaustive.

Biotechnical control - (available for conventional, integrated and organic farming) – Although the availability of an effective pheromone and the possibility of applying mass trapping devices on small surfaces because of the scarce mobility of adults, olive farmers only occasionally apply this control strategy.

Chemical control – (available for conventional and integrated farming) – Due to the scarce economic importance of damage, chemical pesticides and growth regulators (fenoxicarb) should be used with extreme caution because of their side effects on the environment. When infestation is extremely dangerous for production, it is possible to use systemic organophosphates against the carpophagous larval generation.

7.3.3. Black Olive Scale - Saissetia Oleae (Oliv.) - Homoptera Coccidae

Biology: Scale with dorsal fairings H-shaped on the dorsal shield. Polyphagous. It has only one generation per year, hibernating as a neanid. Usually parthenogenetic, the male is very rare. Adults appear in April-July. Females lay 1000 eggs. Young neanids have a high mortality rate during the phase of active dispersion. Sexual maturity is usually reached on branches of two years sucking sap.

Damages: It causes damage due to hardly quantifiable sap subtraction and damage due to sugar dejections on leaves and branches which are the optimal substratum for fungus development [116]. In recent years in Italy this scale has not caused as appreciable damage as in the past.

Control: Biological control - (available for conventional, integrated and organic farming) – The black olive scale has several natural enemies. The females of *Scutellista cyanea* and *Moranila californica* (Hymenoptera Chalcidoidea) lay eggs under the scale body. Other important parasitoids of Chalcidoidea belong to the genus Metaphycus. Conservative biological control is a focal against this insect and can be improved by correct agronomic practices. Frequent and correct pruning strongly reduces the microclimatic conditions favourable to the development of the black olive scale and that an increment of plant diversity increases the population density and diversity of natural antagonists. It is also possible to inoculate entomophagous insects such as *Metaphycus bartletti*, *Metaphycus hevolvolus* and *Metaphycus swirskii* in the olive orchards in order to obtain good results.

Chemical control - (available for conventional and integrated farming) – It must be limited and evaluated with caution because this phytophagous insect only in very occasional cases produces economic damage. White oils, more selective and less polluting than other chemical products, can be utilised against neanids during July-August. The intervention threshold is about 5 to 10 neanids/leaf on a sample of 100 leaves randomly collected from around ten olive trees [117]. Growth regulators which are highly effective against the scale should be used carefully because of their side effects on the antagonists.

7.3.4. Leopard Moth - Zeuzera Pyrina L. - Lepidoptera Cossidae

Biology: White wings with black-bluish dots. Larva is xylophagous on several tree species. Adults live only one week just for mating. Female lays about 1,000 eggs in small groups within a small splitting of bark in order to facilitate the penetration of young larvae within the wood of small branches. During their development larvae feed within larger tree branches. Adults fly in April-October.

Damages: This voracious phytophagous insect can cause a general wasting of the plant because their larvae feed on wooden organs. Branches can desiccate, and when larvae feed within the stem, the plant can die. Tunnel entries are recognisable thanks to the larvae excrement deposited on the outside. Tunnels are the entry point for other insects and fungus.

Control: Control of this species is difficult [118] because larvae are protected by human interventions in tunnels and adults flying through several months can only be controlled by several pesticide treatments (conventional and integrated farming). With the availability of the pheromone, the mass trapping technique turns out to be effective. Mechanical control with wire introduced within tunnels, microbiological control with entomopathogen fungus or nemathodes introduced into tunnels, and a pruning of infested branches seems to be effective on individual trees in conventional, integrated and organic farming.

7.3.5. Goat Moth - Cossus Cossus L. - Lepidoptera Cossidae

Biology: Its cycle is three years long and hibernate as larva. Adults fly from May to September. Young females lay small groups of eggs within splitting of bark, usually on the low part of the stem. Young larvae are gregarious, under the bark, they then dig individual tunnels within the wood. Larvae are mature in the following autumn, and pupate in the spring within a silky cocoon tissue near the exit hole. Adults do not feed.

Damages: The kind of damage is similar to that produced by the leopard moth, but the effects are higher because larger larvae dig larger tunnels.

Control: Control strategies are the same illustrated for the leopard moth.

7.3.6. Jasmine Moth - Palpita Unionalis Hb. - Lepidoptera Pyralidae

Biology: White wings, with nacrcous reflections. Adults fly from March to November with 3 to 5 generations. Females lay about 600 eggs on leaves either individually or in small groups. Eggs hatch in a few days depending on the climatic conditions. Larvae are polyphagous, on the olive tree they feed on leaves or occasionally on drupes. Larvae pupate within a cocoon made from leaves.

Damages: Economic damage is produced within new plantations on small plantlets, because it feeds preferably on the new shoots. It is particularly dangerous in nurseries. It would most likely be extremely dangerous in super-intensive olive orchards.

Control: It is difficult to control due to its long flying period, therefore it is better to control larvae than adults utilising *Bacillus thuringiensis*, available for conventional, integrated and organic farming, or a pesticide with a broad spectrum of action, only under a conventional regime.

7.3.7. Olive Weevil - Otiorrhynchus Cribricollis Gyllenhal - Coleoptera Curculionidae

Biology: Black insect with reddish legs and antenna. It feed on several plants. Adults appear from March-June. They are nocturnal and during the day-time remain under soil. After mating females lay eggs in the soil where larvae feed on the roots of several herbaceous plants. After ten moultings, the larvae pupate in the soil from April-May. It has only one generation per year and hibernates in the soil.

Damages: Economic damage is produced within new plantations on small plantlets, because it feeds preferably on new shoots and leaves. Foliar erosions are semi-circular avoiding main nervure. It is particularly dangerous in nurseries, and it is most likely to be very dangerous in super-intensive olive orchards.

Control: Good results have been obtained by utilising synthetic band traps applied on the trunk which hinder the climbing of adults from the soil to the canopy. Microbiological control by enthomophagous fungus (under conventional, integrated and organic regimes) and poisoned baits of sugary bran (under conventional and integrated regimes) also seem to be effective.

7.3.8. Olive Bark Beetle - Phloeotribus Scarabaeoides Bernard - Coleoptera Scolytidae

Biology: Bark beetle with a small hairy body, black-brownish with reddish antenna. It feeds on the olive and a few other trees and bushes. Adults hibernate within tunnels excavated within small twigs. The exit hole of these tunnels is visible due to the presence of sawdust and faeces. In February-March adults attack unhealthy twigs and pruning residues. Females lay 10 to 100 eggs under the bark. After 10 days the young larvae hatch that then excavate tunnels perpendicularly to those where eggs were laid. At the end of these small tunnels they pupate. This species has 1 to 3 generations per year, but only the first causes appreciable damage.

Damages: Damage is caused by the first generation that feeds within young twigs where they penetrate at the basis of leaves, inflorescences or twig bifurcations. The second generation can only occasionally cause early fruit drop chewing the petiole of the drupes.

Control: It is essentially agronomic, based on the pruning of attacked twigs and on the utilisation of twig-baits (pruning residues on the ground), successively burned at the beginning of May (available for conventional, integrated and organic farming).

7.3.9. Olive Fruit Midge - Prolasioptera Berlesiana (Paoli) - Diptera Cecidomyidae

Biology: Brownish body, head covered by light brown scales. The species is common on *Pistacia lentiscus* where larvae feed on eggs of small insects, sap, fungus, *etc.* On the olive the female searches for oviposition stings of the female olive fly where it lays one or more eggs. After 1 to 2 days the young larvae hatch that first feed on fly eggs, then on the mycelium of the symbiont *Camarosporium dalmaticum* that causes the early fruit drop. It has 4 to 5 generations on the olive per year, at least in Southern Italy.

Damages: It is an antagonist among insects of olive groves as a predator of olive fly eggs, but in table olive cultivation it is a pest because its symbiontic fungus *Camarosporium dalmaticum* causes fruit rot that appreciably decreases the economic value of production.

Control: The control of this Cecidomyidae is usually carried out by utilising the same products utilised against the olive fruit fly. Chemical active agents are utilised under conventional and integrated regimes, whilst vegetal biocides are utilisable in organic farming.

7.3.10. Olive Scale - Parlatoria Oleae (Colvée, 1880) - Homoptera Diaspididae

Biology: Adult females are grey with black exuviae visible on fruits and leaves. It is polyphagous on several plant species feeding on leaves, young twigs and fruits. Usually it is predated by several natural antagonists.

Damages: It causes damage to the drupes deforming them or leaving a dark-brownish spot on the epidermis. Dangerous for table olive cultivation.

Control: Damage from these parasites is very rare because of the efficacy of natural antagonists. It is rarely necessary to control this species utilising white oil in conventional, integrated and organic farming or organophosphates in conventional and integrated farming when more than 1% of drupes are infested or two neanids per leaf are present.

7.4. Control of Main Diseases

Olive diseases depend on favourable climatic conditions, strong nutritional imbalance, and pollution stress, pathogenic activity of bacteria, fungus and viruses.

The olive shows an easy phytopathological situation [119]. In fact, among pathogens only one bacterial species (*Pseudomonas savastanoi*) and four fungus (*Spilocaea oleagina*, *Verticillium dahliae*, *Colletotrichum gloeosporioides*, *Mycocentrospora cladosporioides*) can significantly affect the olive tree and the production of drupes. Furthermore, some diseases due to the contemporaneous activity of several fungus are known (root rot, caries, sooty mold). Among viruses, only SLRV can appreciably affect the olive tree but does not require control.

It is very important to study epidemiology in order to control any individual pathology due to the close links between pathogen biology and environmental conditions and relations between host and parasite. Knowledge regarding these relations offers a good probability for preventing and rationally controlling the disease [120] (Table **7**).

7.4.1. Olive Knot Disease - Pseudomonas Savastanoi pv Savastanoi (Smith)

Biology: Olive knot disease is a widespread disease in all areas where the olive is cultivated. It is caused by the bacterium *Pseudomonas savastanoi*. It is characterised by tubercles that grow on twigs. The main inoculum is the infected plant because aerial organs inhabiting the bacterium are often utilised as propagation material. *P. savastanoi* can only penetrate the host tissues in water through wounds due to meteoric (such as cold) or cultural events (pruning, mechanical harvesting). Climatic conditions such as high humidity rates, prolonged wetness of aerial organs and temperatures of 20 to 25 °C are decisive for bacterium development.

Table 7: Control methods against main diseases

Disease	Damages	Agronomic control	Control with natural products (allowed in organic farming)	Prevention	Chemical Control
Bacteria					
Olive knot disease	Twigs, stem	Cultivar with low susceptibility	Copper	Disinfection of equipment	Copper, Bacticin
Fungi					
Olive leaf spot	Leaves	Cultivar with low susceptibility	Copper	Avoid wet environment	Copper, dodine
Verticillium wilt	Twig wilt	Cultivar with low susceptibility		Avoid consociations with solanacee	Fosetyl-aluminium
Fruit rot	Spot on drupes		Copper	Control olive fruit midge	Copper +pesticide against olive fruit midge
Anthracnose	Drupe mummification	Cultivar with low susceptibility	Copper	Avoid wet environment	Copper
Mycocentrospora cladosporioides	leaves and fruits	Cultivar with low susceptibility	Copper	Avoid wet environment	Copper
Root rot	Wilting	Soil drying	Lime powder	Avoid water stagnation	Systemic fungicides
Sooty mold	Leaves	Periodic pruning	Washing with surfactants	Control the olive black scale	Copper

Damages: Symptoms are tubercles on twigs and other aerial organs caused by virulent strains able to stimulate the production of phytoregulators (auxin, cytokines) promoting cancer formations.

Control: Prevention is the main control method for this disease. Planting cultivars with low susceptibility [121, 122], sterilisation of pruning tools and of wounds after pruning or after unfavourable climatic events are important for reducing the incidence of the disease. Particular attention should be paid to the selection of propagation material and in the disinfection of tools during propagation.

Conventional control against this disease is not easy, because antibacterial products are prohibited in several countries in agricultural practices and because compounds with antibacterial activity are hardly translocated to the plant. The available treatments are those of rameic compounds immediately after predisposing meteorological events or after pruning or removing of cancerous formations. In the latter it is possible to use Bacticin. Control strategies as an alternative to chemical control have low efficacy and are limited to some attempts carried out in California with actinomycetes producing antibiotics, and in Italy on avirulent isolates of the bacterium producing bacteriocines. Some cultivars resistant to rapid temperature decrease (mainly in spring) are very promising for the planting of new orchards.

7.4.2. Olive Leaf Spot - Spilocaea Oleagina (Cast.) Hugh.

Biology: The olive leaf spot is the best known and most widespread disease of the olive. It is recognisable due to the characteristic concentric brown spots on leaves [123]. The fungus preferably infects leaves but also small twigs with germinant conidia forming hyphae within epidermis cells. Two weeks to some months from penetration to the expression of symptoms are necessary depending on environmental conditions. In the Mediterranean olive areas favourable epidemiological conditions are frequent during spring and autumn, when symptoms become evident and leaves drop prematurely [124]. The fungus is dispersed by wind within small water droplets, simply by wind or by insects [125] (*Ectopsocus briggsi*). Susceptibility to fungal infections and their diffusion are strongly affected by thermo-hygrometric conditions and by olive orchard management.

Damages: The leaves drop is the most evident damage of the disease causing strong hormonal and nutritional imbalance and, as consequence, a decrease in production [126].

Control: Copper-based compounds can be utilised in a year with low production because they induce both the elimination of the inoculum and strong phylloptosis. During years with high production it is necessary to sustain production avoiding treatments with copper-based products [127]. In this case it is possible to utilise other fungicides such as dodine, penconazole and bitertanol. The only organic control available is to preserve the biological balance of the canopy where several natural antagonists of the fungus live [128]. In areas where olive orchards are increasing a correct environmental analysis and a right choice of cultivars will be necessary [129, 130]. Some cultivars have shown low susceptibility to fungal infections, probably due to phytoalexins, oleuropein content of leaves, histological and/or physiological characteristics [131]. In Italy, the cultivars Bhardi i Tirana, Carboncella di Pianacce, Cassanese, Dritta di Moscufo, Gentile di Chieti, Kalinjot, Kokermadh i Berat, Leccino, Cipressino, Ottobratica, Zaituna, Pisciottana, Cellina di Nardò and Dolce Agogia showed a low susceptibility. In Israel the cv Maelia 29 has been selected, and used in genetic improvement programs because of its high resistance.

7.4.3. Verticillium wilt - Verticillium Dahliae (Kleb.)

Biology: Symptoms of this disease are the wilting of small twigs or the whole plant, with leaden and crumpled leaves (Fig. **23**) [132, 133]. The fungus attacks the wooden vascular system impeding the water supply of plants. Solanacee are highly susceptible to this fungus. V. dahliae conidia are dispersed by water or animals, penetrating within plant tissues through wounds or lesions where development begins. It produces toxins damaging vascular tissue by occlusions and cell poisoning impeding water supply and causing the wilting of plants or twigs. It is widespread in the Mediterranean Basin. Recent studies have highlighted an expansion in Italy, mainly in Calabria and Sicily [134], due to infected plantlets of highly susceptible cultivars propagated in nurseries, or to infected soil transported by machines, or to infections propagated by Solanacee and other weeds [135].

Damages: Symptoms are the partial or total wilt of plants, usually young plants. Chronic symptoms are present on adult plants with little damage, while young plants can die in the most serious cases. It is possible that the infection is latent in asymptomatic plants with the fungus inhabiting within xylematic tissues.

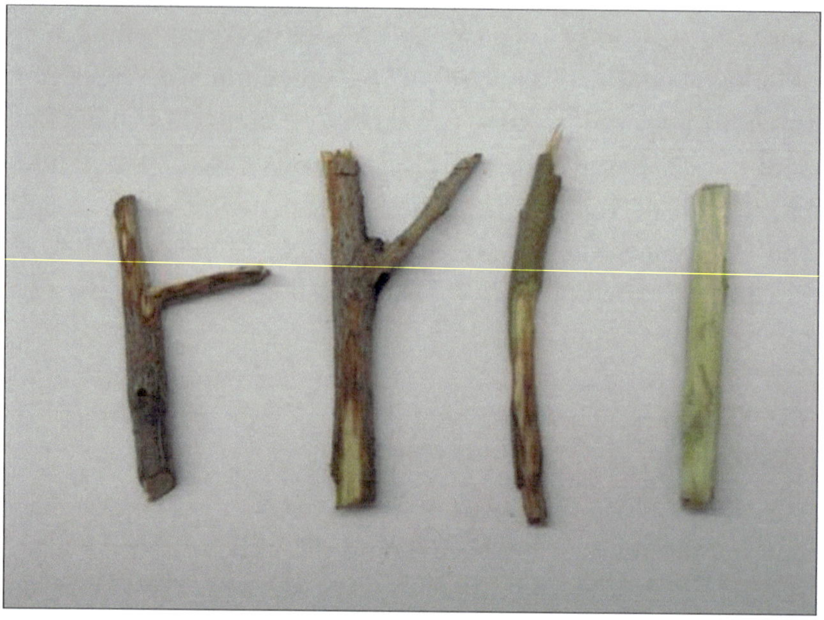

Figure 23: Symptoms of *Verticillium* wilt on twigs.

Control: It is based mainly on prevention, regarding nursery workers that must only propagate certified material and utilise uninfected soils, and farmers that must avoid consociations with Solanacee and Cucurbitacee and pay attention to the use of machines on infected soils and to utilise drop irrigation instead of flow irrigation.

Chemical control, which was ineffective until just few years ago, today can be effective by injecting fosetyl-aluminium [136], or dodine, directly within trunks, permitting stable rehabilitation of infected plants. Olive vegetation waters demonstrated the inhibition capacity of mycelium growing only under laboratory conditions, but with an applicative perspective mainly for reducing the inoculum in the soil. Other available techniques are solarisation and sawdust addition to the soil.

Biological control can be carried out by application of the ascomycetes *Talaromyces flavus* (Klöcker)to the soil, which seems to be effective against microsclerotia of *V. dahliae* acting as parasite or by enzymatic pathways. The control of this disease, like previous ones, can be carried out by searching for sources of genetic resistance. Among studied cultivars some seem to be promising showing appreciable resistance-tolerance and are also utilisable as rootstocks [137]. Recent studies have demonstrated a different behaviour of cultivars towards the inoculation of different *V. dahliae* strands, and the importance of the virulence of tested strands on the response of cultivars. The cultivar Frantoio showed no symptoms and a potential resistance to the pathogen, whilst cv. Ottobratica and Sant'Agostino were highly susceptible. Further studies are necessary in order to define the behaviour of cv. Arbequina, Arbosana and Urano increasingly used in high density plantations (super intensive). Arbequina and Arbosana showed a susceptibility related to the tested verticillium strand, while Urano seems to be resistant becoming very important for an eco-sustainable management of new orchards.

7.4.4. Anthracnose - Colletotrichum Gloeosporioides (Penzig)

Biology: This disease is mainly distributed in some areas of South Italy (Calabria and Apulia), where they cause twig and leaf wilt, and damage to the fruit (mummification), preferring mild and wet environments. The fungus can be preserved as perithecium, mycelium or conidia in rotten fruit and in all attacked vegetal parts, producing symptoms during veraison. It can penetrate within host tissues through stomata and wounds.

Damages: Damage of economic importance is only caused on fruits that mummified before dropping. The mummification of fruits that stay for a long period on the olive is characteristic, representing further pathological evidence.

Control: Agronomic control by planting low susceptible cultivars and carrying out a light and periodic pruning is effective against anthracnose. Cupric products are also effective because of their long persistence and because copper has a detoxifying effect on the aspergillomarasmine B produced by *C. gloeosporioides*. Treatments based on cupric salts (copper oxychloride) should be carried out from the end of summer to the end of October.

7.4.5. Fruit rot - Camarosporium Dalmaticum (Thum)

Biology: It is characteristic for a brown spot on fruits. The pathogen is dispersed by *Prolasioptera berlesiana* Paoli, egg predator of the olive fly. Mycelium growth in the fruit pulp.

Damages: Infected drupes have a characteristic brown spot, depressed and suberified, around the oviposition sting of the olive fly. Infested olives are not utilisable in olive table production and the quality of oil (from both chemical and organoleptic point of view) can strongly decrease if infection involves an appreciable amount of drupes.

Control: Treatments are only necessary for table olive production, utilising copper-based fungicides and pesticides against the olive fly and olive fruit midge, which are usually associated. Other active agents like dodine, carbendazim and systemic tebuconazole-based products are available against this disease.

7.4.6. Mycocentrospora Cladosporioides (Sacc.)

Biology: Symptomatology shows grey spots on the underside of leaves. Sometimes it also attacks fruits on which brown-violet spots appear. The inoculum is on leaves where conidia are preserved. The fungus can also be present on the soil. It actively penetrates host tissues. Mild and wet climate favour the development of the disease.

Damages: Leaf drop follows the appearance of grey spots. Infested drupes show brown-violet spots and are not utilizable as table olives.

Control: Treatments are necessary only for olive table production, utilising copper-based fungicides or other active agents like dodine and carbendazim.

7.5. Control in Organic Farming

Organic farming, as defined by European Union (REG. CEE 2092/91) is a method of agricultural production with sustainable management of the ecosystem, preserving and enhancing the biodiversity and the biological activity of soil, with the goal of reducing pollution due to some agronomic practices. It is based on the reduction of external inputs and on the elimination of synthetic chemical products (pesticides and fertilizers), replacing them with natural products and applying agronomic methods that enhance resistance to diseases. Phytosanitary protection according to European legislation on organic farming is independent from treatments that can cause undesirable effects for the environment and human health, mainly focusing on prevention based on the knowledge of soil-plant-environment-animals relations. For this purpose it is necessary to known the cultivation environment and the biology of fungus and pests threatening them.

An increasing ecological and health sensitivity, stimulated by the European Union, has led several olive farmers towards organic farming. The increasing demand for certified organic products has determined the introduction of several controls by public authorities and certifications along the whole productive process of organic production, paid for by consumers with more expensive costs for organic products. Organic regulations are defined by the European Union that lists the phytosanitary products allowed in organic farming, also successively permitted by regulation of any individual member country.

Organic phytosanitary control in olive farming is hard to apply in some Mediterranean environments favourable to the development of parasites [138, 139]. Olive fly control, also difficult under conventional management, is particularly hard under organic management [140]. Research on alternative control strategies, the recent update of regulations on the employment of deltamethrin and pheromones within traps [106], and the increase of intervention thresholds (level of infestation requiring treatment to reduce economic damage), undoubtedly facilitate the olive fly control. In fact, it is possible to control olive fly infestations in order to obtain high quality olive oil production, except in few difficult areas and for few cultivars with late maturation [141].

Many control strategies of the most dangerous phytophagous insects for the olive (such as the olive fly, olive moth, olive black scale) under organic farming are shared with conventional control, except those referring to the use of chemical pesticides. Control techniques such as agronomic (early harvesting [142], low susceptibility cultivars [143], increase of vegetal diversity), biological [144], bio-technique (mass-trapping) [145], repellents (kaolin) or natural pesticide (azadirachtin, copper) ones, can be usefully utilised under both conventional and organic farming with the same considerations concerning efficacy, costs, and environmental and toxicological impacts [146].

Mass trapping seems to be very promising against the olive fly [147], especially if applied on large surfaces and integrated with other agronomic control methods [148, 149]. The use of natural pesticides, such as azadirachtin and others, can offer new perspectives for olive fly control due to its fagoinhibitor and/or bioregulator activity, causing alimentary and hormonal alterations. The use of treatments covering plant and drupe surfaces against olive fly is also very interesting [150].

The control of fungi is easier than that of insects because copper-based products, which are effective against the main olive diseases, are allowed in organic olive farming. Furthermore, prevention techniques assume a very important role for disease control especially in nurseries from where plantlets must exit accompanied by sanitary certifications [151]. Successively, farmers should pay great attention to environmental field conditions because they can predispose the orchard to disease infection. The choice of cultivars with low susceptibility is a focal point, and the use of correct agronomic techniques (pruning and disinfection of utilised tools, irrigation) can ameliorate the physiological balance of plants preventing field infections. It is also important to avoid consociation with herbaceous plants such as solanacee that can act as inoculum sources.

The utilisation of products of natural origin (vegetation water against verticillium wilt), few cases of biological control (avirulent isolates against olive knot disease, antagonist fungi of olive leaf spot and verticillium wilt), solarisation technique against verticillium wilt can be also cited.

7.6. Control in Integrated Farming

European Agricultural Policy has indicated integrated rural development as the basis for a sustainable development where agriculture should be multifunctional. In European countries with important olive production ecologically sustainable cultivation is also strongly promoted by the European Union.

Integrated farming is an agricultural system producing high quality food, one which is based on resources and natural mechanisms for replacing inputs that are dangerous for the environment ensuring long life for agriculture. In perspective, this model is suitable for a large part of olive orchards, mainly where environmental conditions are favourable to parasite development [152].

In olive farming phytosanitary control is very difficult. The success of an integrated production program is closely linked to an effective control of parasites [153]. The control of pests and fungi should be carried out utilising alternative control methods based on the monitoring of environmental conditions and on the sampling of insect pest populations [154] and fungal infections. Chemical compounds should be only used if absolutely necessary, overcoming intervention thresholds, and should not be found as toxic residues exceeding the allowed limits.

Biological control of disease is very limited, therefore prevention techniques assume an important role [155]. Nursery workers should certify plantlets from a sanitary point of view, and olive farmers should apply agronomical strategies to prevent the development of diseases, such as the choice of low susceptible cultivars, correct fertilisation and pruning procedures, tool disinfection, *etc.*). Chemical control should be carried out only when strictly necessary, with punctual and few treatments, related to the epidemiology of any disease.

7.7. Control of Table Olive Production

Olive oil production is compatible with intervention thresholds of around 15 to 20% of infestation, while table olive production has much more restricted intervention thresholds. In fact, for this kind of production the presence of low density population of phytophagous insects feeding on drupes has important economic consequences because the fruit appearance deteriorates aesthetically. Consequently, the intervention threshold is represented by any initial attack menacing drupe integrity, and any kind of treatment is fundamental mainly where parasites find favourable environmental conditions.

The most dangerous parasite for table olive production is the olive fly (*Bactrocera oleae*). Control methods should be applied starting from the beginning of July, mainly for black table olives which require a long period before harvesting. The control period for green table olives is short because harvesting usually occurs in September. Olive fly control can be carried out by utilising control methods available for the control of olives for oil production, included preventive ones such as the planting of low susceptibility cultivars, biological and biotechnical control.

The timing of treatment can be defined by the field monitoring of preimmaginal stages (analysing a sample of drupes) and of the adult population (utilising chromotropic and pheromone traps). The choice of the active agents depends on the chosen management regime.

The most dangerous fungus for table olive production are *Camarosporium dalmaticum* (fruit rot), *Colletotrichum gloeosporioides* (anthracnose) and *Mycocentrospora cladosporioides*. *Spilocaea oleagina* can only occasionally affect fruits.

7.8. Viruses

The olive hosts several viruses throughout all studied olive areas of the World. To date, 15 viral species belonging to 7 different genus have been isolated from the olive (Table **8**). Among viral species, Arabis Mosaic Virus (ArMV), Strawberry Latent Ringspot Virus (SLRSV), Cherry Leaf Roll Virus (CLRV), Tobacco Mosaic Virus (TMV), Tobacco Necrosis Virus (TNVD), and Cucumber Mosaic Virus (CMV) are ubiquitous, polyphagous and of economic importance for other cultivations (tobacco, peach, strawberry, *etc.*). Other viruses to date have only been isolated from the olive: Olive latent ringspot virus (OLRV), Olive Latent Virus 2 (OLV-2), Olive Vein Yellowing Associated Virus (OVYaV), Olive Semilatent Virus (OSLV), Olive Yellow Mottling and Decline Associated Virus (OYMDaV), Leaf Yellowing-Associated Virus (OLYaV), Olive Mild Mosaic Virus (OMMV), and recently Olive Latent Virus 3 (OLV-3). Olive Latent Virus 1 (OLV-1) is an exception because it has also been isolated from citrus and tulips. Probably, some olive viruses also have other hosts.

Table 8: Viral species isolated from olive

Viral species	Genus	Geographical range
Strawberry latent ringspot virus (SLRSV)	*Nepovirus*	Italy (1979), Portugal, Spain, USA, Egypt
Arabis mosaic virus (ArMV)	*Nepovirus*	Italy (1979), Portugal, USA, Egypt

Table 8: contd...

Cherry leaf roll virus (CLRV)	*Nepovirus*	Italy (1981), Portugal, Spain, USA, Egypt
Cucumber mosaic virus (CMV)	*Cucumovirus*	Italy (1983), Portugal, Spain, USA
Tobacco mosaic virus (TMV)	*Tobamovirus*	Italy (1996)
Tobacco necrosis virus (TNV)	*Necrovirus*	Portugal (2002)
Olive latent virus 1 (OLV - 1)	*Necrovirus*	Italy (1984), Jordan, USA, Turkey, Egypt
Olive latent ringspot virus (OLRV)	*Nepovirus*	Italy (1983), Portugal
Olive latent virus 2 (OLV - 2)	*Oleavirus*	Italy (1984)
Olive vein yellowing associated virus (OVYaV)	*Potexvirus*	Italy (1995)
Olive yellow mottling and decline-associated virus (OYMDaV)	Unknown	Italy (1995)
Olive semilatent virus (OSLV)	Unknown	Italy (1996)
Olive leaf yellowing-associated virus (OLYaV)	*Closterovirus*	Italy (1998), Israel, Lebanon, USA, Egypt
Olive mild mosaic virus (OMMV)	*Necrovirus*	Portugal (2005)
Olive latent virus 3 (OLV-3)	Unknown	Italy (2010)

No data are available for the epidemiology of the olive and it is difficult to hypothesise the way utilized by viruses to reach and colonize plants. However, viruses penetrate the plant through systemic ways, and stay within the propagation material that become the main diffusion vehicle of viral agents. Propagation material has a primary epidemiological role as inoculum source locally for other species and on a larger scale for the diffusion of species representing an obstacle for commercialization of propagation material. In fact, some viruses are the object of quarantine in several countries outwith the European Union.

The high number of viral species on olive is probably due to several factors, and among them a primary role is played by the frequent latency of viral infections concealing epidemiologically healthy plants from infected ones.

Detected virosis are very few compared to the number of described viral species. In fact, only a few symptoms can be attributed to a specific virus, while several diseases have an unknown causal agent. The following symptoms can probably be attributed with certitude to viruses:

- Knobbly fruits: symptoms described in Italy on plants belonging to the cultivar Ascolana tenera infected by SLRSV, and in Portugal on the cv Negrinha. It is characterised by leaf (laciniature, reduction of surface) and fruit (smaller than usual, knobbly, deformed stones) alterations.

- Leaf yellowing complex: foliar yellowing is referred to symptoms characteristic for clear yellowing of leaf and scarce productivity of plants, sometimes accompanied by foliar necrosis and defoliation causing the wasting of plants. To these symptoms the presence of OVYaV, OLYaV and OYMDaV has been associated. Wasting accompanied with yellowing of nervures has been observed in Tuscany on several cultivars, from which TMV and OSLV have been isolated.

It was not possible to associate some symptoms (foliar malformations, hump of fruits, fissures of the bark) to viral agents.

7.8.1. Diagnosis

It is difficult to ascertain the presence of viruses in the olive due to scarce symptomatic reactivity, latency of infections and resemblance of symptoms [156]. Diagnosis through serological analyses is also difficult. Diagnosis of olive viruses is mainly based on molecular techniques (PCR).

7.8.2. Control, Sanitary Selection and Sanitation

The utilisation of health propagation material is important in order to obtain high quality olive production and is mandatory due to phytosanitary regulations. The use of health material for planting new orchards is linked to the market availability certified healthy plants. The contribution of thermotherapy for the sanitation of plants should be optimised. In fact, standardised and validated protocols are available only for other fruit plants and for the grapevine.

Some recent experiments have demonstrated that thermotherapy treatments at 38°C 3 to 4 months long on infected plants eliminate viruses such as CLRV from the vegetative apex, and that the *in vitro* culture of vegetative apices is the most promising technique for sanitation of plants infected by OLYaV.

Plants obtained from sanitary selection and/or sanitation (Primary sources) are the starting point for certified nursery productions. The use of virus-free propagation material is the most effective preventive control against systemic disease, especially viruses.

ACKNOWLEDGEMENTS

This work was supported by the Ministry of Economic Development, (Certificazione della composizione varietale, dell'origine geografica e dell'assenza di prodotti di sintesi negli oli extravergini di oliva – CERTOLIO project) and by the Ministry of Agriculture, Food and Forestry (Trattato internazionale FAO sulle risorse genetiche vegetali, III triennalità 2010–2013, RGV-FAO project).

CONFLICT OF INTEREST

The authors confirm that this chapter contents have no conflict of interest.

REFERENCES

[1] Simmonds N.W. Evolution of crop plants. Longman, London and New York, 1979; pp. 219-221.
[2] FAOSTAT 2012. http://faostat.fao.org/site/567/DesktopDefault.aspx?PageID=567#ancor
[3] Morettini A. Olivicoltura. REDA, Roma, 1972.
[4] Scaramuzzi F. Intensive olive-growing. In: Modern olive growing. FAO, Roma, 1977; pp. 61-84.
[5] Loussert R., Brousse G. L'olivier. G.P. Maison neuve et Larosc Eds., Paris, 1978; pp. 61-63.
[6] Loreti F., Pisani P.L. Lavorazioni del terreno negli arboreti. Riv. di Agron., 1986; 20(2,3): 134-152.
[7] Mikelakis N., Vougioukalou E. Water used, root and top growth of olive trees for different methods of irrigation and levels of soil water potential. Olea, 1988; 19: 17-31.
[8] Fernandez J.E., Moreno F., Martin-Aranda J. Study of root dynamics of olive trees under drip irrigation and dry farming. Acta Hort., 1990; 286: 263-266.
[9] Fernández J.E., Moreno F., Cabrera F., Arrue J.L., Martin-Aranda J. Drip irrigation, soil characteristics and the root distribution and root activity of olive trees. Plant Soil, 1991; 133: 239-251.
[10] Pisanu G., Corrias G. Osservazioni sui sistemi radicali dell'olivo in coltura intensiva. Ann. Fac. Agraria Sassari, 1970; 18: 3-19.

[11] Connel J.H., Catlin P.B. Root physiology and rootstock characteristics. In Olive Production Manual. University of California, 1994; 3353: 43-50.

[12] Tombesi A. Potatura e forme di allevamento nell'olivo. Frutticoltura, 1989; 1: 7-14.

[13] Bongi G., Palliotti A., Rocchi P. Aspetti della fisiologia dell'olivo. Frutticoltura, 1992; 11: 83-87.

[14] Grammatikopoulos G., Karabourniotis G., Kyparissis A., Petropoulou Y., Manetas Y. Leaf hairs of olive (Olea europaea) prevent stomatal closure by ultraviolet-B radiation. Austr. J. Plant Physiol., 1994; 21: 293-301.

[15] Bongi G., Mencuccini M., Fontanazza G. Photosynthesis of olive leaves: effect of light flux density, leaf age, temperature, peltates and H2O vapor pressure deficit on gas exchange. J. Amer. Soc. Hort. Sci., 1987; 112: 143-148.

[16] Lavee S. Biologia e fisiologia dell'olivo. In: Enciclopedia mondiale dell'olivo. COI, Madrid, 1996; pp. 61-110.

[17] Roselli G. Identificazione di cultivar di olivo da alcuni caratteri del polline. Riv. Ortoflorofrutt. It., 1979; 63: 435-445.

[18] Tombesi A. La fertilità nell'olivo. Riv. Ortoflorofrutt. It., 1978; 62: 435-450.

[19] Prevost G., Bartolini G., Messeri C. Cultivar italiane di olivo e loro sinonimi. Menegazzo Ed., Lucca, 1993; pp. 149-152.

[20] Bongi G., Palliotti A. Olive. In: Handbook of Environmental Physiology of Fruit Crops: Temperate Crops, vol. I. CRC Press, Boca Raton, Florida, 1994; pp. 165-187.

[21] Giorio P., Sorrentino G., D'Andria R. Stomatal behaviour, leaf water status and photosynthetic response in field-grown olive trees under water deficit. Environ. Exp. Bot., 1999; 42: 95-104.

[22] Sperry J.S. Evolution of water transport and xylem structure. Int. J. Plant Sci., 2003; 164: 115-127.

[23] Connor D.J. Adaptation of olive (Olea europaea L.) to water-limited environments. Aust. J. Agric. Res., 2005; 56: 1181-1189.

[24] Patumi M., D'Andria R., Fontanazza G., Morelli J., Giorgo P., Sorrentino G. Yield and oil quality on intensively trained trees of three cultivars of olive (Olea europea L.) under different irrigation regimes. J. Hortic. Sci. Biotechnol., 1999; 74: 729-737.

[25] Tognetti R., D'Andria R., Morelli G., Alvino A. The effect of deficit irrigation on seasonal variations of plant water use in Olea europaea L. Plant Soil, 2005; 273: 139-155.

[26] Chaves M.M. Effects of water deficits on carbon assimilation. J. Exp. Bot., 1991; 42: 1-16.

[27] Angelopoulos K., Dichio B., Xiloyannis C. Inhibition of photosynthesis in olive trees (Olea europaea L.) during water stress and rewatering. J. Exp. Bot., 1996; 47: 1093-1100.

[28] Goldhamer D., Dunai J., Ferguson L., Lavee S., Klein I. Irrigation requirements of olives trees and responses to sustained deficit irrigation. Acta Hort. 1994; 356: 172-175.

[29] Moriana A., Pérez-López D., Gómez-Rico A., De los Desamparados Salvador M., Olmedilla, N., Ribas F., Fregapane, G. Irrigation scheduling for traditional, low-density olive orchards: water relations and influence on oil characteristics. Agric. Water Manag., 2007; 87: 171-179.

[30] Lavee S., Woodner, M. Factor affecting the nature of oil accumulation in fruit of olive (Olea europaea L.) cultivars. J. Hort. Sci., 1991; 66: 583-591.

[31] Inglese P., Gullo G., Pace L.S. Summerdrought effects on fruit growth, ripening and accumulation and composition of 'Carolea' olive oil. Acta Hortic., 1999; 474: 269-273.

[32] Briccoli Bati C., Santilli E., Guagliardi I., Toscano P. Cultivation Techniques. In: Olive Germplasm: The Olive Cultivation, Table Olive and Olive Oil Industry in Italy. I. Muzzalupo (Ed.) 2012; InTech, Rijeka, Croatia. pp. 39-70.

[33] Gucci R., Cantini C. Pruning and training system for modern olive growing. CSIRO Publishing, 2000; pp. 93-124.

[34] Kay B. Rates of change of soil structure under different cropping systems. Adv. Soil Sci., 1990; 12: 1-41.

[35] Toscano P. Gestione del suolo: Ciclo delle lavorazioni, inerbimenti e controllo della flora spontanea degli oliveti. Conv. Rassegna olio qualità ambiente. Metodi e sistemi innovativi dell'olivicoltura biologica e sostenibile. Stato della ricerca e della sperimentazione. ISO Rende, 14-16 Apr. 1999, 2000; pp. 73-83.

[36] Dexter A.R. Soil physical quality: Part I. Theory, effects of soil texture, density, and organic matter, and effects on root growth. Geoderma, 2004; 120 (3-4): 201-214.

[37] Pastor M. Incidenza delle male erbe sull'evoluzione del contenuto d'acqua del suolo in oliveto di seccagno. Olivae, 1989; 28: 32-37.

[38] Stein O.R., Neibling W.H., Logan T.J., Moldenhauer W.C. Runoff and soil loss as influenced by tillage and residue cover. Soil Sci. Soc. Am. J., 1986; 50: 1527-1531.

[39] Rasiah V., Kay B. D. Runoff and soil loss as influenced by selected stability parameters and cropping and tillage practices. Geoderma, 1995; 68: 321-329.

[40] Raglione M., Toscano P., Angelini R., Briccoli Bati C., Spadoni M., De Simone C., Lorenzoni P. Olive yield and soil loss in hilly environment of Calabia (Southern Italy). Influence of permanent cover crop and ploughing. VI International meeting on soils with Mediterranean type of climate (IMSMTC). Barcellona (Catalonia), 4-9 Jul 1999. 2000; pp. 1038-1040.

[41] Briccoli-Bati C., Toscano P., Antonuccio S., Failla O. Effetto dell'inerbimento sullo stato vegeto-produttivo e nutrizionale dell'olivo. Inf. Agrario, 2002; 31: 43-46.

[42] Toscano P., Briccoli Bati C., Trombino T. Grass-cover effects on the vegetative and productive state of a young hilly olive-grove. Acta Hortic., 1999; 474 (1): 181-184.

[43] Toscano P., Casacchia T., Briccoli-Bati C., Scazziota B. Inerbimento temporaneo e permanente in olivicoltura. Inf. Agrario, 2006; 21: 36-39.

[44] Toscano P. The controlled grass-cover soil management in southern Italy olive orchard environments. 18th Symposium of the international scientific centre of fertilizers - more sustainability in agriculture: new fertilizers and fertilization management. Rome, 8-12 Nov 2009; pp. 381-386.

[45] Michelakis N.I.C., Vouyoucalou E., Clapaki G. Plant growth and yield response of the olive tree cv Kalamon, to different levels of soil water potential and methods of irrigation. Acta Hortic., 1994; 356: 205–210.

[46] Briccoli Bati C., Tocci C., Toscano P., Gazea F. Influenza di potatura ed irrigazione sullo stato nutrizionale dell'olivo. Conv. "Tecniche, Norme e Qualità in olivicoltura". Potenza (Italy), 15-17 Dec. 1993. 1995; pp. 305-316.

[47] Inglese P., Barone E., Gullo G. The effect of complementary irrigation on fruit growth, ripening pattern and oil characteristic of olive (Olea europaea L.) cv. Carolea. J. Hort. Sci., 1996; 71(2): 257-263.

[48] Pandolfi A., Proietti P., Famiani F., Guelfi P., Farinelli D., Sisani G., Tombesi A. Effetti del tipo di gestione del suolo e dell'irrigazione sull'attività vegeto produttiva dell'olivo in centro Italia. V giornate scientifiche SOI 2000, Sirmione (BS), 28-30 Mar, 2000; pp. 329-330.

[49] Gucci R., Lodolini E. M., Rapoport H.F. Productivity of olive trees with different water status and crop load. J. Hortic. Sci. Biotech., 2007; 82 (4): 648-656.

[50] Xiloyannis C., Dichio B. L'irrigazione sostenibile in frutticoltura. Ital. J. Agron., 2006; 3: 507-517.

[51] Goldhamer D., Dunai J., Ferguson L., Lavee S., Klein I. Irrigation requirements of olives trees and responses to sustained deficit irrigation. Acta Hortic., 1994; 356: 172-175.

[52] Blum A. Drought resistance, water-use efficiency, and yield potential - are they compatible, dissonant, or mutually exclusive? Austr. J. Agric. Res., 2005; 56: 1159-1168.

[53] Peterson G.A., Schlegel A.J., Tanaka D.L., Jones O.R. Precipitation use efficiency as affected by cropping and tillage systems. J. Prod. Agric., 1996; 9 (2): 180-186.

[54] Bacelar E.A., Moutinho-Pereira J.M., Goncalves B.C., Ferreira H.F., Correia C.M. Changes in growth, gas exchange, xylem hydraulic properties and water use efficiency of three olive cultivars under contrasting water availability regimes. Envir. Exp. Botany, 2007; 60 (2): 183-192.

[55] Bini G., Sapia L., Briccoli Bati C., Toscano P. Irrigazione dell'olivo con sistemi a microportata: risultati sulla cv Carolea in Calabria. Inf. Agrario, 1997; 22: 37-44.

[56] Nuzzo, V., Xiloyannis, C., Dichio, B., Montanaro, G., Celano, G. Growth and yield in irrigated and non-irrigated olive trees cultivar Coratina over four years after planting. Acta Hortic., 1997; 449: 75-82.

[57] D'Andria R., Lavini A., Morelli G., Patumi M., Terenziani S., Calandrelli D., Fragnito F. Effects of water regimes on five pickling and double aptitude olive cultivars (Olea europaea L.). J. Hortic. Sci. Biotech, 2004; 79: 18-25.

[58] Basta P., Toscano P., Ciliberti A., Turco D. Osservazioni pluriennali sulle risposte vegeto-produttive di cinque cultivar di olivo irrigate con acqua salmastra a differenti concentrazioni. Conv. Internaz. di Olivicoltura. Centenario I.S.Ol. Spoleto (PG) 22-23 Apr. 2002; pp. 235-240.

[59] Perica S., Goreta S., Selak G.V. Growth, biomass allocation and leaf ion concentration of seven olive (Olea europaea L.) cultivars under increased salinity. Scientia Hort., 2008; 117: 124-129.

[60] Petruccioli G., Parlati M.V. La concimazione dell'olivo. Terra e vita, 1983; 3: 55-59.

[61] Tombesi A.; Michelakis N., Pastor, M. Raccomandazioni del gruppo di lavoro di tecniche di produzione in olivicoltura e produttività. Olivae, 1996; 63: 38-51.

[62] Freeman M., Uriu K., Hartmann H.T. Diagnosing and correcting nutrient problems. in Olive production manual. Ferguson L., Sibbett G.S., Martin G.C. Eds., University of California, 1994; 3353: 77-86.

[63] Toscano P., Madeo A., Godino G. Risposte di tre cultivar di olivo a differenti tipi di concimazione in un ambiente litoraneo del Sud Italia. Conv. Internaz. di Olivicoltura. Centenario I.S.Ol. Spoleto (PG), 22-23 Apr. 2002; pp. 310-315.

[64] Hartmann H.T. Some responses of the olive to nitrogen fertilizers. Proc. Am. Soc. Hort. Sci., 1958; 72: 257-266.

[65] Cimato A., Marranci M., Tattini M. The use of foliar fertilization to modify sinks competition and to increase yield in olive (Olea europaea, cv Frantoio). Acta Hort., 1990; 286: 175-178.

[66] Cimato A., Marranci M., Tattini M., Sani G. La concimazione fogliare come mezzo per incrementare la produzione nell'olivo. Inf. Agrario, 1991; 14: 71-73.

[67] Inglese P., Gullo G., Pace L.S. Fruit growth and olive quality in relation to foliar nutrition and time of application. Acta Hort., 2002; 586: 507-509.

[68] Toscano P., Godino G., Belfiore T., Briccoli-Bati C. Foliar fertilization: a valid alternative for olive cultivar. Acta Hort., 2002; 594: 191-195.

[69] Toscano P. Foliar fertilization on olive growing: first results of a specifics foliar fertilizers applied on some cultivars in different environments of Southern Italy. FAO Land And Water Bulletin, 2005; 10: 127-130.
[70] Toscano P. Risposte di cultivar di olivo alla nutrizione per *via* fogliare e valutazione dell'efficacia nutrizionale di alcuni formulati specifici. Pgt RIOM, Tomo I, Olivicoltura, Dic. 2006. Guido Ed., Rende (CS), 2007; pp. 163-164. ISBN: 978-88-902732-7-8
[71] Toscano, P. Concimazione fogliare insieme a quella al terreno. Olivo&Olio, 2008; 3: 50-53.
[72] Toscano P., Godino G. Concimare le foglie per aumentare la resa. Olivo&Olio, 2010; 3: 40-43.
[73] Tombesi A., Pilli M., Boco M. Prime valutazioni su intensità e periodicità della potatura nell'olivo. Riv. di Frutticoltura, 2002; 10: 71-76.
[74] Tombesi A., Guarella P., Di Vaio C., Toscano P. Innovazioni nella meccanizzazione della raccolta e della potatura e ristrutturazione degli impianti olivicoli. Pgt RIOM, Tomo I, Olivicoltura, Guido Ed. Rende (CS), Dic. 2007; pp. 175-190. ISBN 978-88-902732-5-4.
[75] Tombesi A. Planting systems, canopy management and mechanical harvesting. OliveBioTech: Recent advances in olive industry. Mazara del Vallo (TP), 5-10 Nov 2006; pp. 307-316.
[76] Tombesi, A.; Farinelli, D.; Ruffolo, M., Sforna, A. Primi risultati sulla meccanizzazione agevolata ed integrale della potatura dell'olivo. Conv. COM.SI.OL "Competitività del Sistema Olivo in Italia", Spoleto (PG), 2008; pp. 141-147.
[77] Gucci R., Cantini C. Pruning and training system for modern olive growing. CSIRO Publishing, 2000; pp. 37-90. ISBN 0643-06443-5.
[78] Giametta G., Zimbalatti G. Mechanical pruning in new olive-groves. J. Agr. Eng. Res., 1997; 68: 15-20.
[79] Ferguson L., Krueger W.H., Reyes H., Metheney P. Effect of mechanical pruning on (Olea europea L.) California black ripe cv. 'Manzanillo' table olive. Acta Hort., 2002; 586: 281-284.
[80] Dias A.B., Peca J., Pinheiro A., Santos L., Morais N., Pereira A.G. The influence of mechanical pruning on olive production and shaker efficiency. Acta Hort, 2008; 791: 307-313.
[81] Peca J.O., Dias A.B., Pinheiro A.C., Santos L., Morais N., Pereira A.G., De Souza D.R. Mechanical pruning of olive trees as an alternative to manual pruning. Acta Hort., 2002; 586: 295-299.
[82] Pascuzzi S., Toscano P., Guarella A. La meccanizzazione della potatura di riforma degli oliveti. Atti Conv. Naz. A.I.I.A, III, V, VI Sez., Tecnologie innovative nelle filiere:orticola, vitivinicola, e olivicola-olearia. Pisa-Volterra, 5-7 Sept. 2007, Vol. III; pp. 77-80.
[83] Toscano P. Potatura meccanica solo ad anni alterni. Olivo&Olio, 2010; 3: 44-48.
[84] Farinelli D., Boco M., Ruffolo M., Tombesi A. Ottimizzazione del periodo di raccolta per aspetti agronomici, meccanici e di qualità dell'olio. Conv. Naz. "Maturazione e raccolta delle olive: strategie e tecnologie per aumentare la competitività in olivicoltura". Alanno, PE, 1 Apr. 2006; pp. 175-182.
[85] Toscano P. Come organizzare i cantieri per l'olivicoltura calabrese. Olivo&Olio, 2007; 11-12: 60-64.
[86] Briccoli-Bati C., Toscano P., Perrotta M.L. La raccolta meccanica in diverse tipologie olivicole calabresi. Conv. Naz. "Maturazione e raccolta delle olive: strategie e tecnologie per aumentare la competitività in olivicoltura". Alanno (PE) 1 Apr. 2006; pp. 95-99.
[87] De Simone G., Tombesi A. Influenza delle caratteristiche strutturali della pianta di olivo sulle rese di raccolta con l'uso di vibratori di tronco. Conv. Naz. "Maturazione e raccolta

delle olive: strategie e tecnologie per aumentare la competitività in olivicoltura". Alanno (PE), 1 Apr. 2006; pp. 63-68.

[88] Toscano P., Casacchia T. Un fitoregolatore per facilitare la raccolta meccanica delle olive. Inf. Agrario, 2006; 38: 61-63.

[89] Tombesi A., Tombesi S. Olive harvesting and mechanization. In "Production Techniques in Olive Growing". International Olive Council Ed. Madrid, 2007; pp. 317-346.

[90] Toscano P. Raccolta delle olive efficiente con macchine e cantieri adeguati. Inf. Agrario, 2010; 3: 82-84.

[91] Bellomo F., D'Antonio P. Using a grape harvester in super-intensive olive cultivation. J. of Agricultural Engineering, 2008; 1: 33-39.

[92] Arrivo A., Bellomo F., D'Antonio P. Raccolta meccanica dell'oliveto superintensivo. Inf. Agrario, 2006; 1: 68-71.

[93] Ravetti L. "How Continuous Mechanical Harvesting Transformed the Olive Industry." ISHS International Symposium on Mechanical Harvesting & Handling Systems of Fruits and Nuts. 2-4 April, 2012, Citrus Res. and Education Center. Univ. of Florida/IFAS, Lake Alfred, FL (USA).

[94] Iannotta N, Belfiore T, Brandmayr P, Scalercio S. The effects of treatments against the Bactrocera oleae (Gmel.) on the entomofauna of olive eco-system. II Eur. Meeting of the IOBC "Integrated protection of olive crops", Firenze 2005

[95] Altieri M, Nichols C, Ponti L. Biodiversità e controllo dei fitofagi negli agroecosistemi. Accademia Nazionale Italiana di Entomologia 2003.

[96] Arambourg Y. Entomologie oleicole. Cons. Oleic. Int. Madrid, 1986; pp. 1-360.

[97] Leandri A, Pompi V, Parlati MV, Iannotta N. Residui di fitofarmaci presenti sulle olive e nell'olio dopo trattamenti larvicidi al Dacus oleae (Gmel.). Atti Convegno Tecniche, norme e qualità in olivicoltura, Potenza: Italy 1993 (1995); pp. 815-822.

[98] Parlati MV, Iannotta N. La qualità dell'olio in rapporto alla difesa da Bactrocera oleae (Gmel.). Atti convegno Tecniche, norme e qualità in olivicoltura, Potenza, Italy 1993 (1995); pp. 445-458.

[99] Perri E, Iannotta N, Parlati MV, Zaffina F. Influenza dell'infestazione dacica sulla composizione dei composti fenolici di oli di cultivar calabresi. Atti convegno L'Olivicoltura mediterranea: Stato e prospettive della coltura e della ricerca, Cosenza: Italy 1995 (1996); pp. 555-560.

[100] Iannotta N, Lombardo N. Integrated control to Dacus oleae (Gml.): Influence of infestation level on oil quality. XXIII th. Int. Hort. Cong. Firenze, Italy 1990.

[101] Iannotta N, Perri L, Tocci C, Zaffina F. The behaviour of different olive cultivars following attacks by Bactrocera oleae (Gmel.). Acta Horticulturae 1999; 474: 545-548.

[102] Iannotta N, Monardo D, Perri E, Perri L. Comportamento di diverse cultivar di olivo nei confronti degli attacchi di Bactrocera oleae (Gmel.) e correlazione con la quantità di oleuropeina presente nelle drupe. Atti Conv. Biodiversità e sistemi ecocompatibili, Caserta: Italy 1999.

[103] Delrio G. Spring and summer releases of Opius concolor against Dacus oleae. Atti del Convegno IOBC Fruit Fly Open Meeting, Working Group on Fruit Flies of Economic Importance, Sassari: Italy 1990.

[104] Genduso P, Brancato C. Problems regarding the Opius concolor Szepl. rearing in laboratory. Atti del Convegno IOBC Fruit Fly Open Meeting, Working Group on Fruit Flies of Economic Importance, Sassari: Italy 1990.

[105] Delrio G, Lentini A, Bandino G, Sedda PG. Experiments on the control of olive by masstrapping in Sardinia. Atti del Convegno IOBC Fruit Fly Open Meeting, Working Group on Fruit Flies of Economic Importance, Sassari: Italy 1990

[106] Delrio G, Lentini A. Applicazione della tecnica delle catture massali contro il Dacus oleae in due comprensori olivicoli della Sardegna. Convegno "Olivicoltura", Firenze: Italy 1991 (1993); pp. 53-59.

[107] Haniotakis GE, Kozyrakis M, Bonatsos K. Area-wide management of olive fruit fly by feeding, attractans and sex pheromones on toxic traps. II Intern. Symp. Fruit Flies/Crete 1986; pp. 549-560.

[108] Belcari A. Possibilità di controllo della mosca delle olive mediante l'uso di sostanze ad azione battericida. Atti ARSIA "La sperimentazione nel settore fitosanitario in Toscana", Firenze: Italy 2000.

[109] Perri E, Iannotta N, Muzzalupo I, Rizzuti B, Russo A, Caravita MA, Pellegrino M, Parise A, Tucci P. Kaolin protects olive fruits from Bactrocera oleae Gmelin infestations unaffecting olive oil quality. II Eur. Meeting of the IOBC "Integrated protection of olive crops", Firenze, Italy 2005.

[110] Castoro V, Pucci C. Applicazione di un modello statistico di previsione della gravità dell'infestazione di Bactrocera oleae (Diptera, Tephritidae) nell'ambiente olivicolo materano: esperienze condotte nel biennio 1994-1995. Atti "Giornate Fitopatologiche", Numana, Italy 1996; Vol. 1, pp. 505-512.

[111] Raspi A. L'impiego di un modello previsionale nella difesa antidacica. Atti ARSIA "La sperimentazione nel settore fitosanitario in Toscana". Firenze, Italy 2000.

[112] Iannotta N, Giordano G, Rende G. La tignola dell'olivo in Calabria. Informatore agrario n. (1998) - 26: 69-73.

[113] Niccoli A, Tiberi R, Iannotta N. Andamento dell' infestazione di Prays oleae Bern. in oliveti della Calabria durante il biennio 1983-84. Redia 1984; 67: 483-492.

[114] Niccoli A, Tiberi R, Iannotta N, Lombardo N. Andamento della popolazione ed incidenza dei danni causati da Prays oleae (Bern.) in un biotopo olivicolo calabrese. Annali dell'Istituto Sperimentale per l'Olivicoltura 1984/86; 8: 181-189.

[115] Lentini A, Delrio G. Osservazioni sulla resistenza di alcune cultivar di olivo all'attacco della generazione carpofaga del Prays oleae. Atti Conv. Biodiversità e sistemi ecocompatibili, Caserta, Italy 1999.

[116] Raspi A, Mattedi L, Michelotti F. La cocciniglia mezzo grano di pepe dell'olivo. Terra Trentina 1993; 7: 29-31.

[117] Neuenschwander P. Contribution to the definition of a metodology for the study of the population dynamics of Saissetia oleae (Oliv.). Olea (FAO), 1978; pp. 50-56.

[118] Lombardo N, Rizzuti B. Osservazioni sulla Zeuzera pyrina in un areale olivicolo calabrese e prove di lotta. Annali dell'Istituto Sperimentale per l'Olivicoltura 1987; 9: 39-51.

[119] Prota U. Le malattie dell'olivo e relativa difesa. Atti Conv. L'olivicolura mediterranea, Cosenza, Italy 1996; pp. 451-471

[120] Surico G. Malattie crittogamiche dell'olivo: epidemiologia e prospettive di lotta. International course on olive growing, Scandicci, Italy 1998; pp. 231-237.

[121] Varvaro L, Surico G. Comportamento di diverse cultivars di olivo (Olea europaea L.) alla inoculazione artificiale con Pseudomonas savastanoi (E.F.Smith) Stevens. Phytopath medit 1978; 17: 174-177

[122] Iannotta N, Monardo D, Noce ME, Perri L. Susceptibility of olive genotypes to Pseudomonas savastanoi (Smith). II Eur. Meeting of the IOBC Integrated protection of olive crops, Firenze, Italy 2005.

[123] Frisullo S, Rossi V, Lops F. Spilocaea oleagina: epidemiologia e lotta. Atti Conv. Innovazioni e prospettive nella difesa fitosanitaria, Ferrara, Italy 1994; pp. 233-236.

[124] Scarito G. Influenza dei fattori ambientali sulla sporogenesi di Spilocaea oleagina (Cast.) Hugh. Conv. Innovazioni e prospettive nella difesa fitosanitaria, Ferrara, Italy 1994; pp. 245-246.

[125] De Marzo L, Frisullo S, Lops F, Rossi V. Possible dissemination of Spilocaea oleagina conidia by insects (Ectopsocus briggsi Mac Lachlan). Bull. OEPP/EPPO 1993; 23: 389-392.

[126] Morettini A. Influenza della defogliazione anticipata su la fioritura e la fruttificazione dell'olivo. Ann. Sper. Agraria 1951; N.S. 5,3: 457-479.

[127] Iannotta N, Monardo D, Perri L. Effects on different treatments against Spilocaea oleagina on olive trees. Atti I.S.H.S. "Olive 2000", Valenzano, Italy 2000.

[128] Varvaro L, Balestra GM, Contu G, Pugliese F. Possibilità di lotta contro alcune crittogame dell'olivo mediante l'impiego di batteri epifiti. Atti Conv. Innovazioni e prospettive nella difesa fitosanitaria, Ferrara, Italy 1994; pp. 259-262.

[129] Iannotta N, Fodale AS, Mulè R, Tocci C, Tucci A, Zaffina F. Valutazione della suscettibilità di diverse cultivar di olivo nei confronti di Spilocaea oleagina (Cast.) Hugh. Atti Conv. Biodiversità e sistemi ecocompatibili, Caserta Italy 1999.

[130] Iannotta N, Monardo D. Suscettibilità di cultivar di olivo a Spilocaea oleagina (Cast.) Hugh. e correlazione con il contenuto di oleuropeina nelle foglie. Conv. Germoplasma Olivicoli e tipicità dell'olio, Perugia, Italy 2004: pp. 216-220

[131] Scarito G. Osservazioni sulla suscettibilità all'occhio di pavone di varietà di olivo in vivaio. Tecnica Agricola 1985; 37: 163-167.

[132] Cirulli M, Ciccarese F, Amenduni M. Recenti acquisizioni sulla verticilliosi dell'olivo. Atti Conv. Tecniche, norme e qualità in olivicoltura, Potenza, Italy 1993.

[133] Zazzerini A, Tosi L. Ricerche epidemiologiche su Verticillium dahliae e Phoma incompta su olivo nell'Italia centrale. Atti Conv. Innovazioni e prospettive nella difesa fitosanitaria, Ferrara, Italy 1994; pp. 253-258.

[134] Iannotta N, Fodale AS, Mulè R, Rinaldi R. Indagine sulla diffusione della verticilliosi dell'olivo (Verticillium dahliae Kleb.) in Calabria e Sicilia occidentale. Atti Conv. Tecniche, norme e qualità in olivicoltura, Potenza, Italy 1995; pp. 801-806.

[135] Thanassoulopoulos CC, Biris DA, Tjamos EC. Weed hosts as inoculum source of Verticillium in olive orchards. Phytopath medit 1981; 20: 164-168

[136] Fodale A, Mulè R, Ianotta N, Tucci A. Prove di lotta contro la tracheoverticilliosi (Verticillium dahliae Kleb.) mediante iniezione al tronco di Fosethyl-aluminium. Atti Conv. Tecniche, norme e qualità in olivicoltura, Potenza, Italy 1993; pp. 603-610

[137] Pennisi AM, Cacciola SO, Magnano di San Lio G, Perrotta G. Evaluation of the susceptibility of olive cultivars to verticillium wilt. Bull. OEPP/EPPO 1993; 23: 537-541.

[138] Iannotta N. La difesa fitosanitaria. In: Fiorino P, Ed. Olea Trattato di olivicoltura, ed. Edagricole 2003; pp. 393-407.

[139] Iannotta N. I punti deboli dell'olivicoltura biologica. Informatore Agrario 2005; 42: 107-111.

[140] Cirio U. La mosca delle olive: ecologia e tecniche di lotta. Olivo e olio 1998; 4: 38-49.

[141] Iannotta N. La difesa dell'olivo dai principali fitofagi con metodi conformi al Reg. 2092/91. in "La difesa dai fitofagi in condizioni di olivicoltura biologica". Ed. Accademia Nazionale dell'olivo e dell'Olio, Spoleto, Italy 2004: 49-62.

[142] Iannotta N. La raccolta anticipata contro la mosca delle olive. Informatore Agrario 1996; 34: 73-78.

[143] Iannotta N. Influenza del genotipo sugli attacchi della mosca delle olive (Bactrocera oleae Gmel.). Atti conv. Accademia dei Georgofili Il controllo della mosca delle olive: Stato attuale delle ricerche e nuove strategie, Bosa, Italy 2006.

[144] Raspi A. Lotta biologica in olivicoltura. Atti convegno Tecniche, norme e qualità in olivicoltura, Potenza, Italy 1993 (1995); pp. 483-495.

[145] Rugini E, Caricato G. Biotecnologie in olivicoltura: presente e prospettive per il futuro. Atti Conv. L'olivicoltura mediterranea, Cosenza, Italy 1995; pp. 227-238.

[146] Cirio U. Agrochimici ed impatto ambientale in olivicoltura. Olivae 1997; 65: 32-39

[147] Iannotta N, Perri L. Prova di mass-trapping nel controllo di Dacus oleae (Gmel.) in Calabria. MAF Convegno Olivicoltura, Firenze, Roma, Italy 1991 (1993); pp. 87-92.

[148] Broumas T, Haniotakis G, Yamvrias C, Stavrakis G. Comparative study of a mass trapping method and various bait sprays for the control of the olive fruit fly- First year resuits. In: Casida JE, Ed. Pesticides and alternatives, 1990; pp. 205- 215.

[149] Iannotta N, Perri L, Rinaldi R. Control of the olive-fly by mass-trapping in Calabria. Second International Symposium on Olive Growing. Israele 6-10 settembre 1993, Acta Horticulturae 1994; 356: 411-413.

[150] Iannotta N, Noce ME, Perri E, Perri L, Scalercio S, Vizzarri V. Effetti di trattamenti con caolino contro Bactrocera oleae (Gmel.) sulla maturazione delle olive. Atti conv. Maturazione e raccolta delle olive, Alanno Italy 2006.

[151] Loreti F, Guerriero R, Triolo E, Vitagliano C. Proposta di metodo per la selezione clonale e sanitaria in olivicoltura. Olivae 1993; 47: 60-66.

[152] Viggiani G. La difesa integrata dell'olivo: attualità e prospettive. Informatore fitopatologico 1989; 2: 23-32.

[153] Iannotta N, Parlati MV. Lotta integrata in olivicoltura. Atti P.F. Lotta biologica ed integrata, Ferrara, Italy 1994; pp. 181-184.

[154] Iannotta N. Metodologie per il monitoraggio di adulti di Dacus oleae (Gmel.). Atti del Convegno "Lotta Integrata in Olivicoltura". Firenze, Italy 1991; pp. 25-31.

[155] Frisullo S, Garau R, Laviola C, Lops F, Manici LM, De Marzo L, Nicoletti R, Prota U, Prota VA, Rossi V, Scarito G, Tosi L, Zazzerini A. Problematiche patologiche in olivicoltura. Atti Conv. Lotta integrata in olivicoltura, Firenze, Italy 1991 (1993).

[156] Heriques MIC, Rei FT, Leitao FA, Serrano JF, Potes MF. Virus deseases in Olea europaea L. cultivars. Immunodiagnosis of SLRV. Phytopath medit 1992; 31: 127-132.

Chapter 2: Omics Approaches for the Characterization and Valorisation of Olive Varieties

CHAPTER 2

Omics Approaches for the Characterization and Valorisation of Olive Varieties

Adriana Chiappetta[1,*], Leonardo Bruno[1] and Innocenzo Muzzalupo[2]

[1]Università della Calabria (University of Calabria – DiBEST), 87036 Arcavacata di Rende (CS), Italy and [2]Consiglio per la Ricerca e la sperimentazione in Agricoltura (Agricultural Research Council, Olive growing and olive oil industry research centre, CRA-OLI), C.da Li Rocchi, 87036 Rende (CS), Italy and [2]Agricultural Research Council - Olive growing and Olive Oil Industry Research Centre (CRA-OLI), C.da Li Rocchi, 87036 Rende (CS), Italy

Abstract: The olive (*Olea europaea* L.) is one of the most ancient cultivated fruit trees. Olive trees are considered to be one of the most widely grown fruit crops in the countries of the Mediterranean basin. Olive products, such as olive oil, table olives and olive pastes are staple foods of the Mediterranean diet due to their benefits for human health, as well as other applications such as in cosmetics.

More than 2600 olive plant cultivars have been described using morphological analysis, although many of them might be synonyms, homonyms, ecotypes or the result of crosses between neighbouring olive cultivars. The high number of olive cultivars causes a considerable problem in germplasm collection management and both the traceability and authenticity of olive oils produced, once there is an uncertainty about its correct olive cultivar denomination. Until recent years, cultivar identification was only based on morphological and agronomic traits. However, recognition of olive cultivars based on phenotypic characters was revealed to be problematic, especially in the early stages of tree development. In recent years molecular markers have been applied in olive germplasm to identify cultivars and to determine the relationships between cultivars.

The increasing openness of genetic markers in olive trees allows detailed studies and evaluations of genetic diversity. This will provide a view of what has been attained and what still needs to be done in order to better understand this crop that has lived for centuries and still remains to be fully discovered and understood.

Although current breeding strategies can now benefit from the availability of new polymorphic genetic markers, the characterization of olive germplasm is still far from complete.

A wider gene characterization of loci related to the quality of plant products and adaptive mechanisms could provide new information and tools to support Marker

***Corresponding author Adriana Chiappetta:** University of Calabria – Dep. of Biology, Ecology and Earth Science (DiBEST), Arcavacata di Rende (CS), Italy; Tel: +390984492963; Fax +390984492964;
E-mail: a.chiappetta@unical.it

Innocenzo Muzzalupo & Sabrina Micali (Eds.)

Assisted Selection (MAS) strategies and new biotechnological approaches to develop suitable growing techniques and increase productivity and product quality of this species which is unique in its kind.

Keywords: Abiotic stress, DNA sequencing, *ex situ* collections, expression sequence tags, gene identification, genomic approaches, marker assisted selection, molecular markers, next generation sequencing approaches, olive breeding, olive germplasm, plant breeding.

1. INTRODUCTION TO PLANT BIOLOGY

The major challenge in plant biology is currently the functional elucidation of unknown genes. Generally, the classic approach based on the availability of plant lines in which specific genes are either over expressed or knocked out is useful for elucidating the functions of specific genes through an analysis of the phenotype. However, an evident morphological alteration of the phenotype is always observed for genes involved in cell growth and development while it is not always observed in the case of genes that code for enzymes involved in metabolic pathways [1].

This approach requires knowledge of the entire genome of the organisms studied. The complete genomes of a limited number of organisms are currently known. The sequencing era of genomes started at the end of the last century and *Caenorhabditis elegans* was the first genome to be sequenced in 1998 [2]. This was followed by the sequencing of *Drosophila melanogaster* [3].

In the field of plant biology *Arabidopsis thaliana* was the first plant whose genome was sequenced in 2000 [4] with the objective of identifying and characterizing the structure, function, and regulation of *Arabidopsis* genes and developing technologies for genome studies. After the success of the Multinational *Arabidopsis thaliana* Genome Research Project to complete the sequencing of the *Arabidopsis* genome in 2000, the ambitious goal to determine the function of every *Arabidopsis* gene by the year 2010 was set by a new Project: the Multinational Coordinated *Arabidopsis thaliana* Functional Genomics Project. Numerous international laboratories have taken part in this project and very large datasets and resources have been generated leading to breakthroughs in understanding of the fundamental processes underlying plant growth development and responses to the environment. Studies on *Arabidopsis* have provided an understanding of the importance of plant genome analysis in order to understand the fundamental principles of genome, transcriptome and metabolome [5].

Knowledge of genomes belonging to different organisms allows for the annotation of functionally new and specific genes on the basis of sequence similarities, evaluated by means of specific software analysis programmes. In pluricellular organisms, the revealing of temporal and spatial gene expression provides information on the putative sites of activity of the proteins codified by the cells and contributes to the explanation of the role of specific genes in the organism growth and development processes.

Arabidopsis is a model organism in biology and its genome-related resources including the whole-genome sequence with functional annotations, DNA microarrays, DNA-tagged insertion mutants and metabolic maps (AraCyc; http://www.arabidopsis.org/tools/aracyc) are readily accessible [6]. In such a context, the knowledge acquired about *Arabidopsis* and a limited number of other species such as rice (*Oryza sativa*), poplar (*Populus trichocarpa*) and grapevine (*Vitis vinifera*) represents a starting point to decode the gene functions of other non – model plants owing to the lack of whole-genome sequence information. In particular, genome level studies, integrated with transcriptome, proteome and metabolome approaches, correlating the expression pattern of specific genes with the accumulation pattern of all metabolites involved in a specific pathway, should help in deciphering the functions of specific genes not just in model plants but also in crops such as the olive.

Changes in transcriptome and metabolome profiles can be investigated under different conditions such as different tissue and physiological changes as well as stress conditions.

2. OLIVE TREES

The olive tree (*Olea europaea* L. subsp. *europaea* var. *europaea*), with over 5000 years of history cultivation, and literature devoted, is considered one of the oldest known cultivated tree in the Mediterranean basin. It belongs to the Oleaceae family, which includes 600 species within 25 genera and is widely distributed on all continents.

In addition to the genus *Olea* it also contains the genera *Forestiera*, *Forsythia*, *Fraxinus*, *Ligustrum*, and *Syringa*. The genus *Olea* of the sub-family *Oleideae*, includes two sub-genera, *Olea* and *Paniculatae.* According to recent revisions of the *Olea europaea* taxonomy [7], this species is divided into the following six sub-species based on geographical distribution:

1. Subsp. *europaea*, divided into the two botanical varieties: the wild olive or oleaster (var. *sylvestris*) and the cultivated olive (var. *europaea*), distributed in the Mediterranean Basin.

2. Subsp. *cuspidata*, distributed from South Africa to southern Egypt and from Arabia to northern India and south-west China.

3. Subsp. *maroccana*, present in south-western Morocco.

4. Subsp. *cerasiformis*, present on the Island of Madeira.

5. Subsp. *guanchica*, present in the Canary Islands.

6. Subsp. *laperrinei*, localized in the Sahara region.

Over the last few centuries, cultivation of the olive tree has spread to North and South America, as well as Japan, South Africa, and Australia [1].

Due to the tree's need for a warm but not excessively hot climate, it can be cultivated in both the northern and southern hemispheres between 30 and 45 degrees latitude, with the exception of some equatorial regions where olive trees are grown at high altitude. Nowadays, olives are produced in more than 40 countries spread across all six inhabited continents, and even in exotic places like Hawaii [1].

Native to the Mediterranean regions, *Olea europaea* is the only species within the genus *Olea* able to produce edible fruits [8]. *Olea europaea* is a diploid species

(2n=2x=46). The nuclear DNA content of *Olea europaea* cultivars ranged between 2.90 ± 0.020 pg/2C and 3.07 ± 0.018 pg/2C while the genome size of the wild olive was estimated at 3.19 ± 0.047 pg/2C DNA [9].

Over millennia, new varieties have originated by genetic mutation and by spontaneous crossing with a subsequent natural dissemination of stones. Sexual reproduction involving populations of local wild *Olea* and those selected according to the criteria of local farmers was also an important factor in the development of locally specific varietal populations [1, 10], while many new varieties were established by vegetative means. The longevity of the olive tree and the selection of a large number of varieties have contributed to the conservation of its variability and allowed a large proportion of this genetic diversity to pass on [1, 11]. Another factor that has contributed to increasing biodiversity of this species is the wide genetic variability of olive that has been freely created and distributed without any concern for loyalty to a morphologically defined archetype because the end product is not the whole fruit, as for most other fruit trees, but the result of squeezing the fruit: virgin olive oil. This has led, over time, to the formation of heterogeneous phenotype polyclonal varieties f (varieties–populations) rather than the formation of monoclonal varieties. Intra-varietal polymorphisms in fact, have been reported in the literature [12-14] in which the observed differences within the same variety have been suggested as somatic mutations occurring during vegetative propagation.

2.1. Olive Germplasm: *Ex Situ* Collections

The *ex situ* collections of olive tree germplasm may provide either a source of genes which can be potentially useful as raw material in plant breeding, or plants which are directly valid for sustainable production. With regards to the latter item, reference is made to local varieties that, having evolved for a very long period in a set location, and having developed adaptative traits which are well integrated with the agronomic, environmental, cultural and traditional features of the site and which, more or less recently, have been replaced with new varieties. The needs of modern agriculture, call for the cultivation of a wider range of diverse material that could better respond to the different aspects involved. Specifically, if it is necessary to obtain new varieties with a broader genetic base, which are capable of producing under diverse conditions and of responding to different stresses – *i.e.* pests, drought, low soil fertility, *etc.* –, on the other hand, in some cases, the reintroduction of old local varieties and the safeguarding of traditional farming systems and landscapes, can be very profitable from economic and socio-economic points of view [1]. In general, the lack of information about plant genetic resources has the effect of limiting their use, restricting both the value and the usefulness of a collection even within the owning institute and among other potential users. Hence, the characterisation of the germplasm conserved in a *ex situ* collection is an essential prerequisite to a proper and wide utilization of the conserved plant material and it is the first step toward the definition of the roles that varieties can play in sustainable production, through the direct use or in breeding programmes [8].

Germplasm *ex situ* collections established and maintained by genebanks provide the present and future utilization of plant genetic resources. In the early stages of collection development, the focus was mainly on acquisition *per se*, and less on optimizing collection composition. Many germplasm collections were started from working collections that had been used to support specific purposes, including crop improvement, breeding and taxonomic studies. In many cases, germplasm collections expanded their collections thereafter by including obsolete varieties, research lines or samples obtained from collecting missions to natural distribution areas of crops and their wild relatives. There is still a need for greater rationalization among collections globally [15]. While this approach to the conservation and use of plant genetic resources for food and agriculture is becoming increasingly mainstreamed within national programmes, further efforts are needed in this regard. With the development of molecular markers such microsatellite (SSR) and single nucleotide polymorphisms (SNPs), the amount of data available on genetic diversity has increased dramatically, leading to an improved understanding of issues such as domestication, genetic erosion and genetic vulnerability [15].The largest total numbers of *ex situ* varieties are of wheat, rice, barley and maize accounting for 77% of the total cereal and pseudo-cereal holdings. Other important collections of olive trees are found in several Mediterranean countries: at the Aegean Agricultural Research Institute of Turkey (AARI, Turkey), Agricultural Research Council - Olive growing and Oil Industry research centre *(Consiglio per la Ricerca e la Sperimentazione in Agricoltura - Centro di Ricerca per l'Olivicoltura e l'Industria Olearia*, CRA-OLI, Italy), Horticulture and Subtropical Crops Research Institute (HSCRI, Azerbaijan), *Junta de Andalucía, Instituto Andaluz de Investigación Agroalimentaria y Pesquera, Centro de Investigación y Formación Agroalimentaria Córdoba* (CIFACOR, Spain) and the National Plant Gene Bank of Iran was placed in the Seed and Plant Improvement Institute (NPGBI-SPII, Iran). The largest olive collection (accounting for 17 percent of the total olive trees with more than 500 varieties) is held by CRA-OLI in Italy, followed by the collections of CIFACOR in Spain [16].

The systematic collection of Italian olive varieties for deposit into specific catalogue fields began in Italy in the 1980s. A similar international collection was begun in 1997 by CRA-OLI of Rende, Italy. Collection entailed the following steps: a survey of the territory, individuation, basic characterization, and introduction into the gene bank field. Material identified by other international scientific institutions (International Treaty on Plant Genetic Resources for Food and Agriculture - Plant Genetic Resources RGV-FAO Projects) was also included [16]. The largest *ex-situ* olive germplasm collection consisting of approximately 500 Italian olive varieties, and corresponding to 85% of the total Italian olive germplasm and to more than 18% of the total world olive germplasm, is maintained at the CRA-OLI [16], and this list has been published (web site http://www.certolio.org/data-base-molecolare/).

A useful *ex situ* olive germplasm collection also requires an organizational system devoid of synonymy, homonymy, and mislabelling so that a reliable classification of all varieties can be achieved without unnecessary confusion. Recent research has focused on using biochemical and morphology and molecular markers to characterize olive varieties. The identification of varieties and varieties using molecular markers is a crucial aim of modern horticulture since such a technique would greatly facilitate breeding programmes and germplasm collection management (Fig. **1**) [1, 15].

Figure 1: Olive germplasm collection of Italian olive varieties (CRA-OLI Mirto-Crosia (CS), photo by Godino G.).

2.2. Characterization of the Olive Germplasm

The problem of characterizing the olive tree germplasm is complicated not only by the richness of its genetic patrimony, but also by the absence of reference standards and a well-defined system of nomenclature that is free from synonymy and homonymy [16, 17]. There is still no "standard reference variety" for olive varieties [18] and only recently, some research Italian projects (*i.e.,* "International Treaty on Plant Genetic Resources for Food and Agriculture - Plant Genetic Resources, RGV-FAO", "Improvement and qualification of nursery olive, OLVIVA", "Certification of varietal composition, geographical origin and the absence of synthetic products in the extra virgin olive oils, CERTOLIO") have addressed this issue and are trying to achieve a "standard certificate" for each variety present in different Italian regions [15]. The extent of this diversity has important implications for both the adaptation of varieties to their local environment and for the optimization of the agronomical performance of these varieties under a given set of environmental conditions. For example, each initiative promoting olive cultivation should consider the potential repercussions of such action on any local olive varieties. Every region should preserve its own plant material in order to safeguard both the adaptation and productivity of the species and the unique characteristics of the region's olive oil. However, the study of intra-varietal polymorphisms is important since they may have traits which, although they may not have been considered important in the past, might be important in meeting the challenges of modern olive growing (*i.e.*, salinity tolerance, resistance to low temperatures, *etc.*). The preliminary work performed in olive tree genomics is currently very far from producing results that are useful for selecting new varieties using molecular tools [19]. This combined with the general lack of prior knowledge regarding cultivated and wild olive germplasms, has mainly focused attention on germplasm evaluation. There is a strong need for a reliable means of identifying different olive tree varieties, partly because so many of these varieties are propagated solely *via* vegetative methods. This would also be of substantial benefit to nurserymen and growers, as the cost of plants represents the major investment in establishing new orchards. At the same time, it is also important to improve the *ex situ* plant germplasm collection in order to adequately characterize all varieties, and to develop future breeding programs [15].

Biological and morphological characteristics are widely used for descriptive purposes and are commonly used to distinguish olive varieties [15, 19-22]. Agronomic characterization has also aided in the classification of different olive varieties [15, 19, 21]. Morphological characterization of olive varieties is potentially unreliable because environmental factors strongly influence the plants' morphology. Despite this drawback, the age of trees, their training systems, and the phenological stage of the plants continues to be a key preliminary step in olive tree germplasm description and classification. Improving *ex-situ* olive plant germplasm collections continues to be an important objective, which will ultimately prove useful for characterizing all varieties and for developing future breeding programs [15, 22].

A multiplicity of molecular markers, has been used recently to characterize and distinguish between olive varieties. In light of these efforts, some combinations of enzymatic markers with distinct physiological, morphological, and agronomic characteristics may ultimately provide a method for the reliable and systematic classification of olive tree varieties [15, 23]. Assessments of RAPD profiles, SSR markers, RFLPs and AFLPs, provide direct genotypic information, which has numerous valuable applications in genetic studies. The main advantages of generating RAPD profiles are the technique's simplicity and low cost [15, 24-28]. Nevertheless, RAPD experiments demonstrate poor reproducibility, which hampers comparison between individual studies. Experiments assessing an organism's AFLP markers are more technically demanding than RAPD but are highly effective in detecting DNA polymorphisms [28-30]. In contrast to a plant species' chloroplast DNA (cpDNA), which can occasionally be insufficiently variable for intra-species comparison [31-33], mitochondrial DNA (mtDNA) within a given species varies enormously in terms of organization, structure, size, and gene arrangement [34]. As a result, intraspecies mtDNA variation is common in plants, especially in naturally occurring populations [33]. Taken together, these distinctive features make mtDNA sequencing a powerful tool for analysing a given plant population's genetic structure and phylogenetic relationships [35]. SSR markers are ubiquitous, abundant, and highly dispersed in eukaryotic genomes, but are costly to assess experimentally. Once these markers have been ascertained, the data can be readily shared among laboratories. However, since not all microsatellites are identical [36-42] successful utilization of known microsatellite markers requires prior information regarding the characteristics of a particular genetic locus [15, 36]. Internal transcribed spacer 1 (ITS-1) sequences, RAPD profiles, and inter-SSR (ISSR) markers have been employed to evaluate the colonization history of *Olea europaea* [43]. A number of *Olea europaea* retroelements have also been identified [44], and their copy number has been estimated [45]. Using previously established RAPD profiles [44, 46] SCAR markers linked to leaf peacock spot tolerance were developed. Another method to distinguish inter-variety variability and to characterize clonal variants using single nucleotide polymorphisms (SNPs) in the olive tree genome is also currently being developed [47, 48]. All the aforementioned genetic techniques provide useful information regarding the level of olive tree polymorphism and diversity, which demonstrates their utility for the characterization of olive germplasm varieties [15, 49].

2.3. Olive Edible Products

The olive fruit is a drupe, which is botanically similar to the apricot, almond, peach, nectarine, cherry, and plum. The olive fruit consists of an endocarp, a mesocarp and an exocarp (Fig. **2**). The endocarp is woody and represents 13-24% of the total fruit and encloses the seed (2-4% of the total fruit). The mesocarp represents 70-80% of the total fruit, it is the tissue that is eaten. The exocarp represents 1.5-3.5% of the total fruit; it is free of hairs and contains stomata.

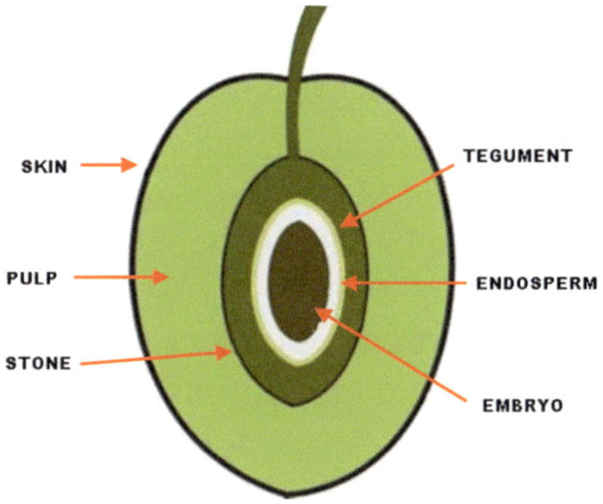

Figure 2: Schematic representation of olive fruit (drupe).

Quantitatively, the largest constituents of the drupe are water (40-70%) and oil (6-25%). The biochemical composition of olive oil consists of a major portion that includes triacylglycerols and represents more than 98% of the total oil weight and a minor one, that is present in a very low amount (about 2% of oil weight), including more than 230 chemical compounds such as sterols, hydrocarbons, aliphatic and triterpenic alcohols, volatile compounds and antioxidants (tocopherols and phenolic compounds).

The quality of its products, olive oil and table olives, is highly dependent on the agronomic and organoleptic characteristics of its drupes, which vary both in relation to the genetic traits, varieties, the stage of ripeness, as well as to the different susceptibility to environmental growth conditions [50, 51].

The genuineness of olive oil is an important aspect within the "Mediterranean diet"; several research and epidemiological studies have linked the healthy aspects of the main components of this diet, and especially olive oil, to the manifested protective effect against vascular disease and the onset of cancer. In the case of olives and olive oil these features correlate to the high percentage of monounsaturated fats as well as to a high content of antioxidant compounds such as phenols and tocopherols, which together with other components characterize the nutraceutical profile of olive products [52-59]. In this respect, phenolic compounds represent a complex mixture in olive derived products, that are responsible for the anti-atherogenic and anti-cancerogenic effects, and for its antioxidant properties [60-63]. Despite the importance and uniqueness of olive products, the current breeding programs for this species are still long-term and linked to the long juvenile developmental phase (20-25 years) and its intrinsic self-incompatibility mechanisms. Although the current breeding strategies can now benefit from the availability of new polymorphic genetic markers, olive germplasm characterization is still far from complete [36, 42, 64].

Therefore it is of prime importance to channel research programs towards innovative improvement strategies to support conventional programs. In particular, a wider gene characterization of *loci* related to plant product quality and adaptive mechanisms could provide new information and tools to support Marker Aided Selection (MAS) strategies and new biotechnological approaches to develop suitable growing techniques and increase productivity and product quality of this unique species.

3. OLIVE TREES AND STRESS

3.1. Abiotic Stress

Environmental *stress* refers to potentially dangerous actions which are connected to extreme climatic and soil conditions which can introduce modifications in the morphology, physiology and metabolism of the plant or its organs. Both adaptive and non-adaptive responses, which occur consequent to a *stress*, tend to alleviate the effects and eventual damage caused to the plant. Although environmental *stress* includes a wide range of potential agents, this chapter will only address those of greater economic importance in olive-growing addressed, such as *stress* caused by water availability and soil salinity, as well as *stress* from temperature and atmospheric pollutants. The olive does not tolerate either low temperatures or root asphyxiation well, but it is resistant to water shortage, to salinity and to high temperatures. These characteristics are the evolutionary result of adaptation processes to the Mediterranean climate, which are marked by mild winters and by a long dry period and high summer temperatures.

There is very little available in literature regarding resistance to cold, to drought and to salinity. The genetic mechanisms which regulate these properties in olive plants are still prevalently unknown, while some information is available on the physiological processes which govern *stress* resistance mechanisms. A better knowledge of *stress* resistance mechanisms, as well as the selection of tolerant genotypes, is fundamental in order to increase these properties in the olive plant.

3.1.1. Drought Stress

Water is the main factor limiting growth and productivity in the typical areas where olive plants are cultivated. The visible symptoms of a lack of water in the olive plant are not easily distinguishable from those caused by other kinds of *stress* and appear rather late, when the water deficit is already at an advanced state. Observation of symptoms on adult leaves is rare at the initial stages of a deficit, while the shoot tips and leaves that have not completely opened can show loss of turgor during the hottest hours of the day. The leaves tend to assume an erect position, reducing the insertion angle on the shoot thus exposing the abaxial surface, which is densely covered by trichomes, to light. Consequently, the risks of both temperature and of photo inhibition for the leaf increase, which are due to lesser transpiration induced by water deficit and to the excess of radiation [65-67].

With an increase in *stress* it is possible to note a loss in turgidity of the adult leaves which, after some days, present a faded colour and then, at more advanced stages of stress, become bronze in colour and bend along the main axis of the leaf. Such symptoms, in adult plants are the prelude of early leaf senescence and phyloptosis in response to long periods of drought.

A limited availability of water can determine an increase in the phenomena of fruit drop and a reduction in the dimensions of the mesocarp cells [68, 69] during fruit development and maturation phases. The effect of water deficit on endocarp growth appears to be more easily reversed compared to that on mesocarp growth [70]. The relation between water availability and the principal compounds which qualitatively characterise the oil are still not clear due to the numerous interactions between factors which control these properties. For example, cultivation in both dry and more southern environments has no influence on either the acidity or composition in fatty acids of the oil in temperate environments, but would appear to increase the drupe phenolic compound content [71, 72].

The olive can reduce the negative effects of water shortage through a series of physiological responses which limit the loss of water whilst maintaining an appropriate functioning level of the main metabolites. One of these regards the dimension of the xylematic veins which with a reduction in dimensions reduces the water transport of the xylem, but represents a useful property for increasing resistance of the plant to drought *stress* significantly reducing the probability of the formation of embolisms in the vascular bundles [73]. The oil, moreover, has a notable capacity of active osmotic adjustment. The synthesis and storage of osmotically active solutions can lower the osmotic potential of the leaves by more than 1.5 MPa [74]. Active osmotic adjustment occurred in olive plants exposed to hydric *stress* and mainly was due to the accumulation of manitol, glucose and organic acids, while mineral elements such as potassium would appear to have no role [75, 76]. The efficiency of the root apparatus in drought *stress* conditions is also dependent on the foliage-root relation. In the olive plant, as in other vegetable species, this relation decreases in condition of drought [67, 77].

3.1.2. Low Temperature Stress

S*tress* from low temperatures, of between 10 and 0°C, are indicated as *stress* from cooling. The optimal temperature for growing olive plants results as being between 20 and 30 °C, but metabolic processes progressively slow down at temperatures lower than 25 °C [78]. Further decreases in temperature diminish both respiration and enzymatic activity with a consequent slowing of the absorption of water and nutrients, a reduction in photosynthetic efficiency and the main cellular processes which lead to a block in growth. Decreases in temperature below 0 °C is referred to as freezing rather than cooling. The effects of freezing temperatures are completely different from those caused by cooling temperatures. One of the most serious manifestations of cold damage is represented by characteristic fissures which are produced longitudinally in the tree cortex. Such damage is the result of the formation of ice in the xylem and of the consequent expansion of internal tissues, as the cortex and other tissues break because they are not sufficiently elastic. The wood beneath the cortex varies from a normal cream colour to a dark brown colour, due to the release of phenolic compounds. The change from the growth phase, which is susceptible to low temperatures, to the phase in which the plants are more resistant, includes a high number of metabolic, biochemical and biophysical adaptations which, on the whole, are known as acclimatisation. During the period of acclimatisation environmental stimuli activate specific gene families which lead to the typical acclimatisation metabolism. These include a variation in hormonal levels and alterations in metabolism which result in an intracellular accumulation of solutions (such as simple sugars, phenolic compounds, polyalcohols and soluble proteins) and in lipid composition variations of the membranes (such as an increase in polyunsaturated fatty acids) [79, 80].

3.1.3. Stress from High Temperatures

The olive plant is more resistant to high temperatures than it is to the cold. The occurrence of damage in the form of burns to the trunk and leaf chlorosis rarely occur and only occur in cases in which plants are exposed to high temperatures and drought or high light intensity. The temperatures at which a species enters heat *stress* are normally defined as temperatures at which irreversible damage occurs. Heat damage can also appear reversible at lower temperatures through a substantial reduction in production. As with other resistances to abiotic *stress*, the genetic factor plays an important role even in heat tolerance. Thus, differences between varieties in the olive plant can be sensitive and in the region of several degrees centigrade. Generally cultivars originating from more northern locations show lesser tolerance to high temperatures, while native varieties from hotter areas are more tolerant. In the olive plant, during summer there is a significant increase of resistance in the region of 3-4 °C compared to winter. Even during the day, in periods characterised by high temperatures (around 35 °C), variations in heat tolerance can occur with a lethal increase in temperature of around 2 °C at midday, compared to late afternoon [81]. Such an increase is probably linked to the synthesis of specific protective proteins (*HSP - Heat Shock Proteins*) which are produced by the plants in a very short period of time in response to high temperature *stress* [81].

3.1.4. Stress from Salinity

Arid and semi-arid cultivation environments where olive growing is widespread are naturally predisposed to salinity problems.

The olive plant is a species with average resistance to soil salinity and is more tolerant than other fruit tree species, such as vines, citrus, pomaceous and stone-fruits. Nevertheless, the threshold levels beyond which there are effects on the growth, productivity, and survival of the olive plant are not easy to establish as they are dependent on several factors such as the genotype, the age of the plant, the growth and environmental conditions. The main visible symptoms of saline *stress* are: chlorosis and necrosis of the foliage, drying of the leaf tips which in the most serious cases becomes drying of the root tips and abscission of the leaves. Olive plants grown in conditions of salinity show a reduction in growth, shortened internodes, small leaves with mesophyll and thickened cellular walls. Furthermore, salinity reduces the vitality and germination of the pollen, the number of perfect flowers for efflorescence, the percentage of development and the growth of the fruits. Salinity has a more marked effect on the production of olives than on the production of oil [77, 82]. An important aspect is the choice of the variety. In olive plants there is a notable genotype difference in the degree of resistance to salinity [83, 84]. Varieties which are considered to be tolerant and which are distinguished for their high qualitative production standard are the Arbequina and the Picual (Spain), the Megaritiki (Greece), the Chemlali (Tunisia) and the Frantoio (Italy).

3.1.5. Stress from Root Asphyxiation

The lack of oxygen in the soil, both partial (hypoxia) and total (anoxia), is a common condition, above all at a certain depth, even in soils which are normally not considered to be subject to root asphyxiation. In conditions of excess water in the soil, water eliminates oxygen from the soil occupying soil pores which had previously been filled with air. In these conditions the oxygen concentration remains high only in the first millimetres of the soil, which remain in direct contact with the air while at a depth of a few tens of centimetres oxygen already disappears. The effects of a lack of oxygen in the olive plant are both dramatic and immediate as the plant does not possess any means of defence against such adversity [85].

3.1.6. Stress from Atmospheric Pollutants

In recent decades there has been a constant and significant increase in water, soil and air pollution caused by the release of substances, in both natural and agricultural environments, which are potentially capable of negatively interfering with plant health and productivity. In olive plants, the first systematic investigations on the physiological effects of O_3 started during the first half of the 1990s, using relatively high doses of O_3 (150 ppb) and very short exposure periods (a few hours) [86]. From these first experiences it was observed that during the phase of exposure to ozone olive plants react to the presence of the pollutant by significantly reducing stomata opening. Although there is currently no direct experimental evidence on the effects of O_3 on the growth and productivity of different olive genotypes, it is however presumable that these physiological limitations may determine negative effects in the long term [87]. The problems associated with SO_2 have been known for a long time and the increase of this pollutant in the atmosphere of industrialised countries reached its highest values around 1970 [88]. In olive plants long term treatment with SO_2 determines physiological and morphological responses which are genotype-dependent, even in the absence of visible symptoms [89].

3.2. Biotic Stress

See part I *Olea europaea* chapter 1 “Botanical and agricultural aspects. Agronomic techniques and orchard management”.

4. GENOMIC APPROACHES

Since olive plants are able to survive for a long time, the tree has retained its germplasm characteristics over thousands of years. Furthermore, the long juvenile phase severely hinders classical breeding and genetic studies [29]. Cultivars were selected over centuries for specific traits and propagated vegetatively. Recently, the need to select molecular markers in order to distinguish, characterize and identify cultivars, emerged. Many important agronomic traits, such as grain quality, disease resistance and stress tolerance are controlled by many genes and are known as "quantitative" 'polygenic,' 'multifactorial' or 'complex' traits. The identification of these desirable traits represents the fundamental basis of plant breeding programmes to select specific genotypes. The use of molecular marker technologies offers optimal opportunities to develop early selection strategies in many fruit crops. DNA-based markers, such as random amplified fragment length DNA (RAPD), amplified fragment length polymorphism (AFLP) and restriction fragments length polymorphism (RFLP) have been widely used, both separately or combined for map construction.

The fist linkage map of the *Olea* genome was constructed by de la Rosa *et al.* [90] using dominant PCR markers such as RAPDs and AFLP and RFLPs and as SSR co-dominant marker. The approach was applied to ninety-five individuals of a cross progeny derived from Leccino and Dolce Agogia cultivars.

The *Olea* breeding programs include important agronomic traits such as oil quantity and quality, olive fruit harvest regime and biotic and abiotic resistance.

Also quantitative trait locus (QTL) markers have been used to select genotypes since the 1980s. One of the main uses of DNA markers in agricultural research has been in the construction of linkage maps for diverse crop species. In particular, the QTLs markers may be used as molecular tools for marker-assisted selection (MAS) in plant breeding [91].

QTLs are very suitable markers in agricultural research in the construction of linkage maps for diverse crop species. Unfortunately, to date QTLs of agronomical interest for *Olea* breeding have not been detected.

Although steps have been taken towards olive genome sequencing [92], the complete sequence of the olive genome has not yet been accomplished. Hence, the lack of olive genome sequence is a big handicap in resolving olive genetic complexity.

4.1. Gene Identification in Crop Species

The main goal of molecular genetics is to isolate and characterize genes involved in specific biochemical and/or physiological pathways. Unfortunately, the number of known gene sequences present in the databases are limited to model species. For this reason in many cases the sequence information from heterologous species may be used to this porpouse. The recent advances in genome sequencing, through the high-throughput sequencing approaches, provid useful opportunities to develop new markers in non-model crop species, as well as to identify genes of agronomic traits. Several sequence information from different species are now becoming available and permits a rapid identification of candidate genes through bioinformatics analysis. Indeed, they are commonly used to deduce consensus motif and to design degenerate primers that, by means of amplification reaction and sequencing, allow to identify genes, gene promoters and polymorphisms in a wide range of agronomically important crop species [15, 93].

cDNA- and oligonucleotide microarray technologies hold great promise to identified candidate genes, to monitor the expression levels of specific mRNAs and to identify polymorphisms. These approaches allow to study the entire gene complement of a genome in a single experiment and represent functional genomic methodology that have revolutionized global gene expression profiling [15, 94, 95]. Microarrays allow to study the global gene expression levels in specific organs and/or tissues of Arabidopsis, rice, maize, strawberry, petunia, ice plants and lima bean under controlled physiological conditions in, for crop species such as wheat, barley and olive, where no reference sequence are available, genes discovery relies on expressed sequence tags (EST).

Concerning olive trees, although its economic importance, genomic little information is still available and predominantly related to pollen allergens and olive fruit characteristics are related to genes involved in fatty acid biosynthesis, fatty acid modification, triacylglycerol synthesis, and fat storage [15, 91, 96, 97].

Furthermore, in recent years, the parallel sequencing of different fruit cDNA collections has allowed to identify genes involved in fruit development and fruit quality traits [15, 98].

4.2. High-Throughput Transcriptomics Approaches: RNA-Seq

With the advent of the RNA sequencing techniques it has made possible the rapid characterization and quantification of entire transcriptomes. The RNA-Seq is a recently developed approach to trascriptome profiling that uses deep-sequencing technologies. It involves direct sequencing of cDNAs by using high-throughput next generation sequencing technologies. This method has clear advantages over existing approach because it is not limited to detecting transcripts that corrispond to existing genomic. The last aspect makes the RNA-Seq very attractive for non model organisms with genomic sequences not yet determined.

RNA-Seq method can help to identify the precise location of transcription boundaries, to a single-base resolution, how the exons are connected, and sequence variations [99]. RNA-Seq does not have an upper limit for quantification and is highly accurate to detect transcripts expression levels, as determined using qPCR and finally it show high levels of reproducibility for both biological and technical replicates.

Next generation sequencing platforms used for RNA-Seq are commercially available from several companies and are continuously improved to increase sequencing speeds, accuracy, and depth at a lower cost. The large amount of data produced by RNA-Seq must be properly processed by bioinformatic programs. The first step is to map the short reads from RNA-seq to the reference genome, or to assemble them into contigs and then allign to genome sequence to determine the transcription structure [99].

Sequencing technology applied to crop species represents the first step to identify genes involved in the control of important agronomic traits. *Arabidopsis thaliana* [4] was the first crop genome to be sequenced [100, 101], followed by rice *Theobroma cacao* [102], apple [103] grape [104] cotton [105] of *Miscanthus* [106] and banana [107].

4.2.1. The Olive Genome as a Crucial Tool for Breeding

In recent years, much attention has turned to the olive fruit. The parallel sequencing of different fruit cDNA collections and miRNA has provided large scale information about the structure and putative function of gene transcripts regulation during the development of leaves and fruit development [98, 108, 109].

A first inventory of sRNAs in the olive has been obtained from juvenile and adult shoots, revealing that the 24-nt class dominates the sRNA transcriptome and atypically accumulates at levels previously unseen in other plant species, suggesting an active role of heterochromatin silencing in the maintenance and integrity of its large genome [110]. A total of 18 known miRNA families were identified in the libraries.

Recently [109] are sequenced six small RNA libraries from fruits (ripe and unripe) and leaves (one year and off year) by using the high-throughput Illumina platform). Bioinformatic analysis of 93,526,915 reads identified 135 conserved miRNA, belonging to 22 miRNA families in the olive. In addition, 38 putative novel miRNAs were discovered in the datasets. Expression of olive tree miRNA varied greatly among the six libraries, indicating the contribution of diverse miRNA in balancing between reproductive and vegetative phases. Predicted miRNA targets were categorized into 108 processes, such as development, hormone – mediated signalling and organ morphogenesis. The KEGG analyses revealed that the miRNA - targeted genes are involved in seven main pathways, belonging to carbohydrate metabolism and hormone signal-transduction pathways. This work demonstrates, for the first time, a comprehensive study on olive miRNA related to bearing.

Regulation of miRNA under different developmental phases and tissues indicated that control of nutrition and hormones, together with flowering processes had a noteworthy impact on olive tree alternate bearing.

Despite its global importance, genomic sequence resources available for the olive are still scarce, though an increasing number of expressed gene functions have been described in recent years through limited Next Generation Sequencing approaches (Fig. **3**).

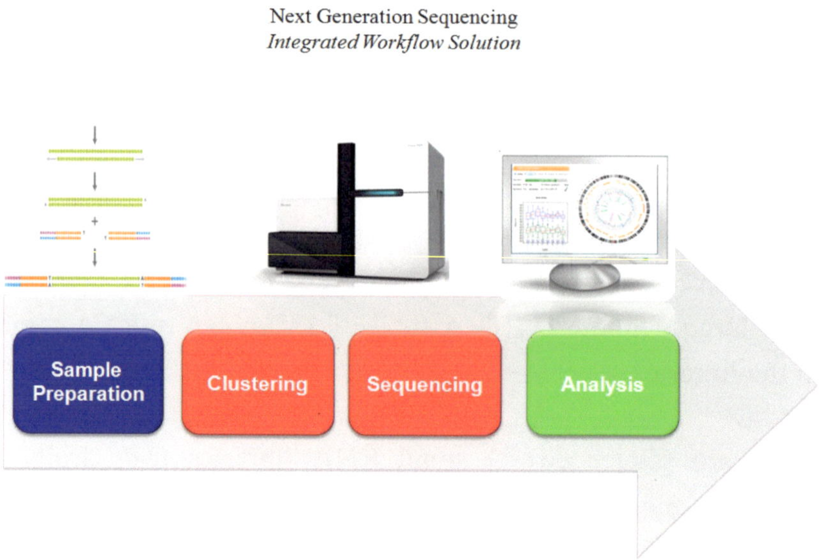

Figure 3: Scheme of next generation sequencing technology for whole genome analysis.

While these studies have highlighted the utility of cDNA sequencing for candidate gene discovery and gene function, a comprehensive description of genes expressed in *Olea europaea* remains unavailable. The choice of the candidate gene in the olive remains the most used, above all due to the studies that regard olive plant physiology.

Recent studies have regarded the isolation and characterisation of genes involved in biochemical pathways which lead to synthesis of chlorophyll and tocopherols [111], to aroma pathways [112, 113] and flavonoid and anthocyanin pathways [114].

It has been known for some time that the quality of fruits and response to stress are two closely related parameters. Taking such presuppositions as a starting point the gene which codifies for genranyl geranyl reductase (*CHLP*) was chosen as the candidate gene. It encodes a chloroplastic enzyme involved in the formation of phytolic side chain of tocopherols chlorophyll, and plastoquinones [15, 111].The choice of such a gene derives from work carried out previously on peach trees, where it was demonstrated that the gene was involved both in the quality and in the response to both biotic and abiotic stress conditions.

In such a context the olive ortholog gene named *OeCHLP* (*Olea europaea GERANYLGERANYL REDUCTASE*) was studied. *OeCHLP* was isolated and characterizated by Bruno *et al*. *In silico* analysis evidenced the high homology of *OeCHLP* with other known homologous genes [115, 116]. Differing from the homologous genes of *Arabidopsis thaliana*, *Glycine maxima* and *Prunus persica*, no introns were present in the genomic sequence of *OeCHLP*. Moreover, two copies of *OeCHLP* gene per haploid genome were detected in the 'Carolea' cultivar used in the present study. Such a feature is common to other Drupoideae species, such *Prunus amygdale*, *Prunus avium* and *Prunus dulcis*, whereas one single copy gene was estimated in both *Prunus persica* and *Arabidopsis thaliana* [115, 116]. *OeCHLP* transcripts were detected in various organs of olive plants. This result is in line with the presence of ubiquitous tocopherols and/or chlorophyll, *OeCHLP* transcripts were present in various organs of plants. This result is in line with the involvement of this gene in the biosynthesis of several compounds, of which chlorophylls, present in plant aereal organs and fruits, and ubiquitous tocopherols [115, 117-120]. Notably, a strong accumulation of *OeCHLP* transcripts was detected in leaf trichomes where many protective products, including terpenes, accumulated [121, 122]. In drupes of olive plants *OeCHLP* gene expression was enhanced in dark fruit most likely in relation to the increase of the level of total tocopherols in mature fruits, suggesting a role in antioxidant synthesis. Change in gene transcript levels occur during ripening phase inolive fruits in relation to genotype considered [123]. In this context, tocopherols confer nutritional value in olive fruits [124] and contribute to product stability and post harvesting shelf life [125] by protecting storage oil from oxidative damage [15, 126]. Increased levels of *OeCHLP* were also detected in fruits damaged by the *Bactrocera oleae* pathogen as well as in fruits wounded by needles (Fig. **4**). These results suggest a role in protection mechanisms related to cell damage and oxidative burst induced by pathogen [15, 114, 127, 128].

All together these results clearly indicate that the modulation of *OeCHLP* gene expression is part of the complex genetic network underlying plant development and stress response.

Analogous studies focused on lipoxygenase pathway and in particular on the molecular characterisation of genes involved in the biochemical pathway due to its importance in the production of an aromatic compost and its function in stabilising fruits.

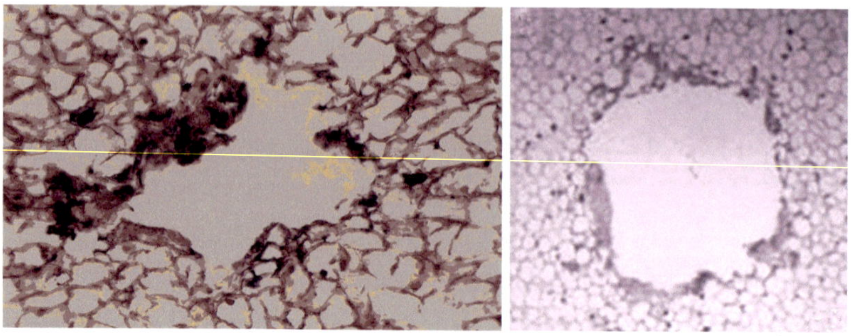

Figure 4: Localisation of *OeCHLP* transcripts in cross section of green olive fruit infected of *Bactrocera oleae* pathogen (captured on an optical microscope by Chiappetta).

The present work adds new information on the LOX pathway in olive plants by demonstrating that both cultivar and maturation stage impact on *OeADH* gene expression. A coordinate modulation with type 1 9/13 LOX gene was also demonstrated. The genotype-dependent differences in *OeADH/LOX* transcriptional pattern between the analysed cultivars appeared to be related to metabolite content variations in crushed fruits, suggesting a role for *OeADH/LOX* expression pattern in marking the aroma profile. Due to the complexity of the LOX pathway, an integrated 'omic' approach should be required to fully dissect the genetic control of LOX cascade and identify molecular markers for flavour characteristics of olive drupes in different cultivars. So far, function and regulatory mechanisms of many genes acting in the pathway have yet to be elucidated. On the basis of our results, the *OeADH* gene features as a suitable component of such a molecular approach, which could provide useful tools for cultivar selection in assisted breeding programs.

More research was performed to further understand the relationship between development, cold response, *FAD* expression and oil composition in the olive tree drupe. The difference in *FAD* transcript levels occurs in response to development and genotype, whereas cold affects *FAD* transcription during oil biogenesis [129].

FUTURE PERSPECTIVES

Genomic sequences for the olive will enable researchers to explore the breadth of genetic diversity present within the species and within the breeding germplasm using high throughput methods of re-sequencing based on NGS technologies. This will give access to all types of variations, namely Single Nucleotide Polymorphisms (SNPs), small insertion/deletions and structural variants (large insertion/deletions). It will allow a better assessment of the relationships among the different accessions, of the geographical patterns of distribution of genetic variation and of the genetic consequences of olive trees domestication. It will finally form the basis for the development of novel molecular marker assays.

They will also allow the analysis of global gene expression and specific gene expression of olive tissues in diverse developmental stages and conditions. The identification and characterization of expression of important genes involved in agronomic and productive traits affecting fruit production and quality, biotic and abiotic stress resistance, important development characters (*e.g.*, juvenility, self-incompatibility, ovary abortion, and chill response), may offer a significant amount of tools and open new opportunities for improvement either through molecular breeding and/or genetic engineering.

Many researches will be focused on gene network activities, using olive microarray and/or qPCR to address the expression patterns of genes, during plant and fruit development and ripening of drupe fruits. Moreover, a new generation of molecular markers will be developed, which will be helpful in localizing genes involved in both monogenic and polygenic agronomic traits, to construct genetic fine-maps. These markers will be also used for marker-assisted selection (MAS) to obtain elite genotypes by allowing the analysis of cross progenies at earlier stages.

ACKNOWLEDGEMENTS

This work was supported by the Ministry of Economic Development, (Certificazione della composizione varietale, dell'origine geografica e dell'assenza di prodotti di sintesi negli oli extravergini di oliva – CERTOLIO project) and by the Ministry of Agriculture, Food and Forestry (Trattato internazionale FAO sulle risorse genetiche vegetali, III triennalità 2010–2013, RGV-FAO project).

CONFLICT OF INTEREST

The authors confirm that this chapter contents have no conflict of interest.

REFERENCES

[1] AA.VV. Olive germplasm – the olive cultivation, table olive and olive oil industry in Italy. Muzzalupo I.(Ed.) 2012. InTech, Rijeka, Croatia.

[2] The *C. elegans* Sequencing Consortium. Genome sequence of the nematode *C. elegans*: a platform for investigating biology. Science, 1998; 282:2012-2018.

[3] Adams M.D., Celniker S.E., Holt R.A., Evans C.A., Gocayne J.D., Amanatides P.G., *et al.* The genome sequence of *Drosophila melanogaster*. Science, 2000; 24:2185-2195.

[4] The *Arabidopsis* Genome Initiative. The *Arabidopsis* is Information Resource (TAIR): a comprehensive database and web-based information retrieval, analysis, and visualization system for a model plant. Nucleic Acids Research. 2001; 29:102–105.

[5] Meinke D.W., Meinke L.K., Showalter T.C., Schissel A.M., Mueller L.A., Tzafrir I. A sequence-based map of *Arabidopsis* genes with mutant phenotypes. Plant Physiology, 2003; 131:409–418.

[6] Mueller L.A., Zhang P., Rhee S.Y. AraCyc. A biochemical pathway database for *Arabidopsis*. Plant Physiology. 2003; 132:453-460.

[7] Green P.S. A revision of *Olea* L (*Oleaceae*). Kew Bullettin, 2002; 140:57-91.

[8] Muzzalupo I., Perri E. Genetic Characterization of Olive Germplasm by Molecular Markers. The European Journal of Plant Science and Biotechnology, 2008; 2:60-68.

[9] Loureiro J., Rodriguez E., Costa A., Santos C. Nuclear DNA content estimations in wild olive (*Olea europaea* L. ssp. *europaea* var. *sylvestris* Brot.) and Portuguese cultivars of *O. europaea* using flow cytometry. Genetic Resources and Crop Evolution. 2006; 54:21–25.

[10] Breton C., Tersac M., Bervillé A. Genetic diversity and gene flow between the wild olive (oleaster, *Olea europaea* L.) and the olive: several Plio-Pleistocene refuge zones in the Mediterranean basin suggested by simple sequence repeats analysis. Journal of Biogeography. 2006; 33:1916–1928.

[11] Rallo P., Dorado G., Martin A. Development of Simple Sequence Repeats (SSRs) in Olive tree (*Olea europaea* L.). Theoretical and Applied Genetics. 2000; 101:984–989.

[12] Lopes T., Capelo A., Brito G., Loureiro J., Santos C. Genetic variability analyses of the somatic embryogenesis induction process in *Olea* spp. using nuclear microsatellites. Trees Structure and Function. 2009; 23:29-36.

[13] Muzzalupo I., Stefanizzi F., Salimonti A., Falabella R., Perri, E. Microsatellite markers for identification of a group of Italian olive accessions. Scientia Agricola. 2009; 66:685-690,

[14] Muzzalupo I., Chiappetta A., Benincasa C., Perri E. Intra-cultivar variability of three major olive cultivars grown in different areas of central-southern Italy and studied using microsatellite markers. Scientia Horticulturae, 2010; 126:324–329.

[15] Muzzalupo I., Perri E., Chiappetta A. Fruit germplasm characterization: genomics approaches for the valorization of genetic diversity. In: Genetic diversity in plants. Caliskan M.(Ed.) 2012. InTech, Rijeka, Croatia. Pp 55-86.

[16] FAO, The second report on the state of the world's plant genetic resources for food and agriculture, Rome, Italy, 2010.

[17] Bartolini G., Petruccelli R. Classification, origin, diffusion and history of the olive. Food and Agriculture Organization of the United Nations, 2002; pp. 2-143, Rome, Italy.

[18] Roselli G., Scaramuzzi F. Definizione dei termini di cultivar e clone. vol. 2. Annali dell'Accademia Nazionale di Agricoltura. 1974; pp. 119–126.

[19] Barranco D., Rallo L. Olive cultivars in Spain. HortTechnology, 2000; 10:107-110.

[20] Cantini C., Cimato A., Sani G. Morphological evaluation of olive germplasm present in Tuscany region. Euphytica. 1999; 109:173-181.

[21] Lombardo N., Godino G., Alessandrino M., Belfiore T., Muzzalupo I. Contributo alla Caratterizzazione del Germoplasma Olivicolo Pugliese, Istituto Sperimentale per l'Olivicoltura (Ed), 2004; pp. 2-112, Cosenza, Italy.

[22] Muzzalupo I., Olive germplasm – Italian catalogue of olive varieties. 2012. InTech, Rijeka, Croatia.

[23] Ouazzani N., Lummaret R., Villemur P. Apport du polymorphisme alloenzymatique à l'identification variétale de l'olivier (*Olea europaea* L). Agronomie. 1995; 15:31-37.

[24] Bogani P., Cavalieri D., Petruccelli R., Roselli G. Identification of olive tree cultivars by using random amplified polymorphic DNA. Acta Horticulturae, 1994; 356:98-101.

[25] Fabbri A., Hormaza J.I., Polito V.S. Random amplified polymorphic DNA analysis of olive (*Olea europaea L.*) cultivars. Journal of the American Society for Horticultural Science. 1995; 120:538–542.

[26] Wiesman Z., Avidan N., Lavee S., Quebedeaux B. Molecular characterization of common olive varieties in Israel and the West bank using randomly amplified polymorphic DNA (RAPD) markers. Journal of the American Society for Horticultural Science. 1998; 123:837–841.

[27] Belaj A., Trujillo I., de la Rosa R., Rallo L., Gimenez M.J. Polymorphism and discrimination capacity of randomly amplified polymorphic markers in an olive germplasm bank. Journal of the American Society for Horticultural Science. 2001; 126:64–71.

[28] Muzzalupo I., Fodale A., Mulè R., Caravita M.A., Salimonti A., Pellegrino M., Perri E. Genetic relationships of Sicilian olive varieties using RAPD markers. Advances in Horticultural Science; 2007; 21:35–40.

[29] Angiolillo A., Mencuccini M., Baldoni L. Olive genetic diversity assessed using amplified fragment length polymorphism. Theoretical and Applied Genetics, 1999; 98:411–421.

[30] Owen C.A., Bita EC., Banilas G., Hajjar SE., Sellianakis V., Aksoy U., Hepaksoy S., Chamoun R., Talhook SN., Metzidakis I., Hatzopoulos P., Kalaitzis P. AFLP reveals structural details of genetic diversity within cultivated olive germplasm from the Eastern Mediterranean. Theoretical and Applied Genetics, 2005; 110:1169-1176.

[31] Amane M., Lumaret R., Hany V., Ouazzani N., Debain C., Vivier G., Deguilloux M.F. Chloroplast-DNA variation in cultivated and wild olive (*Olea europaea* L.). Theoretical and Applied Genetics, 1999; 99:133-139.

[32] Lumaret R., Amane M., Ouazzani N., Baldoni L., Debain C. Chloroplast DNA variation in the cultivated and wild olive taxa of the genus *Olea* L. Theoretical and Applied Genetics, 2000; 101:547-553.

[33] Besnard G., Khadari B., Baradat P., Bervillé A. Combination of chloroplast and mitochondrial DNA polymorphisms to study cytoplasmic genetic differentiation in the olive complex (*Olea europaea* L.). Theoretical and Applied Genetics, 2002; 105:139-144.

[34] Brennicke A., Klein M., Binder S., Knoop V., Grohmann L., Malek O., Marchfelder A., Marienfeld J., Unseld M. Molecular biology of plant mitochondria. Naturwissenschaften, 1996; 83:339-346.

[35] Cavallotti A., Regina TMR., Quagliariello C. New sources of cytoplasmic diversity in the Italian population of *Olea europaea* L. as revealed by RFLP analysis of mitochondrial DNA:characterization of the *cox3* locus and possible relationship with cytoplasmic male sterility. Plant Science, 2003; 164:241-252.

[36] Baldoni L., Cultrera N.G., Mariotti R., Riccioloni C., Arcioni S., Vendramin G.G., Buonamici A., Porceddu A., Sarri V., Ojeda M.A., Trujillo I., Rallo L., Belaj A., Perri E., Salimonti A., Muzzalupo I., Casagrande A., Lain O., Messina R., Testolin R. A consensus list of microsatellite markers for olive genotyping. Molecular Breeding. 2009; 24:213–231.

[37] Rallo P., Dorado G., Martin A. Development of simple sequence repeats (SSRs) in olive tree (*Olea europaea* L.). Theoretical and Applied Genetics, 2000; 101:984–989.

[38] Sefc KM., Lopes MS., Mendonca D., Rodrigues dos Santos M., Laimer da Camara Machado A. Identification of microsatellite loci in olive (*Olea europaea* L.) and their characterization in Italian and Iberian olive trees. Molecular Ecology, 2000; 9:1171-1173.

[39] Carriero F., Fontanazza G., Cellini F., Giorio G., Identification of simple sequence repeats (SSRs) in olive (*Olea europaea* L.). Theoretical and Applied Genetics, 2002; 104:301-307.

[40] Cipriani G., Marrazzo M.T., Marconi R., Cimato A., Testolin R. Microsatellite markers isolated in olive (*Olea europaea* L.) are suitable for individual fingerprinting and reveal polymorphism within ancient cultivars. Theoretical and Applied Genetics, 2002; 104:223-228.

[41] Muzzalupo I., Lombardo N., Musacchio A., Noce M.E., Pellegrino G., Perri E., Sajjad A. DNA sequence analysis of microsatellite markers enhances their efficiency for germplasm management in a Italian olive collection. Journal of the American Society for Horticultural Science, 2006; 131:352–359.

[42] Muzzalupo I., Stefanizzi F., Perri E. Evaluation of olives cultivated in Southern Italy by SSR markers. HortScience. 2009; 44:582-588.

[43] Hess J., Kadereit W., Vargas P. The colonization history of *Olea europaea* L. in Macaronesia based on internal transcribed spacer 1 (ITS-1) sequences, randomly amplified polymorphic DNAs (RAPD) and inter-simple sequence repeats (ISSR). Molecular Ecology, 2000; 9:857–868.

[44] Hernandez P., de la Rosa R., Rallo L., Dorado G., Martin A. Development of SCAR markers in olive (*Olea europaea*) by direct sequencing of RAPD products: applications in olive germplasm evaluation and mapping. Theoretical and Applied Genetics, 2001; 103:788-791.

[45] Stergiou G., Katsiotis A., Hagidimitriou M., Loukas M. Genomic and chromosomal organization of Ty1-*copia*like sequences in *Olea europaea* and evolutionary relationships of *Olea* retroelements. Theoretical and Applied Genetics, 2002; 104:926–933.

[46] Mekuria G., Collins G., Lavee S., Sedgley M. Identification of a RAPD marker linked to tolerance to peacock disease (*Spilocaea oleaginea*) in olives (*Olea europaea*). Acta Horticulturae, 2001; 546:565-567.

[47] Rekik-Hakim I., Grati-Kammoun N., Makhloufi E., Rebai A. Discovery and Potential of SNP Markers in Characterization of Tunisian Olive Germplasm. Diversity, 2010; 2:17-27.

[48] Reale S., Doveri S., Díaz A., Angiolillo A., Lucentini L., Pilla F., Martín A., Donini P., Lee D. SNP-based markers for discriminating olive (*Olea europaea* L.) cultivars. Genome, 2006; 15:1193-1209.

[49] Belaj A., Caballero J.M., Barranco D., Rallo L., Trujillo I. Genetic characterization and identification of new accessions from Syria in an olive germoplasm bank by means of RAPD markers. Euphytica, 2003; 134:261–268.

[50] Loumou A., Giourga C. Olive groves: The life and identity of the Mediterranean. Agriculture and Human Values, 2003; 20:87-95.

[51] Conde C., Delrot S., Geros H. Physiological,biochemical and molecular changes occurring during olive development and ripening. Journal of Plant Physiology, 2008; 165:1545-1562.

[52] Pérez-Jiménez F., Ruano J., Perez-Martinez P., Lopez-Segura F., Lopez-Miranda J. The influence of olive oil on human health: not a question of fat alone. Molecular Nutrition & Food Research. 2007; 51:1199-1208.

[53] Sánchez J. Lipid photosynthesis in olive fruit. Progress in Lipid Research, 1994; 33:97-104.

[54] Hernandez M.L., Mancha M., Martinez-Riva J.M. Molecular cloning and characterization of genes encoding two microsomal oleate desaturases (FAD2) from olive. Phytochemistry, 2005; 66:1417-1426.

[55] Zarrouk M., Marzouk B., Daoud D.B.M., Cherif A. Oil accumulation in olives and effect of salt on their composition. Olivae, 1996; 61:40-45.

[56] Ayton J., Mailer R.J., Robards K., Orchard B., Vonarx M. Oil concentration and composition of olives during fruit maturation in south-western New South Wales. Australian Journal of Experimental Agriculture, 2001; 41:815-821.

[57] Obied H.K., Prenzler P.D., Ryan D., Servili M., Taticchi A., Esposto S., Robards K. Biosynthesis and biotransformations of phenolconjugated oleosidic secoiridoids from *Olea europaea,* L. Natural Product Reports, 2008; 25:1167-1179.

[58] Roca M., Mínguez-Mosquera M.I. Carotenoid levels during the period of growth and ripening in fruits of different olive varieties (Hojiblanca, Picual and Arbequina). Journal of Plant Physiology, 2003; 160:451-459.

[59] Roca M., Mínguez-Mosquera M.I. Involvement of chloropyllase in chlorophyll metabolism in olive varieties with high and low chlorophyll content. Physiology Plantarum. 2003; 117:459-466.

[60] Martinelli F., Tonutti P. Flavonoid metabolism and gene expression in developing olive (*Olea europaea* L.) fruit. Plant Biosystems, 2012; 146:164-170.

[61] Tripoli E., *et al.*, The phenolic compounds of olive oil: structure, biological activity and beneficial effects on human health. Nutrition Research Reviews, 2005; 18:98-112.

[62] Hashim Y.Z.H.Y., *et al.* Inhibitory effects of olive oil phenolics on invasion in human colon adenocarcinoma cells *in vitro*. International Journal of Cancer, 2008; 122:495-500.

[63] Llorente-Cortés V., Estruch R., Mena M.P., Ros E., González M.A., Fitó M., Lamuela-Raventós R.M., Badimon L. Effect of Mediterranean diet on the expression of pro-atherogenic genes in a population at high cardiovascular risk. Atherosclerosis, 2010; 208:442-450.

[64] Baldoni L., Tosti N., Ricciolini C., Belaj A., Arcioni S., Pannelli G., Germana M.A., Mulas M., Porceddu A. Genetic structure of wild and cultivated olives in the central Mediterranean basin. Annuals of Botany London, 2006; 98:935-942.

[65] Faria T., Silvério D., Breia E., Cabral R., Abadia A., Abadia J., Pereira J.S., Chaves M.M. Differences in the response of carbon assimilation to summer stress (water deficits, high light and temperature) in four Mediterranean tree species. Physiologia Plantarum,1998; 102:419–428.

[66] Tattini M., Gravano E., Pinelli P., Mulinacci N., Romani A. Flavonoids accumulate in leaves and glandular trichomes of *Phillyrea latifolia* exposed to excess solar radiation. New Phytologist, 2000; 148:69–77.

[67] Angelopoulos K., Dichio B., Xiloyannis C. Inhibition of photosynthesis in olive trees (*Olea europaea* L.) during water stress and rewatering. Journal of Experimental Botany, 1996; 47:1093–1100.

[68] Rapoport H.F., Costagli G., Gucci R. The effect of water deficit during early fruit development on olive fruit morphogenesis. Journal of the American Society for Horticultural Science, 2004; 129:121-127.

[69] Rosati A., Caporali S., Hammami S.B.M., Moreno-Alìas I., Paoletti A., Rapoport H.F. Differences in ovary size among olive (*Olea europaea* L.) cultivars are mainly related to cell number, not to cell size. Scientia Horticulturae, 2011; 130:185–190.

[70] Gucci R., Lodolini E.M., Rapoport H.F. Productivity of olive trees with different water status and crop load. The journal of horticultural science and biotechnology, 2007; 82:648-656.

[71] Sacco A., Brescia M.A., Liuzzi V., Reniero F., Guillou C., Ghelli S., van der Meer P. Characterization of italian olive oils based on analytical and nuclear magnetic resonance determinations. Journal of the American Oil Chemists' Society, 2000; 77:619-625.

[72] De Nino A., Di Donna L., Mazzotti F., Muzzalupo I., Perri E., Sindona G., Tagarelli A. Absolute method for the assay of oleuropein in olive oils by atmospheric pressure chemical ionization tandem mass spectrometry. Analytical Chemistry, 2005; 77:5961-5964.

[73] Trifilò P., Lo Gullo M.A., Nardini A., Pernice F., Salleo S. Rootstock effects on xylem conduit dimensions and vulnerability to cavitation of *Olea europaea* L. trees. Structure and Function, 2007; 21:549-556.

[74] Gucci R., Lombardini L., Tattini M. Analysis of leaf water relations in leaves of two olive (*Olea europaea*) cultivars differing in tolerance to salinity. Tree Physiology, 1997; 17:13-21.

[75] Rejšková A., Patková L., Stodůlková E., Lipavská H. The effect of abiotic stresses on carbohydrate status of olive shoots (*Olea europaea* L.) under *in vitro* conditions. Journal of Plant Physiology, 2007; 164:174–184.

[76] Munns R. Comparative physiology of salt and water stress. Plant Cell and Environment, 2002; 25:239–250.

[77] Basta P., Toscano P., Ciliberti A., Turco D. Osservazioni pluriennali sulle risposte vegeto-produttive di cinque cultivar di olivo irrigate con acqua salmastra a differenti concentrazioni. International proceeding of olive growing, Centenario I.S.Ol., 2002; 1:235-240.

[78] Galán C., García-Mozo H., Cariñanos P., Alcázar P., Domínguez-Vilches E. The role of temperature in the onset of the *Olea europaea* L. pollen season in southwestern Spain. International Journal of Biometeorology, 2001; 45:8-12.

[79] Famiani F., Proietti P., Micheli M., Boco M., Standardi A., Ferranti F., Reale L. Effects of tree shelters on young olive (*Olea europaea*) tree growth and physiology. New Zealand Journal of Crop and Horticultural Science, 2007; 35:303–312.

[80] Flexas J., Medrano H. Drought-inhibition of photosynthesis in C3 plants: stomatal and non-stomatal limitations revisited. Annals of Botany, 2002; 89:183-189.

[81] Chaves M.M., Pereira J.S., Maroco J., Rodrigues M.L., Ricardo C.P.P., Osório M.L., Carvalho I., Faria T., Pinheiro C. How plants cope with water stress in the field? Photosynthesis and growth. 2002; 89:907-916.[81] Sierra J.M., Zapata J.M. Translational regulation of the heat shock response. Molecular biology reports, 1994; 19:211-220.

[83] Gucci R., Moing A., Gravano E., Gaudillère J.P. Partitioning of photosynthetic carbohydrates in leaves of salt-stressed olive plants. Australian Journal of Plant Physiology, 1998; 25:571-579.

[83] Patumi M., d'Andria R., Marsilio V., Fontanazza G., Morelli G., Lanza B. Olive and olive oil quality after intensive monocone olive growing (*Olea europaea* L., cv. Kalamata) in different irrigation regimes. Food Chemistry, 2002; 77:27–34.

[84] Gucci R., Tattini M. Salinity tolerance in olive. Horticultural Reviews, 1997, 21:177–214.

[85] Sanchez Hernandez M.E., Ruiz Davila A., Perez de Algaba A., Blanco Lopez M.A., Trapero Casas A. Occurrence and etiology of death of young olive trees in southern Spain. European Journal of Plant Pathology, 1998; 104:347–357.

[86] Minnocci A., Panicucci A., Sebastiani L., Lorenzini G., Vitagliano C. Physiological and morphological responses of olive plants to ozone exposure during a growing season. Tree Physiology, 1999; 19:391-397.

[87] Sebastiani L., Minnocci A., Tognetti R. Genotypic differences in the response to elevated CO_2 concentration of one-year-old olive cuttings (*Olea europaea* L. cv. Frantoio and Moraiolo). Plant Biosystems, 2000; 136:99-207.

[88] Sebastiani L., Minnocci A., Tognetti R. Olive (*Olea europaea* L.) plants reactions to atmospheric pollutants and UV-B radiation: current state of the research. Advances in Horticultural Science, 2002; 16:144-154.

[89] Vitagliano C., Minnocci A., Sebastiani L., Panicucci A., Lorenzini G. Physiological response of two olive genotypes to gaseous pollutants. Acta Horticolturae, 1999; 474:431-434.

[90] De la Rosa R., Angiolillo A., Rallo L., Guerrero C., Pellegrini M., Besnard G., Bervillé A., Martin A., Baldoni L. A first genetic linkagemap of olive (*Olea europaea* L.) cultivars using RAPD and AFLP markers. Theoretical and Applied Genetics, 2003; 106:1273-1282.

[91] Ribaut J.M., Hoisington D. Marker-assisted selection: new tools and strategies. Trends Plant Science, 1998; 3:236–239.

[92] Cattonaro C., Del Fabbro C., The French-Italian Public Consortium for Grapevine Genome Characterization. Application of illumina genome analysis platform and whole transcriptome analysis for grape genome annotation. In Illumina Usermeeting, 2008, Rottach-Egern, Germany.

[93] Bracci T., Busconi M., Sebastiani L., Fogher C. Molecular studies in olive (*Olea europaea* L.): overview on DNA markers applications and recent advances in genome analysis. Plant Cell Report, 2011; 30:449–462.

[94] Duggan D.J., Bittner M., Chen Y., Meltzer P., Trent J,M. Expression profiling using cDNA microarrays. Nature Genetics, 1999; 21:10-21.

[95] Li Y., Huang Y., Bergelson J., Nordborg M., Borevitz J. Association mapping of local climate-sensitive quantitative trait loci in *Arabidopsis thaliana*. Proceedings of the National Academy of Sciences of the United States of America, 2010; 107:21119-21204.

[96] Hatzopoulos P., Banilas G., Giannoulia K., Gazis F., Nikoloudakis N., Milioni D., Haralampidis K. Breeding, molecular markers and molecular biology of the olive tree. European Journal of Lipid Science and Technology, 2002; 104:574-586.

[97] Banilas G., Moressis A., Nikoloudakis N., Hatzopoulos P. Spatial and temporal expressions of two distinct oleate desaturases from olive (*Olea europaea* L.). Plant Science; 2005; 168:547-555.

[98] Alagna F., D'Agostino N., Torchia L., Servili M., Rao R., Pietrella M., Giuliano G., Chiusano ML., Baldoni L., Perrotta G. Comparative 454 pyrosequencing of transcripts from two olive genotypes during fruit development. BMC Genomics, 2009; 10:349-353.

[99] Wang Z., Gerstein M., Snyder M. RNA-Seq: a revolutionary tool for transcriptomics. Nature Review Genetics, 2009; 10: 57–63.

[100] Yu J., Hu S.N., Wang J., Wong G.K.S., Li S.G., Liu B., Deng Y.J., Dai L., Zhou Y., Zhang X.Q., Cao M.L., Liu J., Sun J.D. A draft sequence of the rice genome (*Oryza sativa* L. ssp *indica*). Science, 2002; 296:79-92.

[101] Matsumoto T., Wu J.Z., Kanamori H., Katayose Y., Fujisawa M., Namiki N. The map-based sequence of the rice genome. Nature, 2005; 436:793-800.

[102] Scheffler B.E., Kuhn D.N., Motamayor J.C., Schnell R.J. Efforts towards sequencing the Cacao genome (*Theobroma cacao*). In: Plant and Animal Genomes XVII, 2009. San Diego, CA, USA.

[103] Velasco R., Zharkikh A., Troggio M., Salvi S., Pindo M. Apple genome sequencing and post-genomic program at IASMA research center. In: Plant and Animal Genomes XVII, 2009. San Diego, CA, USA.

[104] Velasco R., Zharkikh A., Troggio M., *et al.* A high quality draft consensus sequence of the genome of a heterozygous grapevine variety. PLoS ONE. 2007; 2:e1326.

[105] Wilkins T.A., Mudge J., Abidi N., Allen R., Auld D. The sequencing and resequencing of cotton. In Plant and Animal Genomes XVII. 2009. San Diego, CA.

[106] Swaminathan K., Varala K., Moose S.P., Rokhsar D., Ming R., Hudson M.E. A genome survey of *Miscanthus Giganteus*. In Plant and Animal Genomes XVII. 2009. San Diego, CA

[107] Hribova E., Neumann P., Macas J., Dolezel J. Analysis of genome structure and organization in banana (*Musa acuminata*) using 454 sequencing. In: Plant and Animal Genomes XVII, 2009. San Diego, CA, USA.

[108] Munoz-Merida A., Gonzalez-Plaza J.J., Canada A., Blanco A.M., Garcia-Lopez M.C., *et al.* *De novo* assembly and functional annotation of the olive (*Olea europaea*) transcriptome. DNA Reserch, 2013; 20:93–108.

[109] Yanik H., Turktas M., Dundar E., Hernandez P., Dorado G., Unver T. Genome-wide identification of alternate bearing-associated microRNAs (miRNAs) in olive (*Olea europaea* L.). BMC Plant Biology, 2013; 13:10.

[110] Donaire L., Pedrola L., Rosa Rde L., Llave C. High-throughput sequencing of RNA silencing-associated small RNAs in olive (*Olea europaea* L.). PLoS One, 2011; 6:e27916.

[111] Bruno L., Chiappetta A., Muzzalupo I., Gagliardi C., Iaria D., Bruno A., Greco M., Donato G., Perri E., Bitonti MB. Role of geranylgeranyl reductase gene in organ development and stress response in olive (*Olea europaea*) plants. Functional Plant Biology, 2009; 36:370–381.

[112] Muzzalupo I., Macchione B., Bucci C., Stefanizzi F., Perri E., Chiappetta A., Tagarelli A., Sindona G. LOX gene transcript accumulation in olive (*Olea europaea* L.) fruits at different stages of maturation: relationship between volatile compounds, environmental factors, and technological treatments for oil extraction. The Scientific World Journal, 2012; 2012:1-9.

[113] Iaria D.L., Bruno L., Macchione B., Tagarelli A., Sindona G., Giannino D., Bitonti M.B., Chiappetta A. The aroma biogenesis-related *Olea europaea* alcohol dehydrogenase gene is

developmentally regulated in the fruits of two *Olea europaea* L. cultivars, Food Research International, 2012; 49:720-727.

[114] Iaria D.L., Forgione I., Chiappetta A., Muzzalupo I. A *de novo* transcriptomic approach to identify flavonoids and anthocyanins "switch-off" in olive (*Olea europaea* L.) drupes at different stages of maturation. PLoS ONE 2014 (in press).

[115] Keller Y., Bouvier F.D., Harlingue A. Metabolic compartmentation of plastid prenyllipid biosynthesis: evidence for the involvement of a multifunctional geranylgeranyl reductase. European Journal of Biochemistry, 1998; 251:413–417.

[116] Giannino D., Condello E., Bruno L., Testone G., Tartarini A., Cozza R., Innocenti A.M., Bitonti M.B., Mariotti D. The gene geranylgeranyl reductase of peach [*Prunus persica* (L.) Batsch] is regulated during leaf development and responds differentially to distinct stress factors. Journal of Experimental Botany, 2004; 55:2063–2073.

[117] Threlfall D.R., Whistance G.R. Biosynthesis of phytoquinones. Homogentisic acid: a precursor of plastoquinones, tocopherols and alpha-tocopherolquinone in higher plants, green algae and blue-green algae. The Biochemical Journal, 1970; 117:593–600.

[118] Bollivar D.W., Wang S.J., Allen J.P., Bauer C.E. Molecular genetic analysis of terminal steps in bacteriochlorophyll a biosynthesis: characterization of a *Rhodobacter capsulatus* strain that synthesizes geranylgeraniol-esterified bacteriochlorophyll *a*. Biochemistry, 1994; 33:12763–12768.

[119] Tanaka R., Oster U., Kruse E., Rudiger W., Grimm B. Reduced activity of geranylgeranyl reductase leads to loss of chlorophyll and tocopherol and to partially geranylgeranylated chlorophyll in transgenic tobacco plants expressing antisense RNA for geranylgeranyl reductase. Plant Physiology, 1999; 120:695–704.

[120] Munnè-Bosch S., Alegre L. The function of tocopherols and tocotrienols in plants. Critical Reviews in Plant Sciences, 2002; 21:31–57.

[121] McCaskill D., Croteau R. Monoterpene and sesquiterpene biosynthesis in glandular trichomes of peppermint (*Mentha piperita*) rely exclusively on plastid-derived isopentenyl diphosphate. Planta, 1995; 197:49–56.

[122] Aziz N., Paiva N.L., May G.D., Dixon R.A. Transcriptome analysis of alfalfa glandular trichomes. Planta, 2005; 221:28–38.

[123] Muzzalupo I., Stefanizzi F., Perri E., Chiappetta A. Transcript Levels of *CHL P* gene, antioxidants and chlorophylls contents in olive (*Olea europaea* L.) pericarps: a comparative study on eleven olive cultivars harvested in two ripening stages. Plant Foods for Human Nutrition, 2011; 66:1-10.

[124] Valk E.E., Hornstra G. Relationship between vitamin E requirement and polyunsaturated fatty acid intake in man. International Journal for Vitamin and Nutrition Research, 2000; 70:31–42.

[125] Goffman F.D., Bohme T. Relationship between fatty acid profile and vitamin E content in maize hybrids (*Zea mays* L.). Journal of Agricultural and Food Chemistry, 2001; 49:4990–4994.

[126] Sattler S.E., Gilliland L.U., Magallanes-Lundback M., Pollard M., DellaPenna D. Vitamin E is essential for seed longevity and for preventing lipid peroxidation during germination. The Plant Cell, 2004; 16:1419–1432.

[127] Ebel J. Oligoglucoside elicitor-mediated activation of plant defense. BioEssays, 1998; 20:569–576.

[128] Klessig D.F., Durner J., Noad R., Navarre D.A., Wendehenne D., *et al.* Nitric oxide and salicylic acid signaling in plant defense. Proceeding of the National Academy of Sciences of the United States of America, 2000; 97:8849–8855.

[129] Matteucci M., D'Angeli S., Errico S., La Manna R., Perrotta G., Altamura, M.M. Cold affects the transcription of fatty acid desaturases and oil quality in the fruit of *Olea europaea* L. genotypes with different cold hardiness, Journal of Experimental Botany, 2011; 62:3403-3420.

CHAPTER 3

Technological Aspects: Table Olive and Olive Oil Processing

Flora Valeria Romeo[1,*] and Innocenzo Muzzalupo[2]

[1]Consiglio per la Ricerca e la sperimentazione in Agricoltura (Agricultural Research Council, Citriculture and Mediterranean Crops Research Center, CRA-ACM), Corso Savoia, 95024 Acireale (CT), Italy and [2]Consiglio per la Ricerca e la sperimentazione in Agricoltura (Agricultural Research Council, Olive Growing and Olive Oil Industry Research Centre, CRA-OLI), C.da Li Rocchi, 87036 Rende (CS), Italy

Abstract: The technological aspects related to table olives and olive oil are summarized and discussed in this chapter. There are three main trade preparations of table olives: Spanish-style olives, Californian-style olives and naturally black or turning colour olives. There are also many other traditional table olive recipes that are less known in the international market but which are specially linked with a given territory as the product of natural and cultural resources. Some traditional process will also be described and the use of starter cultures in table olive fermentation will also be discussed. Interest in the development of and use of starters for table olive fermentation is increasing as appropriate bacterial inoculation can help to achieve a more controlled process, reduce processing time and improve the organoleptic and hygienic quality of the final product. Suitable hygienic conditions to adopt during olive processing and their alterations are then considered. Moreover, the traditional and the most innovative technologies for mechanical olive oil extraction are described with particular attention given to the discussion of technical aspects and analysis of the repercussions of each processing step on the most important quality markers are subsequently discussed.

Keywords: Innovative technologies, *Lactobacillus plantarum,* olive fermentation, olive oil extraction systems, olive quality, olive varieties, packaging technologies, preparation of table olives, product management, sensory analysis, starter culture.

1. TABLE OLIVES

1.1. Introduction to Table Olives

Fermentation is an economical means for the temporary preservation of different kinds of vegetables, including table olives. The cultivation of table olives in Italy, even if of limited importance in relation to olives destined for the production of

***Corresponding author Flora Valeria Romeo:** Agricultural Research Council - Citriculture and Mediterranean Crops Research Center (CRA-ACM), Corso Savoia, 190, 95024 Acireale (CT), Italy; Tel: +390957653136; Fax: +390957653113; E-mail: floravaleria.romeo@entecra.it

Innocenzo Muzzalupo & Sabrina Micali (Eds.)

oil, is characterised by some areali where the spread of prized cultivars has a longstanding tradition which is also due to favourable pedoclimatic and social conditions. On the whole, an overview of the processing of olives produced in southern Italian regions is noticeably lacking. Limiting factors can be attributed to the lack of knowledge regarding the potential of olive varieties, which are usually exploited for the production of oil, whilst they are not used for the production of table olives; to the lack of knowledge regarding the parameters on which to base choice and the evaluation of the vocation of an olive variety in the transformation into table olives, on the scant knowledge of biological, chemical and physical processes which regulate the production of vegetable conserves, and finally the difficulty of maintaining transformation standards in compliance with actual regulations and the current market. The regions of Southern Italy have a remarkable olive growing heritage, mainly in the existing varieties, among which only some have shown themselves to be suitable for table olives. The choice of appropriate variety mainly depends on some requirements which the olives must possess in order to meet the requirements of both the consumers and the productive cycle. Such requirements are mainly linked to the carpological properties of the drupes produced (weight, flesh/stone ratio) and by the flesh composition properties, such as a limited content of oil and, above all, good consistency of the flesh.

1.1.1. World Consumption and Production

In recent years, world consumption of table olives has undergone a constant increase which explains both the increase in production which, in producing countries, has determined an increase of both domestic consumption and the spreading of product information with better presentation of its quality. The main countries which produce table olives are: Spain, Turkey, Egypt, Algeria, Argentina, Morocco, Syria, Italy, Greece and the USA. Spain, Italy and Greece are the largest European producers. It is clear that productive data and the position of various countries are subject to variation in relation to production alternance and climatic trends which can be more or less favourable to olive growing.

The largest consumers of table olives are also the greatest producers, and the EU is always in first place. The EU results as being the area with the greatest consumption on a global scale followed by Turkey, the USA and Egypt (Table **1**). In Europe, the main consuming countries are Spain, followed by Italy, Germany and France.

Table 1: Total world production and consumption of table olives

	PRODUCTION	CONSUMPTION
	Data 2010/11	Data 2010/11
EU	809	574
Turkey	330	260
Syria	142	119
Morocco	110	32
Egypt	200	200
Algeria	128	129
Argentina	250	35
USA	154	240
Total world production	2,440	2,205

Data are expressed as 1,000 tons [1].

The International Olive Council (IOC) has estimated that world production has undergone a net increase to 2,565 million tonnes (+5%) with a market which is also in a phase of strong growth, estimations indicate around 8%, which is mainly attributable to an acceleration of exports (+28%). In Europe, Spain has predicted a downturn in both olive production and olive oil production, while Greece has predicted a rise. Other members of the International Olive Council, such as Turkey (410,000 t) foresee a steep increase in their production of table olives, which would position Turkey in second place behind Spain and ahead of Egypt, which has an assessed production of 300,000 t. Turkey has created a national initiative in order to promote an increase in the consumption of both olives and olive oil. Around 85% of domestic table olive production is consumed locally while the remaining 15% is destined for export [1].

From an analysis of the main producer nations' consumption, it is possible to notice that the greatest consumption occurs in the Mediterranean basin.

Nationally produced olives are concentrated in Puglia, Calabria and Sicily (Table **2**). It is also where there is the greatest concentration of companies which transform olives (more than 50). These businesses, on the whole, are modest in size or even very small (artisan level).

The increase in table olive consumption, recorded in recent campaigns is attributable to the priority given to marketing campaigns, by producers, which have focussed on the introduction of new products (bread, snacks, seasoned olives, olive pâté, *etc.*) and research into new packaging forms which are always more capable of transmitting reassuring messages to the consumer.

Table 2: Olive production in Italy

Region	**Total Production**	**Harvested Production**	**Table Olives:**	**Oil olives:**
			Total Production	**Total Production**
Piemonte	82	81	7	74
Valle d'Aosta	-	-	-	-
Lombardia	4,520	4,520	-	4,520
Liguria	31,440	19,600	850	18,750
Trentino-Alto Adige	1,280	1,280	-	1,280
Bolzano	-	-	-	-
Trento	1,280	1,280	-	1,280
Veneto	7,820	7,775	33	7,773
Friuli-Venezia Giulia	204	204	21	183
Emilia-Romagna	5,403	5,403	-	5,403
Toscana	110,570	107,797	1,482	106,325
Umbria	40,291	40,291	6	40,285
Marche	28,979	28,352	882	27,470
Lazio	139,968	132,015	2,697	129,318
Abruzzo	129,644	128,402	1,407	126,995
Molise	36,750	36,750	277	36,473
Campania	247,959	246,137	1,415	244,721
Puglia	1,179,120	1,107,429	18,661	1,088,767
Basilicata	36,603	36,393	113	36,280
Calabria	990,687	924,043	13,651	910,392
Sicilia	333,891	314,010	31,059	282,951
Sardegna	28,012	27,800	3,135	24,665
ITALY	**3,353,225**	**3,168,283**	**75,696**	**3,092,628**

Data are expressed as tons [2].

Only a few olive varieties were unaffected by the 2009 olive price crisis in both the Italian and international markets. The average international price at origin of fresh table olives in 2009 reached a minimum of 0.40 €/Kg for Hojblanca (Spain) and a maximum of 1.68 €/Kg for round black Greek olives [3]. The context is characterised as being a sector with wide margins for improvement, also with regards to pricing policies, which should have the aim of recognising the excellence of Italian products.

1.1.2. Olive Cultivar and Optimal Characteristics as Table Olives

The olive is a species that is characterised by a high number of varieties, more than 1,200 have been described in the world. In Italy there are more than 500 varities which, unfortunately, have more than treble that number of synonyms. The classification of cultivars results as being extremely complex also because the germoplasm has been further enriched with new cultivars and clones deriving from genetic improvement processes. The identification procedure of varieties which was initially based on morphological properties and phenotypes has now been furthered by molecular techniques such as DNA analysis, which allows clarification of doubts about the formation process of the species. The use of table drupes results as being widespread and is important, even though it is necessary to stress the commercial weakness of this in Italy, which is mainly due to an insufficient quantity for wide scale distribution, to the heterogeneity of production due to the high number of cultivars and to the reduced number of transformation establishments with suitable working capacities. In this kind of company it is possible to apply modern technologies capable of guaranteeing a high quality product which is able to satisfy export requirements.

With regards to the best known table olives, the most important varieties in Spain are Gordal (or Sevigliana), Manzanilla and Hojiblanca; in Portogallo Galega; the Conservolia, Kalamata and Chalkidiki varieties in Greece; Domat, Gemlit, Memecik and Halkidiki varieties in Turkey and the Picholine Marocaine in Morocco. While the following varieities are widespread in the USA: Manzanilla, Mission, Gordal, Ascolana and Barouni [4]; Meski, Marsaline and Chemchali in Tunisia; Dan, Djlt and Sorani in Syria; Sigoise e Azeradj in Algeria; Manzanilla and Pigeon Egg in Israel; and Arauco, Gordal, Manzanilla, Ascolana and Bella di Cerignola (both from Italy) in Argentina.

The situation in Italy is complex as production is supplied by many varieties which can be found all over the peninsula. The most common varieties in Sicily are: Nocellara del Belice, Nocellara Etnea, Tonda Iblea, Ogliarola Messinese, Moresca and Giarraffa; in Calabria: Carolea, Grossa di Cassano, Grossa di Gerace and Dolce di Rossano; in Puglia: Bella di Cerignola, S. Agostino, Coratina, Termite di Bitetto and Provenzale; in Basilicata: Majatica and Dolce di Melfi; in Lazio: Itrana; in Campania: Ortice, Ortolana and Caiazzana; in Abruzzo the Intosso variety; in Umbria the Dolce Agogia; in Marche the Ascolana; in Toscana it is possible to find the S. Caterina; in Liguria the Taggiasca; in Sardinia: Bosana, Tonda di Cagliari, and the Pizz'e Carroga [5, 6].

The different varieties are classified according to the use of the drupes:

- Oil variety: small fruits <3.5 grams, oil yield >20%;
- Dual attitude: average sized fruits 3.5-4.5 grams, oil yield >17-18%;
- Table olive: large fruits >4.0 grams, oil yield <16%.

However, it is often the case that local traditions in each region have lead to the use of a cultivar which is not strictly for table use, but a local. Even the transformation techniques used today are rooted in traditions which have been perfected thanks to science and productive technology.

The properties which destine the use of a variety for direct consumption are mainly to be found in the morphological properties and in the physical-chemical properties of the fruit. The choice is guided by the form and dimension of the olives, by the uniformity of the calibre, the pulp/stone ratio, the ease with which the flesh can be removed from the stone, the integrity of the epicarp and the content in terms of oil and reducing sugars. Other agronomic factors are important such as high and constant productivity, resistance to biotic and abiotic adversity, reduced and known sensitivity to different parasites, plant dimensions, type of cultivation (irrigated or not), *etc*. Moreover, each olive processing style necessitates the choice of suitable cultivars. An in-depth knowledge of the physical and chemical characteristic of the cultivar could lead the producer to adopt the most suitable process.

After processing the olives should have a healthy aspect with good pulp consistency which should neither be soft nor woody nor wrinkled; furthermore, the olives should be odourless and without an unpleasant taste.

The optimal characteristics which should be possessed by a good table olive variety [7], can be summarised as follows:

- Olive pulp ≥80%;
- Olive pulp / stone weight ratio >4;
- Pulp: which is easily removable from the stone;
- Appearance: absence of lesions, marks and deformations;
- Consistency: for green olives (full pulp, compact, crunchy, not swollen by irrigation water and unwrinkled); for black olives (firm texture, unwrinkled);
- Calibre: weighting uniformity of olives within the same calibre.

Some types of olives are exempt from these recommendations as, despite the reduced drupe size, they fare well on the table olive market for their flavour which is greatly appreciated by the consumer (*e.g.* Sinopolese in Calabria).

According to the IOC trade standards applicable to table olives [8], table olives are classified in one of the following three trade categories:

- "Extra" or "Fancy": High quality olives endowed to the maximum extent with characteristics specific to the variety and trade preparation are considered as belonging to this category.
- "First choice" or "Select": This category covers good quality olives with a suitable degree of ripeness and endowed with the characteristics specific to the variety and trade preparation. All the types, preparations and styles of table olives may be classified in this category, except for chopped or broken olives and olive pastes.

- "Second" or "Standard": This category includes good quality olives which, although they cannot be classified in the two previous categories, comply with the general quality conditions defined for table olives by the IOC standard.

The IOC standard also clearly defines the definitions and tolerances of defects for each category.

1.2. Trade Standard Applying to Table Olives

1.2.1. About International Olive Council

The International Olive Council is the world's only international intergovernmental organization in the field of olive oil and table olives. It was founded in Madrid, Spain, in 1959, under the auspices of the United Nations. It used to be known as the International Olive Oil Council or IOOC until 2006, when its name was changed to International Olive Council (IOC). The official website www.internationaloliveoil.org is a resource to learn more about this institution and members, news, publications, economic data, related documents and areas of activity regarding the olive world.

1.2.2. Content of Trade Standard IOC/OT/NC no. 1 December 2004

The IOC/OT/NC no. 1 December 2004 established a unified qualitative standard applying to table olives in international trade. The document has its bases in Codex Stan STAN 66-1981 Rev. 1-1987 elaborated by the Codex Alimentarius and IOC.

Table olives are classified according to types, trade preparations, styles and size.

There are three types:

- Green olives: Fruits harvested during the ripening period, prior to colouring and when they have reached normal size;
- Olives turning colour: Fruits harvested before the stage of complete ripeness has been attained, at colour change;

- Black olives: Fruits harvested when fully ripe or slightly before full ripeness has been reached.

The trade preparations with the aim of removing olive bitterness and of preserving the product are:

- Treated olives: (green, turning colour and black olives) alkaline treated and packed in brine for partial or complete fermentation.

- Natural olives: (green, turning colour and black olives) fermented in brine without a chemical debittering agent.

- Dehydrated and/or shrivelled olives: (green, turning colour and black olives) alkaline treated or not, preserved in brine or partially dehydrated in dry salt and/or by heating or by any other technological process.

- Olives darkened by oxidation: (green and turning colour olives which become total black) darkened by oxidation in an alkaline medium and sterilized with brine in hermetically sealed containers; the ferrous salts included in the list of additives of the trade standard are used for this preparation.

- Specialties: such specialties retain the name "olive" as long as the fruit used complies with the general definitions and all relevant requirements of IOC standard, including requirements relating to limits for defects. The IOC standard also provides a list of the principal styles of presentation.

The olives are size-graded according to the number of fruits per kg indicated in the IOC standard. Size-grading is compulsory for olives in the whole, stoned (pitted) and stuffed styles.

Physico-chemical characteristics of the packing brine are well defined in the composition and quality section of the IOC standard. A minimum sodium chloride content, minimum acidity (expressed as % lactic acid) and maximum pH values are recommended for treated and natural olives, depending on the type of preparation as shown in paragraph 3.1.2.1. of the trade standard. These values change due to the addition of preservatives, heat treatment or refrigeration applied. A minimum of sodium chloride is also necessary for dehydrated and/or shriveled olives obtained without heat treatment.

The olives darkened by oxidation are regulated only by Good Manufacturing Practices because the pH, chlorides and acidity of the final product is dependent on the kind of processes used to obtain and to preserve it. Moreover, this product is usually sterilised.

The IOC standard gives indications about the characteristics of the pasteurisation and sterilisation treatments to be applied to table olives in order to obtain a safe product.

The list of additives are approximate to those described by the Codex Stan STAN 66-1981 Rev. 1-1987 [9], but in the Codex standard the maximum level is referred to a weight of every additive per weight of olives including brine, while in the IOC the maximum level is measured in weight of additive per weight of flesh.

According to the IOC standard, the containers used may be made of metal, tin, glass, plastic materials or of any other material, except wood, which complies with existing technical and health requirements. The regulation explains how to calculate the minimum fill for each kind of container, when a container should be considered defective and the net drained weight tolerances. The IOC regulation also describes the compulsory inscriptions of pack and containers and point-of-sale displays.

1.3. Main Trade Preparations of Table Olives in the World

Olive oil and table olive production has increased dramatically over recent decades, which has more than doubled for olive oils and 2.7 times for table olives. This increase is partly due to the establishment of intensive olive plantations using production systems which are very different from traditional systems, and incorporating techniques which have allowed high yields and a large degree of mechanisation [10].

There are three main trade preparations of table olives: Spanish-style olives, Californian-style olives and naturally black or turning colour olives (Greek-style). The Spanish processing method includes treatment with sodium hydroxide solution, washing, brining, fermentation and packaging. The Greek-style method is milder and includes washing, natural fermentation in brine, air-oxidation for colour improvement, and packing. The Californian method includes lye treatment, washing, iron-salt treatment and air-oxidation, canning and sterilisation. This last method includes a final sterilisation, so it is usually a safe product [4].

The three main techniques for table olive treatment used in Italy concern 82% green olives,16% black olives and 2% processed at the cherry-ripened stage [11].

1.3.1. Spanish-Style Green Olives

The fruits are picked when they are still green or a yellowish green, they are fully grown and the pulp maintains a good consistency. They must be manually picked in order to avoid any mechanical damage, the olives should be preventatively calibrated so as to obtain uniformity in the characteristics of the product. The fruits are treated with a diluted sodium hydroxide solution to eliminate the greatest amount of oleuropeinic glucosides. Sodium hydroxide cleaves oleuropein, resulting in an increase of concentrations of oleoside-11-methyl ester and hydroxytyrosol. Tyrosol, a hydrolysis product of ligstroside, is also formed during this step [12]. The NaOH concentration usually used may vary from 1.2 to 2.5% (w/v). Lye concentration is variable depending on olive cultivar, its degree of ripeness and temperature of the process. The lye treatment lasts until the lye solution penetrates the olive pulp to a depth of two-third to three-quarters of the pulp thickness.

The length of the treatment time can vary between 5 and 7 hours which is necessary for the olive variety usually used, while olives with particular drupe dimension and pulp consistency can require longer treatment times which, in some cases, can even exceed 10 hours. When the olives undergo longer treatment a more diluted alkaline solution is used so as not to alter the pulp structure in layers attacked by NaOH.

Following treatment with an alkaline solution, the successive phase consists of water washes which eliminates most of the soda which remains in the pulp. The duration and number of washes is another characteristic of the process which is very important. An excessive number of washes can lead to the loss of a high quantity fermentable material, including nutrients for microorganisms which cause fermentation. Moreover, excessive rinsing can lead to a depletion of the organic acids which are naturally occurring in olive pulp and therefore can lead to a loss of the buffering capacity with a consequent difficulty in maintaining a low pH value. The washing sequence applied lasts between 24-48 hours and can be described as follows:

- First wash carried out with vigorous water spraying;
- Immersion wash in water for 2-3 hours;
- Successive 3-4 immersion washes, for a longer period of time than preceding phase.

The inconveniences that can appear with alkaline treatments are to be found when the fruits have been treated with a low concentration sodium hydroxide solution for long periods of time, in which a successive poor fermentation can be encountered. When fruits are treated with a high concentration of NaOH, it is possible to obtain a pulp that is not very consistent and a high loss of fermentable materials following the excessive permeability of the drupe during the washing phase.

After washes which lead to a lowering of the pH value, the olives are put into brine, obtained by dissolving sodium chloride in water. A correct initial saline concentration and a correct pH value permit the development of strains of bacteria which produce lactic acid. The best strategy would appear to be that of maintaining a saline concentration of between 5-6% during the first fermentation phase, increasing to 7% at the end of the fermentation process. The brine concentration should be further increased during the storage phase in order to avoid the growth of altering or pathogen microorganisms. Monitoring of the pH of the brine, during the first fermentation phases and the eventual acidification, is a very important control procedure. At this stage the main goal is that of preventing an excessive development of Gram-negative bacteria. The attainment of optimal acidity conditions and brine concentration ensures that pasteurisation of the product will not be required (Fig. **1**). In any case, it is possible to apply heat treatment for the correct stabilisation of the product.

Figure 1: Spanish-style green olives (photo by Romeo).

1.3.2. Californian-Style Black Olives

The olives suitable for this process can be harvested as green-yellow, turning colour or totally pigmented. Fresh olives or olives that have fermented (for 2-6 months) can be used as a starting material in this kind of process. Brining for Californian-style black ripe olives is different from the Spanish-style green olive processing methods in that the olives are not treated with lye prior to brining. Therefore, the diffusion rate of fermentation substrates from olives in storage brine is slow in the California-style black ripe processing method [12].

The process for ripe olive production consists of successive olive treatments with a dilute lye. The olives are subjected to successive treatments in which the olives are immersed in diluted sodium hydroxide solutions for preset periods of time, so as to permit a gradual penetration of the alkaline solution within the pulp. The number of treatments varies from three to five. The lye concentration depends on the olive variety, on the state of ripeness, on the treatment temperature, which is generally lesser than that used for the Seville type debittering process, and varies from 1 to 2%. The alkaline concentration and the treatment time are both heavily influenced by the conditions used in the preliminary phase of storing in brine. The highest alkaline concentration is used in the first treatment phase while more diluted concentrations are used in successive phases. During the intervals between lye treatments, the fruits are suspended in water through which air is bubbled. Throughout this operation the fruits darken progressively. The colour formed is not stable and fades progressively after oxidation, therefore in order to prevent this loss, fruits are immersed in ferrous lactate or gluconate solution for several hours during the last washing step. The product has a final pH above 4.6, usually between 5.8 and 7.8, so its preservation is only achieved by sterilisation. The quality of these processed olives depends on final product colour stability and homogeneous firmness. For this purpose, an adequate protection from high temperature during storage seems necessary to comply with the current shelf life of this product [13].

The California-style processing method results in the lowest concentrations of phenolic compounds, especially hydroxytyrosol. Greek- and Spanish style processing methods provide more appreciable levels of phenolic compounds in table olives [12].

1.3.3. Greek-Style Black Olives

The fruits are harvested as ripe olives or just before the full pigmented state. The colour of the olives may be reddish black, violet black, deep violet, greenish black, or deep chestnut [12]. Olives are processed according to a traditional anaerobic method in which, after harvest, sorting and washing, the drupes are immersed in 8-10% (w/v) NaCl brine, where they undergo spontaneous fermentation by a mixed microbiota of Gram-negative bacteria, lactic acid bacteria and yeasts. Anaerobic conditions are necessary to avoid film-yeast and mould growth, which could affect the texture, aroma and taste of the final product. After an initial stage of vigorous fermentation, where the diverse microbial groups compete for nutrients, the process is dominated by lactic acid bacteria and yeasts that coexist throughout the process [14].

Due to the risk of a high pH value and low acidity the brine saline concentration is usually increased which frequently, at the end of the fermentation process, reaches and exceeds 10%. At the end of fermentation, with the aim of increasing colour, the olives can be exposed to air to favour the darkening process. The problems which can arise during this traditionally slow fermentation period, can be due to organoleptic decline and decline of nutrients following anomalous fermentation. At the end of the active fermentation phase it is necessary to stabilise and protect the product from further fermentation by means of the addition of a fermentation inhibitor, such as potassium sorbate, or by means of heat treatment. The pH is often controlled through the addition of acetic acid, and brought to a value of around 4.5. Black olives obtained in this way can be packaged in different types of containers according to form, volume and material. The brine used as the preserving liquid, which is often different from that where fermentation occurred, must have a minimum salt content of 7%. It is possible to use aromatisation (Fig. **2**), addition of vinegar, or transformation in pâté, obtained by pureeing of the pulp among the conditioning treatments which are used on the product following fermentation.

Figure 2: Greek-style black olives.

1.4. Other Widespread Methods Linked to Local Traditions

1.4.1. Natural Green Olives

This preserving method is applied, with variations, according to the location of production. The fermented olives can be whole (Fig. **3**), pitted or crushed. Pitting and, above all, crushing can be carried out mechanically. The olives must not be over ripe in order to be processed mechanically thus avoiding pulping of the olives which would lead to an unsuitable product. After pre-treatment the olives are then placed in barrels in brine with 6-7% salt concentration which, in some preparations, can reach higher concentrations which compromises successive fermentation. The sodium chloride brine content, in specialised firms, is rigorously controlled and gradually increased until final values of around 10-12% are reached. The salt which is added to the brine compensates for the salt absorbed by the drupes, and for the meeting of conditions which allow for the preservation of the product and development of microorganisms which are useful in the fermentation process. The olives acquire properties which are typical of the finished product and, above all, a slow debittering process occurs following the enzymatic resolution of oleuropein during this phase. Microorganism growth depends on the olives and on all the conditions linked to ripening, picking, and drupe treatment. Spontaneous fermentation agents, as well as adhering to the olive external surface, derive from different types of contamination. Characteristics and fermentation phases can be very similar to those which take place in the natural black olive debittering system (Greek-style).

Lactic bacteria and yeasts are the most important microorganisms for the primary fermentation of olives. *Lactobacillus plantarum* and *L. pentosus* are regarded as the main species leading to this process and are often used as a starter in guided olive fermentation [15, 16]. Lactic acid bacteria are part of the indigenous microbial community of olives, and species belonging to *Lactobacillus* genus are predominant during olive fermentation, whereas *Leuconostoc*, *Streptococcus, Enterococcus*, and *Pediococcus* are present in lower concentrations [17]. The main roles of yeasts in the processing of fermented olives, are associated with the production of alcohols, ethyl acetate, acetaldehyde and organic acids, compounds that are relevant for the development of taste and aroma and for the preservation of the typical characteristics of this fermented food [18]. *L. plantarum* generally coexists with the yeast population until the end of the fermentation process and during storage.

Spontaneous fermentation with a consequent lowering of the pH value to safe values (below 4.5), which are sufficient to protect the product from contamination by pathogen agents, does not usually require stabilisation treatments except, in the case of commercialisation, when packing with brine (preserving liquid) in containers is used followed by pasteurisation. The olive which is obtained maintains a hint of bitterness in its taste which is often appreciated by the consumers along with pulp crunchiness.

Figure 3: Natural fermented green olives (photo by Romeo).

1.4.2. Green Semi-Fermented Olives or Picholine Method

The variety most commonly used for this method, and after which the method is named, is the Picholine. It is a processing system which has spread throughout the south of France and successively to some areas of north Africa. This method, after having undergone some changes, and given a different name, is also applied in some olive growing areas of central Italy.

After preliminary sorting and cleaning treatments, olive debittering is carried out by means of 1.8-2.2% soda solution treatments for a variable length of time ranging from 8 to 72 hours. This period is strongly influenced by drupe characteristics, and as in the Seville system, the alkaline solution is made to penetrate the pulp to a depth of ¾ of its thickness. The olives are then placed in 5-6% salt brine which is replaced with fresh 7% salt brine after a period of two days. The pH is corrected by the addition of citric acid until safety values (<4.5) are met. After a maximum of 10 days the olives are ready but, given the absence of true fermentation and salt concentration, it is necessary to stabilise the product. The olives must be refrigerated, pasteurised or have conservatives added.

1.4.3. Salted Black Olives

Over-ripened olives are used which, after picking and preliminary treatments, are vigorously washed in water and placed in appropriate containers in layers alternated with a layer of rock salt until a percentage of 15-20% salt is reached. After a period of about 20-60 days, the olives acquire their typical wrinkled aspect which is due to dehydration caused by the extremely high salt concentration. They are extremely salty and maintain a slight taste of bitterness due to the incomplete degradation of oleuropein. This processing method originated in Greece where the Megaritiki variety presents optimal characteristics for its application.

When ready, the olives can be seasoned, pasteurised and packaged. Given the high salt concentration, heat treatment is rarely applied, unless the product is being prepared for markets which are further afield.

1.4.4. Baked Black Olives

The method was initially created for the preparation of the Maiatica and Ferrandina (Matera, Italy) olives, which then spread to other regions and other varieties [6]. The method consists of blanching the olives in water of a temperature of around 90°C for a maximum of three minutes, so as to encourage drupe permeability. Then, in purposely built barrels equipped with a drainage outlet, salting takes place in alternating layers of dry salt with a ratio of about 1/10, for a period of 2-3 days. The olives lose a large amount of vegetation water during this phase with consequent further sweetening due to the contemporary migration and hydrolysis of bitter compounds.

The olives, which are now partially dehydrated, are placed on wooden frames or plastic nets in a single layer, and the dehydration process takes place in a forced air circulation oven. Drying of the product is generally completed within 48 hours at an initial temperature of 30-40°C and a final temperature of 50°C. Once the olives are dehydrated and have a humidity content that meets safety values, they are piled on grates where, over a period of 24 hours, they reacquire a certain elasticity.

It is necessary to have olives which are completely ripe, which have a thin and extremely permeable skin, a good pulp/stone ratio, pulp which is black and intense reddish in all its thickness for this type of preparation. The residual humidity of the final product, compared to fresh pulp, progressively decreases during the processing phases, until it reaches a final value of around 13-15% at the end of the drying process. Dried olives can develop a rancid taste because of the oxidation of olive oil caused by the high temperature. Olives processed in this way are usually sold in food-safe flexible polylaminate packaging or packed in a modified atmosphere.

1.5. Traditional Preparations

Besides these most prominent preparations there are also many other traditional table olive elaboration recipes that are less well known internationally.

In the traditional process of natural brining which is widespread in the South of Italy, olives are washed, put into containers and then filled with freshly prepared brine. In order to preserve the acidity of these olives, the practice requires a pH of 4.5 or lower. Olives are handled in order to favour the growth of *L. plantarum* in brines, which is thought to be essential to produce the lactic acid needed for preservation and typical flavor. Traditional olive productions are often unpacked but are frequently sold in local markets in glass jars or plastic pouches. In general, the final packing is achieved with fresh brine, but the re-use of fermentation brine is also possible for this operation but leads to different sensorial characteristics [19].

Unfortunately the majority of traditional products is affected by a low level of standardization, leading to a high heterogeneity in physico-chemical and microbiological characteristics.

1.5.1. Castelvetrano System

The method was developed in the area of Castelvetrano (Trapani, Italy) for the Nocellare del Belice variety, which is the main cultivar used in the area. The "Castelvetrano" system (also called "dolcificata" in the local language) is typical of the western area of Sicily. Olives are harvested when green at the end of dimensional growing. Fruits are mechanically graded by size in order to select only olives with a diameter >19 mm. The olives are then placed in plastic drums, treated with lye (1.5-2.2%) overnight and after 12 hours 5-8 kg of salt is added to obtain a 6-7% salt concentration. Treatments are usually carried out in 220 litre PVC barrels which can hold around 130-145 Kg of olives depending on the drupe dimension. The containers must be kept absolutely full with successive addition of brine. This period corresponds to the time necessary so that a balance between the olives and the brine constituents is reached created by the alkaline solution used for the debittering process. After 15 days of storage the product is ready, the olives are then washed and sold in local markets before the temperature increases with the spring season [20].

The conserve obtained in this way has a very limited life and is closely linked to the conservation temperature. Residual alkalinity within the barrels, with a temperature increase, causes anomalous fermentation which, as well as altering the organoleptic properties of the olives, does not create an environment for correct and safe food hygiene. Often, in order to improve preservability, the brine pH is adjusted with the addition of lactic acid so as to bring it to a value of less than 4.5, which allows the product to be preserved without requiring sterilisation which can alter the organoleptic properties. Given the instability of the product, it is necessary to ensure preservation using one of the following methods: pasteurisation, addition of preserving agents or refrigeration.

1.5.2. Itrana Olives

This is a production that is typical of the region of Lazio where the Itrana cultivar takes its name from the town of Itri. The olives are picked once ripening is complete and left to ferment directly in drinking water for less than a month. After this initial period, in which a first fermentation occurs due to bacteria present on the fruits, salt is added to create brine (7%) in which fermentation continues. The olives from this cultivar are ready in six months and sold unpackaged or in glass containers. The olive from the Itrana cultivar is prized for the ease with which the pulp detaches from the stone.

1.5.3. Cracked Olives

These green olives are very common in southern Italy, in particular in Sicily and Calabria. In Calabria the Carolea and Imperiale cultivars are used, and are picked when green and still hard. The olives are cracked manually or mechanically and washed several times in cold water. They are then immersed in boiling water for a few minutes for quick debittering, and then cooled in cold water. They are left in running cold water or in brine for 3-5 days. At the end they are drained, seasoned with garlic, oregano and sometimes red chilli pepper, and consumed alone or as an accompaniment.

1.5.4. "Singate" Olives

Singate olives are picked when green but are riper than cracked olives. Each olive is longitudinally cut with a knife 4-5 times. *Singate* olives are placed in a container into which boiled water is poured and left to act for 5-10 minutes (lightening the colour). Then the olives are debittered in running cold water for 5-8 days. *Singate* olives can be seasoned in the same way as cracked olives.

1.5.5. "Passulune" Olives

In the western area of Sicily region, the *passulune* olives are natural over-ripe olives, left on the tree and harvested in December-January. These olives are washed with hot water and left to air-dry. As reported by several authors [21, 22], these olives have undergone debittering consequent to an attack of the *Camarosporium dalmaticum* fungus, introduced in the drupe by a parasite from *Bactrocera oleae* eggs. The phenomenon is encountered in rather a consistent manner in years in which there is a large presence flies.

1.5.6. "Mpassulute" Olives

In Calabria, *'mpassulute* olives are natural ripe olives harvested in December-January and left to wilt in the sun for 7-8 days. They are then washed in hot water (about 50-55°C), then drained and placed in a cotton sack and are abundantly salted in layers. The sack is placed on a shelf and stones are placed on top of the sack to press the olives. The olives are mixed at least twice and in the meantime lose liquid due to salting. After around 10 days the olives can be seasoned.

1.5.7. Smoked Olives

This traditional technique is found in Lazio for the Itrana cultivar. The olives are hung in baskets at the fireplace hood so that they are smoked for a period of 20 days [5]. The thermal effect of the smoke causes rapid debittering, whilst endowing the olives with aromatic tastes which are very characteristic. The olives are very dry at the end of the process and are rehydrated in hot water and then seasoned.

1.5.8. Oven-Dried Olives

This type of olive is typical of southern regions, above all Basilicata, Sicily and Calabria. Ripe olives are cut and blanched and then left in cold water which is changed at least twice a day, for around 10 days. The olives are then dried, covered in salt and placed in an oven at a low temperature (45-50°C) for one or more hours (depending to the olive calibre) in order to obtain drying. The best known example of this type of olive is the black Maiatica of Ferrandina which is one of the most important typical products of Lucania.

1.6. Biological Method: Starter Culture Addition

Foods and drinks that today are products of fermentation, which represent the edible foundations of every society, are those which derive from microbial or enzymatic activity. This is exploited in order to cause biochemical modifications in the substrate to improve organoleptic and hygienic properties of the final product, which consequently becomes more nutritious, more digestible and safer. The nutritional values of fermented foods is of great importance in developing countries in which the inhabitants suffer from nutritional deficiencies, while in developed countries attention is focussed on improving control, and on the safety and stability of these types of food [23].

The fermentation of vegetables is mainly carried out by lactic bacteria, even though yeasts and other microganisms can also be involved, dependent on the percentage of salt added and on other factors. Lactic acid bacteria (LAB) have long been employed in fermentation as a food preservation technique owing to their progressive acidification of the fermenting brine with a consequent pH decrease [24]. The species of the genus *Lactobacillus* is widely occurring in many natural environments often playing important roles in fermentation processes and in the regulation of relationships among species of complex ecosystems [15, 16, 25, 26]. The role of these bacteria is carried out through the production of lactic acid which, acidifying the product, consents a longer preservation time.

Lactic bacteria compete with the multitude of microorganisms present on the olive epicarp, and since they represent a minor part of microflora, it is necessary to create an environment which is suitable for their development, disfavouring competitive bacteria. The addition of salt and the natural breathing process of olives and microorganisms, which consume oxygen in the brine, contribute to this process. Acidification, along with CO_2 that may also be produced, create an even more stringent environment for competitors. As soon as these conditions occur, lactic bacteria will find themselves in a favourable condition compared to competitive microorganisms.

It is evident that the most delicate phase of fermentation is the first one as just one error is sufficient for the product to be subjected to consequences which are, at times, irreversible.

In table olive processing, the starter culture addition should lead to the complete utilization of fermentable sugars and rapid acidification. LAB starters are selected for their good tolerance against high levels of salt, olive phenols and low pH values. Moreover, some strains show an inhibitory effect against undesirable organisms due to the production of bacteriocin, peptides that were found to be active against a number of natural competitors of *L. plantarum* in the fermentation brines and also against bacteria that can cause olive spoilage [25].

L. pentosus and *L. plantarum* are the most frequently isolated species in table olives; the other species used as inocula, with little exception, have always been studied in conjunction with them [26, 27].

The development of starter culture use in table olive fermentation is due to achieving a more controlled process, reducing debittering time and improving the sensorial and hygienic quality of the final product [17, 28, 29].

After the selection of the most appropriate species, *L. plantarum, L. pentosus* or others, the starters are grown overnight at 32°C in a broth medium suitable for lactobacilli (MRS broth). This fresh culture may also be inoculated a second time in broth supplemented with 4 or 5% (w/v) of NaCl to allow adaptation of strain cultures to the saline environment of the brine. Then, the cells are harvested, washed and re-suspended in physiological water, 0.9% (w/v) NaCl, and added to brines. The starter culture should have a final concentration of at least 10^8 CFU/mL in order to withstand the competitive mechanisms. The better choice is contemporary inoculation of two different LAB species because in vegetable fermentations, *L. plantarum* bacteriophages could be present and destroy all starter cells. Bacteriophages that infect *L. plantarum* have been isolated from fermented vegetables, meat and cheese [30, 31].

1.7. Hygiene and Microbiology of Fermented Olives

The processing phases, in a company that produces table olives, are the following: receiving of the raw materials or receiving of the product (in the case in which the company works with semi-finished products or imported raw materials), storage, sorting, washing, fermentation and debittering phases, acidification, packing, eventual heat treatment and storing of the finished product. It is necessary to be aware of the hazards, the relative preventative measures to be adopted and relative critical limits, and the corrective actions for each of these phases. In the initial receiving phase, hazards are represented by physical, chemical and microbiological pollutants, which are preventable through careful selection or control of suppliers, avoiding relations built exclusively on trust, but rather through direct control of the quality of raw materials. If a semi-finished product has been purchased, it is necessary to check the pH, therefore corrective action could be acidification or heat treatment.

Instead, with regards to storing, a hazard is represented by microbial proliferation which can be prevented by working according to hygiene practices and regulations and by carrying out temperature and humidity checks on the spaces and storage areas. The same type of danger also exists for the sorting phase, where the handling of the product by staff, if not properly trained in hygiene, represents a contamination risk of staphylococci and coliforms.

Water quality used for washing the olives is extremely important. The water used must be drinkable otherwise there is a risk of it being a vehicle for foreign bodies and chemical and/or microbiological contaminants. Monitoring of the pH value in the phase of acidification and fermentation, which follows washing, is very important as it must not be too close to the critical limit of 4.6, as it is necessary to consider that such a value could undergo slight variations, within a short period of time, which depend on factors linked to the transformation method used. In that case the sole corrective action consists of acidifying the product once more.

Various hazards exist in packaging: the material used, mechanical damage or glass fragments, correct sealing of the containers, air-tightness of containers or heat sealing when dealing with plastic containers, recontamination of the packed product as a result of damaged containers or containers which have not been sealed correctly.

Checking the heat process of acidic foods such as table olives, is simply represented by the verification of the time-temperature treatment binomial employed. Temperature control during sterilisation is extremely important for olives with a $pH > 4.6$.

Finally, when storing the finished product, the only hazard is represented by the deterioration and alteration of the product consequent to an improper conservation at temperatures that are too high or in places that are effected by solar radiation. Moreover, it is necessary to remove all damaged containers, or those in which the product shows a colour change. Transfer risks from containers nowadays are improbable if undamaged and if appropriate containers are used which are suitable for acidic foods.

Fermentation of vegetables in brine is carried out by lactic bacteria, even if yeast and other microorganisms can be involved, according to the percentage of salt and other factors. Salt is used as it conditions the type and entity of microbial activity. It is important to understand fermentation mechanisms in order to correctly dose salt since it is possible to use brine that is so concentrated that it greatly delays, or even completely precludes, fermentation. The salt concentration to be used depends on the type of vegetable to be preserved and on its tendency for cellular decline during storage in brine. Salt aids in inhibiting pectinolytic enzymes and consequently prevents softening.

When a high concentration of organic acids is used in order to prolong preservation, the resulting product has a strong taste while market tendencies for fermented vegetables focuses more and more on products with barely perceptible acidity.

1.7.1. Environment Hygiene

Legislative decree n. 155/97 in Italy provides for the application of company self-regulation under the HACCP system. The Business Hygiene Self-Regulation Manual contains an analysis of the risks in relation to dangers with their relative control and surveillance procedures. On the other hand, the Self-Regulation System must also activate all the general preventative measures such as hygiene of the staff, of the environment, of the equipment, and the suitability of the water. The points identified as being critical must be placed under control and documentation must be produced.

The order of the spaces and equipment must provide for a "processing path" so as to prevent cross contamination of the food from the moment it enters the premises until it is packed or distributed. A one directional flow of the produce must be created, preventing the finished foods (olives) coming into contact with raw materials and unprocessed materials, the latter having an extremely high microbial load. It is possible to distinguish separate areas in the "processing path" for the storage, preparation and distribution of goods, for the cleaning of equipment and an area for waste disposal.

Aeration must be sufficient to guarantee adequate air change and eliminate odours and vapours which always form during processing. The uptake point of the air intake must be situated far from polluting sources, and in any case it is necessary to ensure that it does not pollute the premises. Aspiration and air-conditioning systems must be well maintained and efficient and if the hoods are equipped with filters they must be washed and cleaned weekly. The appropriateness of construction materials is fundamental: the walls of the kitchens and places where foods are handled or packaged must be completely covered in waterproof materials, the floor should be built with a slight slope for the draining of water from the washing phase. The spaces used for storage must be dry, washable, aerated directly from outside, and have anti-insect screens on the windows.

Bottle stores must be ready-made in protected areas, avoiding contamination during storage or storing in uncovered places where infestation by rats, birds, or insects can occur or where dirt can gather, or even where overheating can occur due to sunshine.

In the vicinities of the premises and their annexes rubbish, waste or other solid or liquid materials, capable of generating noxious emissions, cannot be stored.

1.7.2. Staff Hygiene

One of the most important moments in prevention is to ensure that the staff have an in-depth understanding of the importance of hygienic behaviour. It is opportune to organise a training programme which provides information on general personal and environmental hygiene, the obligations and responsibilities of food processing industries and hygiene risks linked to the production, handling and packing of food products, preventative measures to be adopted and relations with controlling bodies. Training should be updated at least each year or when there are changes in regulations.

Food handling staff are an important source of contamination: skin and above all hands, eyes, nose, mouth and ears. Moreover, when hygiene is not paid attention to after using the bathroom, there is a risk of contamination which is typical of the gastrointestinal tract. Microorganisms which can be transferred from man to food, and which are potentially pathogenic are coliform bacteria and positive-coagulase staphylococci.

1.8. Table Olive Deterioration

When control of the pH and saline concentration is inadequate, olives can present anomalous alterations. Ensuring hygienic practices throughout processing and packaging avoid most of the possible olive spoilages. Alterations can be identified which affect both green and black table olives, during processing and in the storage phase of the finished product which, obviously, depreciate the product. Levels of spoilage organisms can increase depending on temperature, so fermentation brines should be maintained at about 20°C for the entire duration of the process, but above all during fermentation.

1.8.1. Putrid Fermentation

This alteration usually occurs in green Spanish olives during the first fermentation phase, when the high concentration of fermentable substances and the high pH values allow for the development of bacteria (especially belonging to the

Clostridium genus) which are responsible for putrid fermentation. The alteration is usually found in dirty containers and with a pH which is not adequate for the limited action of lactic bacteria. The affected produce have an aroma of decomposing organic matter even though subjected to washing or repeated brine substitution. The precautions are of a hygienic-sanitary nature: containers and utensils must be hygienically suitable, rendered such by the use of opportune disinfectant products. It is important to control brine saline concentration during the first fermentation phases.

1.8.2. Butyric Fermentation

This alteration also regards green Spanish olives during the first fermentation phase when, following alkaline deposits (residual soda), high pH values can occur which together with low saline concentration and higher than usual temperatures, favour the development of anaerobic butyric bacteria of the *Clostridium butyricum* species. These can be present in dirty barrels, in equipment and in the environment. Consequently, even in this case, a correct hygiene of the barrels and equipment, as well as the monitoring of fermentation parameters (pH and brine salt concentration) allow for the alteration control and prevention. The olives affected present a typical aroma and taste of rancid butter which will be impossible to eliminate from the drupes. The alteration can only be avoided through preventative measures: control of the pH trend and saline concentration, occasional homogenised brine recycling.

1.8.3. Zapateria Spoilage, Propionic-Butyric Fermentation

This alteration usually strikes green olives processed with the Spanish system. This spoilage occurs late in processing. The olives irreversibly smell of old leather [32] of such an intensity that the products effected are absolutely inedible. This type of secondary fermentation is determined by sporigenic bacteria of the *Clostridium* genus which are common in the environment and could easily contaminate olives and processing utensils. Some authors have also attributed this alteration to the presence of bacteria of the *Propionibacterium* genus [33]. Usually alteration develops at the end of lactic fermentation if the salt concentration is not controlled effectively. Propionibacteria develop in these conditions which, following consumption of lactic acid and successive pH value increase, allow growth of

Clostridium bacteria. Prevention, which is the only means of avoiding alteration, is carried out through simultaneously controlling the pH and saline concentration. With a pH of less than 4.7 all *Clostridium* bacteria are inhibited by a salt concentration greater than 3%, whereas, propionic bacteria are inhibited by a pH of 4.5 and by 5% salt concentration. It is possible to note that both the pH value and the brine concentration are determinant to a correctly conducted olive fermentation in brine.

1.8.4. Gassy Alterations:Gas-Pocket or Fish-Eye, Alambrado

Gassy alterations are characterised by the formation of gas pockets either under the cuticle or in the pulp. These alerations affect both green and black olives and are caused by gram-negative bacteria of the *Escherichia, Enterobacter, Klebsiella* and *Aeromonas* genera which are able to produce carbon dioxide. These bacteria grow well in high pH conditions or where the salt concentration is less than 6%, above all during the first fermentation phase. The gassy alterations can occur early in processing, but with directly brined olives which are turning colour or naturally black olives. It can occur at anytime depending on conditions [32]. The formation of blisters, characterized by fissures under the skin is a spoilage known as *Alambrado*. Vaughn *et al.* [34] and Durán Quitana *et al.* [35] related this alteration to the presence of *Saccharomyces cerevisiae* and *Wickerhamomyces anomalus* strains.

This alteration can be prevented through pH correction to values around 4 with the addition of acetic or lactic acid, by ensuring strict anaerobic conditions and scrupulous hygiene standards, by increasing salt concentration of the brine so that NaCl $\geq$8% w/v, by supporting the use of starter cultures. Finally, good hygiene practices are necessary to reduce the bacteria load and, consequently, the Gram-negative bacilli, which are the cause of gas development.

1.8.5. Softening

This alteration causes softening of the pulp, until the complete pulping occurs. Olive consistency is linked to the pectin substances present in the pulp. These polymers are attacked and modified by enzymes which are present both in the fruit and, above all, are produced by microorganisms such as moulds of the following genera *Fusarium, Penicillium, Aspergillus*, *Geotrichum,* but also Gram-negative and *Bacillus* spp. bacteria and yeasts. Some yeast strains as *Rhodotorula minuta*, *Wickerhamomyces anomalus*, *Debaryomyces hansenii* and *Pichia galeiformis* are able to produce enzymes that could cause the softening of fruits such as proteases, xylanases and pectinases. Certain strains of pigmented yeasts, *R. glutinis*, *R. minuta* and *R. rubra*, can grow and form pellicles in olive brines and produce polygalacturonases which also cause softening. Under the usual anaerobic conditions in which these products are packed, olives should be stable even in the presence of these microorganisms [36].

This alteration occurs during fermentation and during storage due to a pH>4.8 and/or a low saline concentration. Moreover, for olives that are treated with the Spanish system, it is good practise to avoid excess water soaking and maintain anaerobic conditions above all with the aim of protecting olives from this alteration. The phenomenon is irreversible. It can, however, be avoided only through precautionary measures such as: control of the pH value, of the saline concentration and by preventing the formation of yeast and mould films on the brine surface and by ensuring that the containers are kept completely full. During fermentation this alteration is prevented by encouraging a rapid lowering of the brine pH during the first phases.

1.8.6. Shriveling

This alteration mainly occurs in brine with a salt concentration which is too high and which consequently leads to osmotic dehydration of the olive pulp. Such a phenomenon is reversible and, acting promptly with opportune brine dilution, the olives reacquire their initial swelling.

1.8.7. Darkening

This alteration affects natural fermented black olives, or olives which are exposed to air to envourage the blackening process. The drupes assume a bluish-black colouring (also known as *cyanosis*). With green Spanish-style olives, darkening reduces their visual appeal because they turn grey-green in colour. The alteration is often accompanied by softening of the pulp and by an unpleasant odour. This alteration is irreversible. It can be prevented through the use of brine with a saline concentraion that is greater than 8%, avoiding excessive temperatures (30-32°C). In some cases a correlation between irrigated olives and darkening onset has been identified [4]. Product pasteurisation and the addition of antioxidants prevents cyanosis in packed olives.

1.8.8. Brine Surface Films

Yeasts and moulds that can utilize oxygen from the air above the brine surface can grow in the upper levels of the fermentation container, lowering brine acidity and thus raising the pH [32]. The increasing pH levels could promote olive spoilage and growth of microorganisms which are potentially pathogens, so this situation must be prevented through brine surface monitoring and, where necessary, correction of the pH value. The problem can be prevented in fermentation tanks with well-sealed lids and the air layer between the liquid and the top of tank must be reduced as much as possible to produce anaerobic conditions. The prevention of this alteration on the surface of brines in packed olives can be realized by adding a preservative such as potassium sorbate, as permitted by the list of IOC. This alteration could occur both early and late during fermentation.

1.9. Sensory Analysis of Table Olives

Sensory analysis allows for the identification of the organoleptic properties of a product by means of sensory organs. It is carried out by a panel, a group of persons who have been selected, trained and equipped with the necessary skills by means of scientific methods to perform sensory analysis in accordance with international sensory analysis standards.

Several years ago, the International Olive Council of Madrid (IOC) formed a study group consisting of experts from the main olive-growing nations in order to discipline sensory analysis of table olives.

On November 2011 the International Olive Council adopted the revised version of the official method COI/OT/MO/Doc. No 1 (2008) and it is well documented throughout COI/OT/MO No 1/Rev. 2 document [37] which establishes the necessary criteria and procedures for the sensory analysis of the odour, taste and texture of table olives. The purpose of this method is to perform the sensory classification of table olives according to the intensity of any defects as determined by a group of 8–10 trained tasters. The test room and equipment are ruled by specific COI standards: standard COI/T.20/Doc. No 6/Rev. 1, guide for the installation of a test room (or ISO 8589:2007) and glasses according to standard COI/T.20/Doc. No 5 (glass for oil tasting). All the other equipment and accessories are described in COI/OT/MO No 1/Rev.2 document [37].

The olive samples stored under refrigerated conditions are removed sufficiently in advance to allow them to reach the test room temperature when they are to be tasted. The sample presentation is well described in COI/OT/MO No 1/Rev.2 document (Fig. **4**), available from the COI web site (see section 2.1).

The panel group is trained to know the specific vocabulary for table olives sensory test, in particular the descriptive gustatory attributes: salty, bitter and acid, which involve distinct areas of the tongue: the region affected by the perception of salty taste is the lateral-anterior, the region affected by the perception of acid taste is the posterior and the region affected by the perception of bitter taste is the base of the tongue.

The negative attributes:

Abnormal fermentation: olfactory sensation perceived directly or retronasally, that may be:

- Putrid: odour of decomposing organic matter.

- Butyric: sensation reminiscent of butter or cheese.

- Zapateria: sensation of rotten leather (caused by the combination of volatile fatty acids).

Musty: olfactory-gustatory sensation perceived directly or retronasally, characteristic of olives attacked by mould.

Rancid: olfactory sensation perceived directly or retronasally, characteristic of olives that have undergone a process of rancidity.

Cooking effect: olfactory sensation perceived directly or retronasally, characteristic of olives that have undergone excessive heating in terms of temperature and/or duration during pasteurization or sterilization.

Soapy: olfactory–gustatory sensation of soap.

Metallic: olfactory–gustatory sensation reminiscent of metals.

Earthy: olfactory-gustatory sensation reminiscent of soil or dust.

Winey–vinegary: olfactory–gustatory sensation reminiscent of wine or vinegar.

Texture is defined as the set of rheological (related to the flow and deformation of matter) and structural (geometrical and surface) properties of a product perceptible to the mechanical receptors, tactile receptors and, in some cases the visual and auditory, receptors.

The kinaesthetic sensations are the follows:

Hardness: attribute relating to the force required to attain the deformation or penetration of a product. It is evaluated by compressing the product between the teeth or between the tongue and palate.

Fibrousness: attribute relating to the elongated conformation of the particles, oriented in the same direction. It is evaluated by perceiving the fibers between the tongue and palate when chewing the olive.

Crunchiness: attribute relating to the friction or fracture between two surfaces. It is related to the force required to fracture a product with the teeth and is determined by compressing the fruit between the molars.

The tasters bend the glass gently to help the sample aromas to be released. Then they remove the watch-glass and smell the sample for no more than 20 seconds. The tasters then place one of the olives contained in the glass in their mouth to assess the other sensations cited above. They chew the olive after removing the stone. They concentrate on the salty, bitter and acid stimuli, the retronasal olfactory sensations and the kinaesthetic sensations (in this sequence). Finally they assess the intensity of each of these sensations by making the corresponding mark on the intensity scale of the tasting sheet. After rinsing out their mouth with water, they repeat the test for each of the olives contained in the glass.

TABLE OLIVE PROFILE SHEET

	INTENSITY →
PERCEPTION OF NEGATIVE SENSATIONS	
Abnormal fermentation (type)	________________
Other defects (specify)	________________
PERCEPTION OF GUSTATORY SENSATIONS	
Salty	________________
Bitter	________________
Acid	________________
PERCEPTION OF KINAESTHETIC SENSATIONS	
Hardness	________________
Fibrousness	________________
Crunchiness	________________

Sample code:
Name of taster:
Date:

Figure 4: Table olive profile sheet (COI/OT/MO No 1/Rev.2 document).

1.10. Main Packaging Technologies of Table Olives

The packaging operation of whole or cracked fermented table olives, performed without atmosphere modification, led to instability and spoilage of the final product. This instability is mainly displayed by swelling and browning due to microbiological and physicochemical changes which occur during shelf life [38]. The main problem for the stabilization of the product, if sugars are not completely fermented, would be related to the microbial population present at the time of packaging and to the CO_2 produced by the fruits' respiration during the first days of packing. Oxygen consumption and CO_2 production change the environment selecting anaerobic bacteria. The obtained final product could have sensory alterations or hygienic risks.

1.10.1. Pasteurisation and Sterilisation

Pasteurisation is carried out at temperatures lower than 100°C and permits the obtainment of products that can be defined as partially preserved. It determines the destruction of vegetable forms of the majority of microorganisms, while it allows the survival of bacterial spores. The pasteurised product is free from pathogen forms, but a small part of microorganisms is still present in it, such as those which are thermoduric which are able to survive this treatment and serve as indicators of correct thermal treatment.

Olive production factories mainly use tunnel type pasteurisers. Containers, filled with olives and the preserving liquid, are sprinkled with hot water during the heating phase until the operating temperature is met. The pasteurising tunnel is formed by a conveyor belt which moves the containers into different heating, operating and cooling sections, set at different temperatures.

The temperatures applied in sterilisation are greater than 100°C. Theoretically, sterilisation should destroy all microorganisms, including spores. In practice, technology is defined as sterilisation when it regards a heat treatment that consents the reduction of *Clostridium botulinum* spores by 12 logarithmic cycles.

Sterilisation (or in this case "appertisation"), is carried out with the use of autoclaves in which the product to be treated is placed. When the risk of botulism is extremely high, the food must reach very high temperatures (from 110 to 120°C) for extended periods of time. Naturally acidic foods, such as vegetable preserves, are the only food for which it is possible to use a lower temperature, immersing the containers in boiling water and reaching temperatures of around 100°C. The necessary heat is insufflated in the chamber by means of superheated steam so that the product reaches the desired temperature for the necessary time.

The olive total polyphenol content undergoes a diminishment in all the samples subjected to treatment both following a partial release of the pulp to the brine and in the oxidisation process of the same compounds, causing a browning of the olives, due to polyphenol oxidase (PPO) activity [39]. Bibliographic research reports that it is possible to observe an increase in luminosity in food subjected to heat treatment, which can be attributed to the loss of PPO enzyme activity, while a diminishment of such chromaticism is associated with a higher degree of browning [40].

1.10.2. Post-Sterilisation of Packaged Olives

The advantages of this method consist in the optimisation of thermal exchange during treatment due to the possibility of varying the form of the container, obtaining a less drastic sterilisation treatment, the containment of energy costs, greater rapidity, and the use of lighter and more economic packaging. The technique consists of the sterilisation of the foodstuff in a flexible package, and for liquid or semi-liquid foods in flexible heat resistant packaging. The product obtained with this heat treatment is of average quality that does not maintain the aromatic qualities of the fermented olive.

1.10.3. Aseptic Packaging

Aseptic packaging consists of the sterilisation of packaging materials preceding the food heat treatment. Filling and sealing operations must occur in sterile conditions. This is the critical point of the process since maintaining asepsis during filling leads to the disadvantage of having sophisticated and costly systems, and in having a non-resealable package. The advantages include a less drastic sterilisation treatment, in which the organoleptic and nutritional properties of the olives are protected. Moreover, it leads to containment of energy costs, greater rapidity, and use of lighter packaging materials. The most common decontamination method of materials used are superheated steam, hydrogen peroxide with a concentration of 15-35%, hydrogen peroxide+UV, alcohol+UV, peracetic acid, ozone and chlorine in various forms.

1.10.4. Vacuum Packaging

This packaging method is very common for dried olives, or for those which do not require a preserving liquid. The objective of vacuum packaging is that of removing air and therefore oxygen in order to confer stability on the product, impeding the proliferation of aerobic moulds and bacteria, browning, and colour modifications, even by means of the activation of enzymes such as polyphenol oxidase. Furthermore, the fat fraction which is high in olives, is protected from rancidification. The product maintains its aromatic properties almost intact since the volatilisation of low molecular weight molecules is avoided.

1.10.5. Modified Atmosphere Packaging (MAP)

The MAP technique is not a simple protection from degrading agents, as with traditional packaging, nor does it have the "passive" role of removing air from the packaging, but rather the possibility of intervening in the control of some degradation phenomena.

The use of modified atmospheres, in any case, should not be considered as a means of qualitative improvement of an inferior quality food product, but rather it should be considered as a technological operation which, together with other quality control interventions, can extend the shelf-life of the product. However, the positive effects of modified atmosphere packaging can be rendered null by the inappropriate choice of packaging materials, in terms of barrier and selectivity to gas and water vapour, of the delivery system and of the gas mix.

1.10.5.1. Nitrogen Atmosphere Packaging

Nitrogen is a gas which does not interact with the organic substance which it comes into contact with, therefore it results as being totally inert. Nitrogen is odourless, tasteless, slightly soluble and slightly permeable through plastic materials and does not alter the sensorial properties of foods. An atmosphere formed exclusively by nitrogen is to be considered very stable due to the low solubility of the gas in foodstuffs and due to its low permeability through packaging materials. For these reasons it is used to avoid collapsing of the packaging materials on the product. An atmosphere of this type can be used for products in which there is no risk of development of anaerobic microflora during preserving. The lack of oxygen avoids oxidisation of fats as well as the possibility of proliferation of aerobic microorganisms, such as moulds.

1.10.5.2. High Carbon Dioxide Content Atmosphere

Carbon dioxide inhibits vegetable respiration and slows ripening, inhibits hydrolysis of pectins (avoiding fluidification), reduces vegetable tissue damage from the cold. It melts in the food water giving rise to an acid reaction. An anaereobic environment in contact with food is created by substituting air with carbon dioxide, favouring the multiplication of anaerobic or microaerophilic species, such as lactic bacteria. The long preservability of these food products is due to the proliferation of the lactic bacteria. In the use of atmospheres rich in carbon dioxide it is necessary to prevent possible inconveniences mainly deriving from the reaction of the gas with food components and its solubility in liquids. High concentrations of the gas can lead to changes in sensorial properties and, more specifically, the product can result as being "acidic" or "fizzy" rendering it unpleasant for the consumer. Moreover, this type of atmosphere lacking in oxygen, should be applied to acid foods since there could be a risk of growth of bacteria of *Clostridium* genus, which is a potential producer of highly toxic molecules. Finally, according to Arroyo-López *et al.* [41] storage under CO_2 atmosphere led to a better preservation of polyphenols, a delay and control of microbial growth, and a contribution to the retention of the fresh appearance of the fruits.

1.11. Virgin Olive Oil

1.11.1. Morphological Characterisation of Olives

The identification of the cultivars present in the company is the first requisite for the production of high quality extra virgin olive oil. This step is recommended in order to identify the optimal harvesting time which is typical for each cultivar (Fig. **5**), to adapt the operative transformation conditions to the chemical and physical olives composition properties, and therefore to optimise the entire process which transforms olives into olive oil. At the end of monocultivar processing the oils can be mixed to form blends.

1.11.2. Technological Ripening of Oil Olives

From a bibliographical analysis it is possible to deduce how the quality of an extra-virgin olive oil is dependent, above all, on the properties of the raw materials, intended as the degree of ripening and healthiness of the olives [42]. Studies in literature show how sugars in olives are the precursors for biosynthesis of the oil [43-47] and during the oil extraction process, olives with a high sugar content can give origin to defective final products, due to fermentative phenomena due to the sugar component [48]. Simple sugar concentration can be considered as an index of the technological ripening of the olives, capable of indicating the correct level of ripening for their processing. Studies carried out have demonstrated that the correct degree of ripening of olives corresponds to the attainment of a minimum and constant value of sugar concentration and to a constant and maximum oil content. The minimal sugar value permits the obtainment of olive paste with a reduced risk of fermentative activity in the oil mill. The kinetics of the decrease in sugars over time resulted as being variable depending on years and productive areas. The phenomena which are responsible for this variability are unknown, even if it would appear logical to presume that they are linked to pedoclimatic and varietal aspects. It was, however, possible to observe that a constant environmental temperature ≤ 10°C (recorded at 8 a.m.) corresponded to a safe minimum constant value of sugar concentration.

Figure 5: Olive drupes at different stage of ripening.

1.11.3. Harvest

Harvesting can be carried out using different systems: manual harvesting, shaking or mechanically by raking, or with mechanical shakers.

The most used technique is manual harvesting, but mechanical harvesting is gaining popularity due to the high labour costs involved in manual harvesting. It is essential to avoid lifting olives from the ground, because some elements which are naturally occurring in the soil, such as microorganisms, facilitate contamination of the fruits by moulds; or even metals such as iron or copper which compromise conservation of the oil as they accelerate the oxidisation process.

1.11.4. Transport and Olive Fruit Storage

It is a simple operation yet, if conducted incorrectly, has the potential to cause a negative effect on the final quality of the oil. After harvesting, it is recommended that olives are transported to the oil mill in perforated plastic crates with a maximum weight of 25 kg. They should then be processed within a few hours of arriving in the mill (about 8 hours), the storing environment must be sheltered from outdoors, be cool and aerated (with a temperature of 14-18 °C), and be clean and odourless.

The olives must be kept in cool aerated places and preferably protected from light and heat sources. Particular attention must be paid to this phase in order to avoid problems of overheating, of mould or anomalous fermentation due to an extended lack of aeration of the fruits or due to having lifted them from the ground. For this reason, intact olives must not be stored along with olives which are at an advanced stage of ripening, nor with those which have been lifted from the ground and which have surface damage and evident surface blemishes. This advice allows olives to be kept intact as long as possible, containing the action of endogenous enzymes (lipase) which are responsible for the increase in oil acidity as well as external microbial proliferation which is responsible for the decaying processes such as the fermentation phenomena which determines oil defects such as the oil being winey or fusty [48].

1.11.5. Leaf Removal

This operation, which is carried out using an oscillating screen, often together with aspirators, is necessary to avoid the accumulation of large quantities of leaves or other vegetable waste during the productive process, but it also serves to remove foreign bodies such as earth, stones, wood residue, *etc*.

Nevertheless, the common practice of leaving some leaves in the working process to influence the final oil colour does not, contrary to common belief, significantly change the total chlorophyll value present and consequently the intensity of the green colour of the oil. Such a value is entirely dependent on the degree of ripeness of the fruits.

1.11.6. Olive Washing

Before milling, a good rule is to wash the olives generously in drinkable water. Washing allows for the removal of stones and twigs thus lowering the risk of damage to the milling system, as well as removing any damaging exogenous microorganisms and/or residues of phystosanitary treatments so that olives suitable to be transformed into a quality product are obtained.

1.11.7. Olive Crushing

Olive malaxation has the objective of obtaining a homogenous paste, the consistency of which depends on the degree of ripeness of the olives and on their quality. It must be carried out for about 20 minutes, using typical molazze or disk or hammer mills.

The traditional olive crusher which was used for many centuries was the stone crusher. The stone crusher consists of a basin formed by a plinth and a stainless steel edge with an opening for the unloading of olive paste at the end of milling. Two or four granite wheels rotate and revolve on a rough granite base at different distances from the centre of the tank. The rotation speed is normally 12- 15 rpm.

The popularity of the stone crusher extraction system using pressure gradually declined. In comparison with continuous crushers, this apparatus shows significant limitations in terms of olive oil quality. In particular, it reduces the phenolic concentration as the olive pastes are in long, extensive contact with air during processing. Contact with the air stimulates polyphenol oxidase and peroxidase, producing a high oxidation of phenolic compounds. Other weaknesses of the stone crusher are its low working capacity, the high hourly machine footprint, and its low ability to release the chlorophyll found in the olive skin, responsible for the green colour of extra-virgin olive oil. This aspect is particularly relevant when the stone crusher is combined with a solid-liquid centrifugal separation. The crushing operation in oil extraction by centrifugation is generally replaced by the use of continuous crushers.

On the contrary, with olive presses the pulp and stone are instantly cracked by a ring with a nut in the middle where the olives are violently broken into pieces.

Thanks to its rapidness, this system is recommended for all the cooperatives which need to process large quantities of product within a short time frame. The bitter and spicy taste of the oil is almost certainly due to the use of this method. Consequently, the highlighting/exaltation of a particular class of antioxidant substances (polyphenols) increases. The level of fruit ripening also influences the use of this system, which is more appreciated when the olives are picked when not completely ripened [49].

1.11.8. Malaxation Process

The mixing and heating (25-35 °C) of the olive pastes during malaxation causes the breakdown of water-oil emulsion, allowing oil droplets to form larger droplets, which separate easily from the aqueous phase during the solid-liquid and liquid-liquid separation processes.

The olive paste, obtained through pressing, is characterised by the presence of numerous enzymes. Some of these enzymes which influence the quality of the final oil are: polyphenol oxidase; peroxidase, lypoxygenase and β-glucosidase.

Such enzymes are endogenous and compartmentalised in whole olives but, as with β-glucosidase,they can even be of an exogenous nature, deriving from contamination of the paste by microorganisms. Polyphenol oxidase and peroxidase have a decaying action on the phenolic compounds of the olive paste, reducing its antioxidant power [50]. Lypoxygenase is the enzyme that activates the enzymatic path of the degradation of free fatty acids, lineolic acid and lineolenic acid, in order to form aldehydes, alcohols and esters, which are responsible for the typical oil aromas such as fruttiness [51]. β-glucosidase is the enzyme which hydrolises the main phenolic compounds of olives, oleuropein and ligostride, in the corrisponding aglycones, so that they are rendered more soluble in the oil [50].

The activity of such enzymes is conditioned by time, by temperature, by water activity and, in the case of polyphen oxidases, peroxidases and lipoxygenases, by the atmospheric oxygen content level when kneading.

Simulations carried out in the "Laboratory Enzymatic Bioreactor" and pilot tests conducted in the experimental olive mill have demonstrated the role of enzymatic activities in olive paste and created the basis for a complete phenomenological description of kneading which is indispensable for planning and control of the operation [52].

Kneading is the setting for the combination of transport phenomena (*e.g.* breaking of the oil-water emulsion and coalescence of drops of oil, migration of olive components either in the oily phase of aqueous phase), and of transformation phenomena which are mainly of a enzymatic nature and concern the phenolic and triglyceride compounds present in the olive paste. Kneading is configured as a unitary chemical-enzymatic transformation operation of the olive oil. Evidence has demonstrated how the following are evident in kneading:

- The effect of time on the phenolic compound content. For a kneading system with a reduced oxidative impact characterised by kneaders with slightly depressed vertical axes: the greater the kneading time, greater is the phenolic compound content present in the oil, due to the coalescence phenomena and due to the action of β-glucoxidase on the phenolic compounds.

- The direct effect of oxygen on the polyphenol oxidase and peroxidase activity and therefore on the phenolic compound content of the extracted extra-virgin olive oil: a greater oxygen content in kneading determines a higher activity of polyphenol oxidase and peroxidase compounds and therefore a lowering of the oil phenolic compound content.

Studies conducted have therefore exhaustively demonstrated that the transformation role of polyphenol oxidase, peroxidase and β-glucosidase enzymes in olive paste on the flavour of the oil, while the transformation role of the lipoxygenase enzyme mainly regards volatile compounds of virgin olive oil.

The C6 and C5 compounds, especially C6 linear unsaturated and saturated aldehydes, alcohols and esters represent the most important fraction of volatile compounds found in high quality virgin olive oils. The C6 and C5 compounds, produced from polyunsaturated fatty acids by the enzymatic activities exerted by the lipoxygenase (LOX) pathway and their concentrations, depend on the level and the activity of each enzyme involved in this LOX pathway (Fig. **6**).

LOX proteins constitute an important class of lipidhydrolysing enzymes that catalyse the oxygenation of polyunsaturated fatty acids such as linolenic (LnA) and linoleic acids (LA) [53, 54].

In olive (*Olea europaea* L.) fruits, the LOX pathway is responsible for the production of desirable organoleptic properties that differentiate virgin olive oil from other vegetable oils. Hexanal (E)-2-hexenal, (E)-2-hexen-1-ol, 1- hexanol, and (*Z*)-3-hexen-1-yl acetate are five biomarkers produced as a consequence of lipid degradation following tissue disruption, and they are among the most important volatile compounds in olive oil aroma [56-58]. Dhifi *et al.* [59] has reported that the qualitative and quantitative composition of the olive oil aroma is highly dependent on the enzymatic store, involved in the LOX pathway, which is linked to fruit ripening. Olive fruit growth and development takes place in 5 months after flowering, depending on the variety and climatic conditions, and it includes different phases such as cell division, cell expansion, and storage of metabolites [56]. The quality of olive oil is influenced by genetic and environmental factors and also by the maturation state of drupes, but it is equally affected by technological treatments such as malaxation [60, 61].

Malaxation for 20 to 40 minutes allows small oil droplets to combine into bigger ones which can be removed by centrifugation. Centrifugation is an absolutely necessary step for effective extraction of the oil [60]. Longer mixing increases the oil yield and allows the formation of minor components that enhance its flavour, but it produces more oxidation products which make the oil acidity and peroxide values higher, shortening its shelf life [62]. It has been stated that an enzymatic system is present in olive fruit, which is genetically determined, including acylhydrolase (AH), LOX, fatty acid hydroperoxide lyase (FAHL), alcohol dehydrogenase (ADH), and alcohol acyltransferase (AAT). It becomes quickly active upon cell disruption and is involved in the formation of green sensory notes, covering the range between sweet-fruity-green to bitter-powerful-green [56]. Thus, the process of obtaining olive oil can be considered a good example of a system that produces secondary green volatiles.

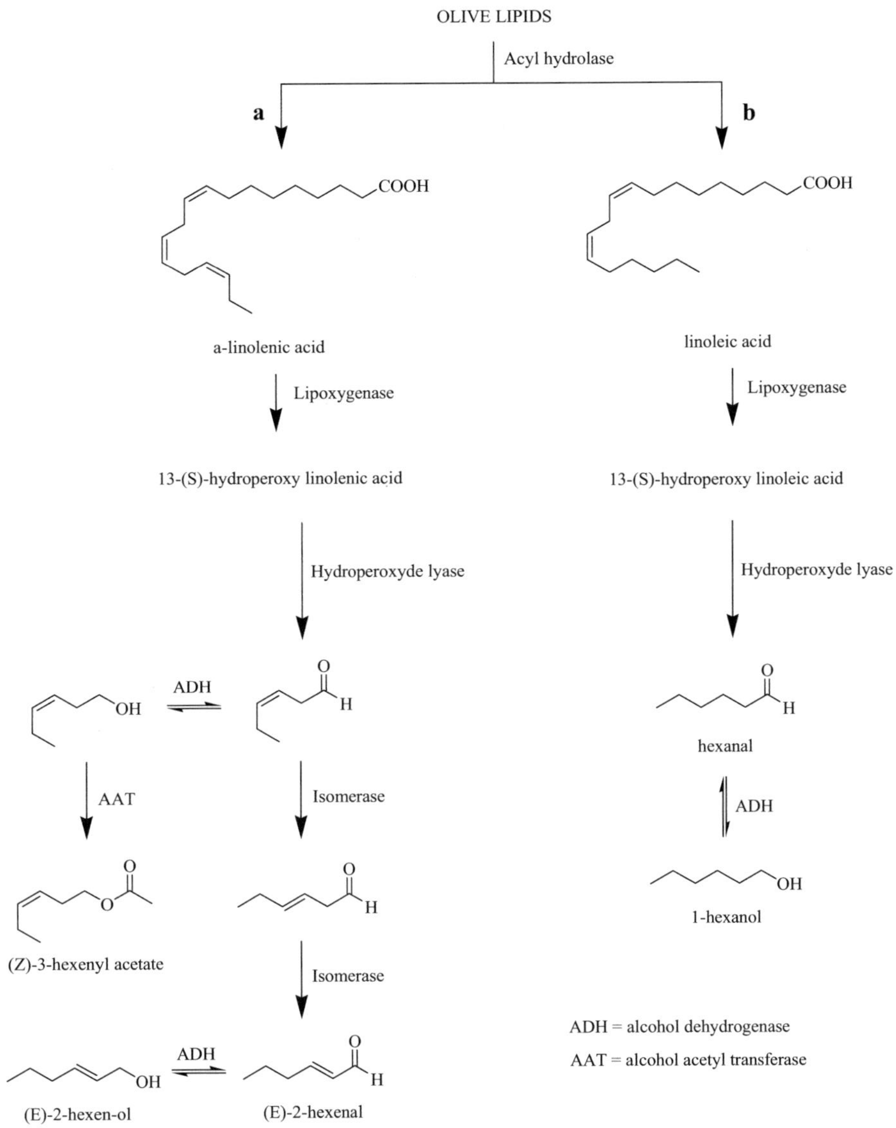

Figure 6: Scheme of the LOX metabolic pathway [55].

These considerations necessarily lead to a rethinking of the extra-virgin olive oil production process. In particular, they have led to the belief that extra-virgin olive oil can no longer be considered as the product of simple mechanical extraction, as specified in current regulations, but rather it is the product of complex biochemical transformations of the nutritional components, present in the fruits, which is not yet completely known.

1.11.9. Olive Oil Extraction Systems

Different extraction technologies, such as pressure and centrifugation and selective filtration (*i.e.* "surface tension" or "percolation") enabling the separation of oily must from the olive paste can be used [49].

1.11.9.1. Pressure Extraction System

Pressing is one of the oldest methods of oil extraction and has evolved considerably over centuries. In olive oil mills equipped with this system the press separation of the oil from the paste is currently carried out using open hydraulic presses, whereas close cage presses have almost disappeared not only due to high purchase prices, but also to their maintenance costs. The previously malaxed paste is subsequently stratified on stacked filter mats, each covered with approximately 1.25 cm of paste and interposed with metal disks. This operation is carried out mechanically thanks to a dispenser, which takes the paste from the malaxer and stores it on the nylon and/or polypropylene filter mats. Both types of filter mats have a central hole to allow the expressed oil and water (olive juice) to exit in both directions. From a theoretical point of view, this system guarantees intrinsic oil quality. However, its use presents a few problems, mainly due not only to its low working capacity per hour, in which case the storage of olives lengthens, but also to the proper use of the filter mats and to the types of materials used to build the equipment. The critical aspects of the process regarding the use of the press, which impacts on the quality of the oil, are concerned with both the proper management of the filter mats and the use of construction materials made of stainless steel. It is important to highlight that the filter mats can represent a source of contamination, due to fact that they may introduce fermentation and an oxidation defect into the oil, causing sensorial defects [63].

1.11.9.2. Extraction by Centrifugation

The majority of VOO is currently extracted by centrifugation in Mediterranean countries. The decanter consists of a drum containing a cylindrical and a conical part with a horizontal axis, inside which an additional cylinder worm is placed, which acts as a screw conveyor. The differential speed of the latter is slower than that of the outer drum in order to discharge the solid part. In recent years, this extraction system has evolved considerably in order to reduce the amount of water used during the process. In fact, the decanters can be classified as follows:

1. Traditional three-phase decanters.
2. Two-phase decanters.

In this regard we can affirm that in traditional three-phase decanters which require the addition of olive paste to a large quantity of water (water addition ranging from 0.5 to 1 m^3/ton) in order to reduce the vicousity and increase oil separation, there is a modification of the phenolic compound dsitribution, not only hydrophilics, which are lost during the aqueous phase. Two-phase decanters, which do not require water addition, allow limitation of this negative effect [48].

Over the years, numerous improvements have been introduced leading to the use of the two-phase decanter. Fundamentally, these are changes which have consented the centrifuging of olive paste without water. In this way the decanter releases only two phases: the oil and olive residue while it does not emit vegetable water, thus totally eliminating the contaminating effect of the discharge.

1.11.10. Separation of the Oil from Vegetation Water

Prior studies have demonstrated the importance of filtration on the quality of extra-virgin olive oil. The extracted oil (Fig. 7) is cloudy with relative stability formed not only by the oil, but also by traces of water and solid substances. This composition makes the product a seat of chemical and enzymatic degradation, above all affecting the triglyceride and phenolic components and therefore filtration is necessary to stabilise the oil [48].

1.11.11. Olive Oil Storage

During storage, the phenolic composition of extra virgin olive oil is modified by the endogenous enzymatic activities contained in the cloudy phase. These enzymes may reduce the "pungent" and "bitter" sensory notes, the intensity of which is strictly linked to the content of aglycon secoiridoids, and, at the same time, produce olfactory and taste defects.

Figure 7: Oil extraction.

Oil filtration partially removes the water and enzymes from virgin olive oils, and enables the phenolic content to stabilize during its storage. The filtration process of virgin olive oil is a procedure carried out in two steps: first, the suspended solids are removed, and second, the elimination of humidity gives the oil a brilliant aspect. Normally, organic or inorganic materials are used in conjunction with a variety of filtration equipment to enhance or enable the separation of suspended solids and water-oil [64].

After good conservation, olive oil can be sold in different sized containers: usually 3 or 5 litre cans are used, but it is preferable to store it in dark glass bottles, as olive oil is light sensitive.

ACKNOWLEDGEMENTS

This work was supported by the Ministry of Economic Development, (Certificazione della composizione varietale, dell'origine geografica e dell'assenza di prodotti di sintesi negli oli extravergini di oliva – CERTOLIO project) and by the Ministry of Agriculture, Food and Forestry (Trattato internazionale FAO sulle risorse genetiche vegetali, III triennalità 2010–2013, RGV-FAO project).

CONFLICT OF INTEREST

The authors confirm that this chapter contents have no conflict of interest.

REFERENCES

[1] IOC-International Olive Council, 2012. Accessed 07/12/2012: http://www.internationaloliveoil.org/estaticos/view/132-world-table-olive-figures.

[2] ISTAT, 2012. Tavola C27 - Superficie (ettari) e produzione (quintali): olivo, olive da tavola, olive da olio, olio di pressione. Dettaglio per regione – anno 2011. Accessed 20/11/2012: http://agri.istat.it/jsp/dawinci.jsp?q=plC270000010000012000&an=2011&ig=1&ct=311&id=21A|15A|32A.

[3] Agea & Unaprol, 2010. Accessed 19/11/2012: http://www.grimoli.it/PDF/SCENARIO%20ECONOMICO%20OLIVE%20DA%20TAVOLA%202010.pdf.

[4] Garrido-Fernández A., Fernández-Díaz, M.J., Adams R.M. Table Olives. Production and Processing. Chapman & Hall, London, UK, 1997.

[5] Lanza B., Corsetti A. Olive da mensa. In: L'ulivo e l'olio. Ed. ART S.p.A., Bologna, Italia, 2009; pp. 635-655.

[6] Cappello A., Poiana M., Romeo F.V. Monografia: Le olive da tavola in Sicilia, produzioni di qualità. Castelvetrano, Regione Siciliana Assessorato Agricoltura e Foreste, 2005; pp. 197.

[7] Brighigna A. La drupa e i suoi costituenti. In: Le Olive da Tavola: Tecnologie, Marketing e Impieghi. Ed agricole, 1984; pp. 1-5.

[8] IOC-International Olive Council. COI/OT/NC no. 1. Resolution no. RES-2/91-IV/04. Trade standard applying to table olives, 2004; pp. 1-19.

[9] Codex Alimentarius Commission. CODEX STAN 66-1981 (Rev. 1-1987), Codex standard for table olives, 1987; pp. 1-19.

[10] IOC-International Olive Council, 2012. Accessed 20/11/2012: http://www.international-oliveoil.org/news/view/660-year-2012-news/316-international-seminar-on-present-and-future-of-mediterranean-olive-sector.

[11] Romeo F.V., Piscopo A., Poiana M. Effect of drying, chemical and natural processing methods on black Biancolilla olives. Grasas Y Aceites, 2012; 63 (2): 223-230.

[12] Charoenprasert S., Mitchell A. Factors Influencing Phenolic Compounds in Table Olives (*Olea europaea*). Journal of Agricultural and Food Chemistry, 2012; 60: 7081-7095.

[13] García-García P., López-López A., Garrido-Fernández A. Study of the shelf life of ripe olives using an accelerated test approach. Journal of Food Engineering, 2008; 84: 569-575.

[14] Nisiotou A.A., Chorianopoulos N., Nychas G.-J.E., Panagou E.Z. Yeast heterogeneity during spontaneous fermentation of black Conservolea olives in different brine solutions. Journal of Applied Microbiology, 2010; 108: 396-405.

[15] Sánchez A.H., Rejano L., Montaño A., de Castro A. Utilization at high pH of starter cultures of lactobacilli for Spanish-style green olive fermentation. International journal of Food Microbiology, 2001; 67 (1-2): 115-122.

[16] Hurtado A., Reguant C., Bordons A., Rozès N. Influence of fruit ripeness and salt concentration on the microbial processing of Arbequina table olives. Food Microbiology, 2009; 26 (8): 827-833.

[17] Randazzo C.L., Ribbera A., Pitino I., Romeo F.V., Caggia C. Diversity of bacterial population of table olives assessed by PCR-DGGE analysis. Food Microbiology, 2012; 32: 87-96.

[18] Alves M., Gonçalves T., Quintas C. Microbial quality and yeast population dynamics in cracked green table olives' fermentations. Food Control, 2012; 23: 363-368.

[19] Romeo F.V., De Luca S., Piscopo A., Perri E., Poiana M. Effects of post-fermentation processing on the stabilisation of naturally fermented green table olives (cv Nocellara Etnea). Food Chemistry, 2009; 116: 873-878.

[20] Romeo F.V., Piscopo A., Mincione A., Poiana M. Quality evaluation of different typical table olive preparations (cv Nocellara del Belice). Grasas Y Aceites, 2012; 63 (1): 19-25.

[21] Iannotta N., Belfiore T., Noce M.E., Scalercio S., Vizzarri V. Correlation between *Bactrocera oleae* infestation and *Camarosporium dalmaticum* infection in an olive area of southern Italy. Acta Horticulturae, 2012; 949: 309-316.

[22] Lanza B. Nutritional and Sensory Quality of Table Olives. In: Olive Germplasm – The Olive Cultivation, Table Olive and Olive Oil Industry in Italy. Ed. Muzzalupo I., InTech, 2012; pp. 343-372. Available on: http://dx.doi.org/10.5772/51723.

[23] Smith J. E. Biotecnologie. Zanichelli Editore, 1998; pp. 235.

[24] Marsilio V., Seghetti L., Iannucci E., Russi F., Lanza B., Felicioni M. Use of a lactic acid bacteria starter culture during green olive (*Olea europaea* L cv Ascolana tenera) processing. Journal of the Science of Food and Agriculture, 2005; 85 (7): 1084-1090.

[25] Leal-Sánchez M.V., Ruiz-Barba J.L., Sánchez A.H., Rejano L., Jiménez-Díaz R., Garrido, A. Fermentation profile and optimization of green olive fermentation using *Lactobacillus plantarum* LPCO10 as a starter culture. Food Microbiology, 2003, 20: 421-430.

[26] Romeo F.V. Microbiological Aspects of Table Olives. In: Olive Germplasm – The Olive Cultivation, Table Olive and Olive Oil Industry in Italy. Ed. Muzzalupo I., InTech, 2012; pp. 321-342. Available on: http://dx.doi.org/10.5772/51479.

[27] Hurtado A., Reguant C., Bordons A., Rozès N. Lactic acid bacteria from fermented table olives. Food microbiology, 2012; 31 (1): 1-8.

[28] Servili M., Settanni L., Veneziani G., Esposto S., Massitti O., Taticchi A., Urbani S., Montedoro G.F., Corsetti A. The use of *Lactobacillus pentosus* 1MO to shorten the debittering process time of black table olives (Cv. Itrana and Leccino): a pilot-scale application. Journal of Agricultural and Food Chemistry, 2006; 54: 3869-3875.

[29] Panagou E.Z., Schillinger U., Franz C.M.A.P., Nychas G.-J.E. Microbiological and biochemical profile of cv. Conservolea naturally black olives during controlled fermentation with selected strains of lactic acid bacteria. Food microbiology, 2008; 25 (2): 348-358.

[30] Lanza B., Zago M., Carminati D., Rossetti L., Meucci A., Marfisi P., Russi F., Iannucci E., Di Serio M.G., Giraffa G. Isolation and preliminary characterization of *Lactobacillus plantarum* bacteriophages from table olive fermentation. Annals of Microbiology, 2012; 62 (4): 1467-1472.

[31] Zago M., Lanza B., Rossetti L., Muzzalupo I., Carminati D., Giraffa G. Selection of *Lactobacillus plantarum* strains to use as starters in fermented table olives: Oleuropeinase activity and phage sensitivity. Food Microbiology, 2013, 34: 81-87.

[32] Kailis S., Harris D. Producing table olives. Landlinks Press, Australia, 2007; pp. 275-276.
[33] Hutkins R.W. Fermented vegetables, In: Microbiology and Technology of fermented foods, Blackwell Publishing (Ed.), Oxford, UK, 2006; pp. 233-260, ISBN-13: 978-0-8138-0018-9.
[34] Vaughn R.H., Stevenson K.E., Davé B.A., Park H.C. Fermenting yeasts associated with softening and gas-pocket formation in olives. Applied Microbiology, 1972; 23: 316-320.
[35] Durán-Quitana M.C., González F., Garrido A., Aceitunas negras al natural en salmuera. Ensayos de producción de alambrado. Inoculación de diversos microorganismos aislados de salmueras de fermentación. Grasas y Aceites, 1979; 30: 361-367.
[36] Arroyo-López F.N., Romero-Gil V., Bautista-Gallego J., Rodríguez-Gómez F., Jiménez-Díaz R., García-García P., Querol A., Garrido-Fernández A. Yeasts in table olive processing: Desirable or spoilage microorganisms? International Journal of Food Microbiology, 2012; 160: 42-49.
[37] IOC-International Olive Council. COI/OT/MO No 1/Rev.2. Method: sensory analysis of table olives, 2011; pp. 1-17.
[38] Arroyo-López F.N., Bautista-Gallego J., Segovia-Bravo K.A., García-García P., Durán-Quitana M.C., Romero C., Rodríguez-Gómez F., Garrido-Fernández A. Instability profile of fresh packed ''seasoned'' Manzanilla-Aloreña table olives. LWT - Food Science and Technology, 2009; 42: 1629-1639.
[39] Marsilio V., Campestre C., Lanza B. Phenolic compounds change during California-style ripe olive processing. Food Chemistry, 2001; 74: 55-60.
[40] King A.D., Magnuson J.A., Torok T, Godman N. Microflora and storage quality of partially processed lettuce. Journal of Food Science, 1991; 56: 459-461.
[41] Arroyo-López F.N., Bautista-Gallego J., Durán-Quintana M.C., Rodríguez-Gómez F., Romero-Barranco C., Garrido-Fernández A. Improvement of the storage process for cracked table olives. Journal of Food Engineering, 2008; 89: 479-487.
[42] Di Giovacchino L., Sestili S., Di Vincenzo D. Influence of olive processing on virgin olive oil qualità. European Journal of Lipid Science Technology, 2002; 104: 587-601.
[43] Lavee S. Olive. In: Monselise, S.P. (Ed.), Handbook of fruit set and development. CRC Press, Florida, 1986; 261-276.
[44] Rangel B., Platt K. Thomson W.W. Ultastructural aspects of the cytoplasmic origin and accumulation of oil in olive fruit (*Olea europaea*). Physiologia Plantarum. 1997, 101: 109-114.
[45] Nergiz C., Engez Y. Compositional varation of olive during ripening. Food Chemistry. 2000, 69: 55-59.
[46] Salas J.J., Sanchez J., Ramli U.S., Manaf A.M., Williams M., Harood J.L. Biochemistry of lipid metabolism in olive and other oil fruit. Progress in Lipid Research. 2002; 39: 151-180.
[47] Taiz L., Zeiger L. 2002. Fisiologia Vegetale. Piccin, Padova.
[48] Mugelli M., Migliorini M., Viti P., Cherubini C., Cini E., Zanoni B. 2005. Olio extravergine di oliva ricerche e innovazioni per il miglioramento della qualità. Camera di Commercio di Firenze, Firenze.
[49] Servili M., Taticchi A., Esposto S., Sordini B., Urbani S. 2012. Technological aspects of olive oil production. In: Olive germplasm – the olive cultivation, table olive and olive oil industry in Italy (Ed. I. Muzzalupo), InTech, Rijeka, Croatia.
[50] Servili M., Selvaggini R., Esposto S., Taticchi A., Montedoro G.F., Morozzi G. Health and sensory properties of virgin olive oil hydrophilic phenols: agronomic and technological aspects of production that affect their occurence in the oil. Journal of Chromatography, 2004; 1054: 113-127.

[51] Angerosa F., Mostallino R., Basti C., Vito R. Influence of malaxation temperature and time on the quality of virgin olive oils. Food Chemistry, 2001; 72:19-28.

[52] Migliorini M., Zanoni B., Berti A., Cherubini C., Cini E., Daou M., Mugelli M. 2008. Protocolli innovativi per la produzione di olio extra vergine nella realtà aziendale Toscana. Camera di Commercio di Firenze, Firenze.

[53] LorenziV., Maury J., Casanova J., Berti L. Purification, product characterization and kinetic properties of lipoxygenase from olive fruit (*Olea europaea* L.), Plant Physiology and Biochemistry, 2006; 44: 450–454.

[54] Palmieri-Thiers C., Canaan S., Brunini V. *et al.* A lipoxygenase with dual positional specificity is expressed in olives (*Olea europaea* L.) during ripening, Biochimica et Biophysica Acta, 2009; 1791: 339–346.

[55] Muzzalupo I., Macchione B., Bucci C., Stefanizzi F., Perri E., Chiappetta A., Tagarelli A., Sindona G. LOX gene transcript accumulation in olive (*Olea europaea* L.) fruits at different stages of maturation: relationship between volatile compounds, environmental factors, and technological treatments for oil extraction. The Scientific World Journal, 2012; 2012:1-9.

[56] Padilla M.N., Hernandez M.L., Sanz C., Martınez-Rivas J.M. Functional characterization of two 13-lipoxygenase genes from olive fruit in relation to the biosynthesis of volatile compounds of virgin olive oil, Journal of Agricultural and Food Chemistry, 2009; 57: 9097–9107.

[57] Olias J.M., Perez A.G., Rıos J.J., Sanz L.C. Aroma of virgin olive oil: biogenesis of the "green" odor notes, Journal of Agricultural and Food Chemistry, 1993; 41: 2368–2373.

[58] Benincasa C., de Nino A., Lombardo N., Perri E., Sindona G., Tagarelli A. Assay of aroma active components of virgin olive oils from Southern Italian regions by SPME-GC/ion trap mass spectrometry, Journal of Agricultural and Food Chemistry, 2003: 51: 733–741.

[59] Cavaliere B., de Nino A., Hayet F. *et al.* A metabolomic approach to the evaluation of the origin of extra virgin olive oil: a convenient statistical treatment of mass spectrometric analytical data, Journal of Agricultural and Food Chemistry, 2007, 55: 1454–1462.

[60] Dhifi W., Angerosa F., Serraiocco A., Oumar I., Hamrouni I., Marzouk B. Virgin olive oil aroma: characterization of some Tunisian cultivars, Food Chemistry, 2005; 93: 697–701.

[61] Servili M., Taticchi A., Esposto S., Urbani S., Selvaggini R., Montedoro G. Influence of the decrease in oxygen during malaxation of olive paste on the composition of volatiles and phenolic compounds in virgin olive oil, Journal of Agricultural and Food Chemistry, 2008; 56: 10048–10055.

[62] GoMez-Rico A., Inarejos-Garcpia A.M., Salvador M.D., Fregapane G. Effect of malaxation conditions on phenol and volatile profiles in olive paste and the corresponding virgin olive oils (*Olea europaea* L. Cv. cornicabra), Journal of Agricultural and Food Chemistry, 2009; 57: 3587–3595.

[63] Muzzalupo I., Pellegrino M., Perri E. 2012. Sensory Analysis of Virgin Olive Oils. In: Olive germplasm – the olive cultivation, table olive and olive oil industry in Italy (Ed. I. Muzzalupo), InTech, Rijeka, Croatia.

[64] Piscopo A., Poiana M., 2012. Packaging and Storage of Olive Oil. In: Olive germplasm – the olive cultivation, table olive and olive oil industry in Italy (Ed. I. Muzzalupo), InTech, Rijeka, Croatia.

CHAPTER 4

Botanical and Pomological Aspects of Stone Fruits Physiology, Agronomy and Orchard Management

Daniela Giovannini[1], Alessandro Liverani[1], Alisea Sartori[2] and Guido Cipriani[2,3,*]

[1]Consiglio per la Ricerca e la sperimentazione in Agricoltura (Agricultural Research Council), Fruit Tree Research Unit, CRA-FRF, 47121 Forlì, Italy; [2]Consiglio per la Ricerca e la sperimentazione in Agricoltura (Agricultural Research Council), Fruit Tree Research Centre, CRA-FRU, 00134 Rome, Italy and [3]Dipartimento di Scienze Agrarie e Ambientali, University of Udine, 33100 Udine, Italy

Abstract: *Prunus* is a genus of about 230 species, distributed primarily in north temperate regions. *Prunus* has been historically divided into a number of genera by various botanists. Three subgenera correspond to broad categories of stone fruits. Main domesticated species are peach, apricot, cherry and plum. Different characters have been used to uniquely identify genotypes and cultivars. In the present chapter the main pomological and phenological characteristics are presented. Features and ripening dates are described for 72 peach cultivars: 19 with white flesh, 28 with yellow flesh, 13 canning peaches and 12 with flat shape. Forty-one different nectarines, 46 apricots, 48 cherries and 18 Japanese and 5 European plums cultivars are also described. A second part of the chapter is also devoted to the description of the physiological phenomena related to production and to the main agricultural techniques and best practices for orchard management.

Keywords: *Prunus*, botany, rootstock, stress, dormancy, chilling requirement, pollination, fruit growth, fruit ripening, planting system, training system, pruning, thinning.

1. BOTANICAL AND POMOLOGICAL ASPECTS

1.1. General Description

Prunus is a genus of about 230 species, distributed throughout the world, with the greatest native diversity in Asia. The group is considered to have a primarily north

***Corresponding author Guido Cipriani:** Dipartimento di Scienze Agrarie e Ambientali – University of Udine, Italy; Tel: +390432558635; Fax: +390432558603; E-mail: guido.cipriani@uniud.it

Innocenzo Muzzalupo & Sabrina Micali (Eds.)

temperate distribution because three quarters of the described species are native to temperate regions in Asia, Europe, or North America, but there is significant representation of the genus in tropical regions. *Prunus* also exhibits considerable ecological diversity, with different species occurring from lowlands to alpine zones and in wet and dry forests as well as deserts [1].

Species of *Prunus* are woody plants, ranging from small shrubs to very large trees, which may be either deciduous or evergreen.

The leaves are alternate and simple, pinnately veined, and usually glabrous but pubescent in a few species; stipules are present but fall early. The leaf margin range from serrate to entire; the leaf teeth are generally glandular, and glands also occur on the petioles or leaf blade bases of most species. The inflorescence ranges from a solitary flower to an umbel-like cluster or a raceme. The flowers are radially symmetric and have well-developed hypanthia, as is characteristic of all Rosaceae. There are five petals that vary in color from white to pink or red. The flowers are usually perfect, with ten or more stamens and a single, uni-carpellate pistil with a superior ovary that matures into a drupe [1].

The mesocarp varies from fleshy and juicy to dry or leathery; the endocarp is hard and contains a single seed.

As in many other Rosaceae, the primary transport carbohydrate in species of *Prunus* is the sugar alcohol sorbitol and the plants produce cyanogenic glycosides that result in a characteristic acrid odor when the plant is wounded [1].

1.2. Taxonomy and Phylogeny

Based on its base chromosome number (x=8) and fruit type, both of which are quite rare in Rosaceae, *Prunus* has been often classified in the subfamily Prunoideae or Amygdaloideae, with the few other genera in the family exhibiting one or both of those characters. Most recent intrafamilial classification of Rosaceae [2] placed *Prunus* in the tribe Amigdaleae, included in the subfamily Spiraeoideae, recently renamed Amygdaloideae [1].

Prunus has been historically divided into a variable number of genera, from two to six, by various botanists. Molecular phylogenetic studies, however, do not provide support for these divisions, and instead indicate that a broad circumscription of *Prunus* is most appropriate.

Rehder [3] divided *Prunus* into five subgenera, several of which he further divided into sections. Three subgenera correspond to broad categories of stone fruits. *Prunus* is characterized by sulcate fruits with bloomy surfaces and stems with solitary axillary buds and no terminal buds, and includes plums and apricots (section *Armeniaca*), the latter ones characterized by usually pubescent fruits. Species of the genus *Amygdalus* (peaches and almonds) also have sulcate fruits but their stems have three axillary buds and terminal buds are also present. In the subgenus *Cerasus* (cherries) the fruits are neither sulcate nor bloomy. These subgenera have flowers that are either solitary or born in umbel-like clusters. No major fruit crops are included in either of the two subgenera *Padus* and *Laurocerasus*, though the fruits of *P. serotina* are consumed in Mexico and *P. serotina* is also used for timber [1].

Molecular phylogenetic studies of *Prunus* conducted over the last decade have led to several conclusions about the relationships across the genus. Neither the subgenus *Padus* nor the subgenus *Laurocerasus* are monophyletic, but these two subgenera form a well-supported clade, within which several geographically distinct lineages are resolved. Most of the subgenus *Cerasus* forms a strongly supported clade. Members of the remaining subgenera together form a moderately well-supported monophyletic group within which the subgenera *Prunus* and *Amygdalus* are strongly supported clades. The position of the *Cerasus* group is variable among.

Analyses based on chloroplast DNA markers, and combined data from nuclear and chloroplastic DNA and morphology, have supported a sister relationship between the *Cerasus* and the *Amygdalus-Prunus* clade, while the analysis based on nuclear internal transcribed spacer (ITS) sequences alone has supported a sister relationship between *Cerasus* and the *Padus-Laurocerasus* clades. These results suggest the possibility that hybridizations were involved in early diversification of the genus [1].

1.3. Peaches

The peach, and its smooth skinned mutant, the nectarine, are primarily grown in temperate zones, between latitudes 30 °C and 45 °C North and South. The peach flower bud is hardy from about -23 °C to -26 °C and this limits its cultivation at higher latitudes. Most peach cultivars require from 100 to 1000 hours of chilling below 7 °C and they are highly susceptible to early spring frosts.

Fruit shapes vary from beaked, round, to flat, while colors vary from yellow, white, to red, and the flesh can be melting or non-melting. Moreover they can be clingstone or freestone. The cultivated peach is diploid and has a chromosome number of 2n=2x=16.

Five species can be termed as peach: *P. persica*, *P.* (Carr.) Franch, *P. mira* Koehne, *P. kansuensis* Rehd. and *P. ferganensis* (Kost. & Rjab) Kov. & Kost.

The domesticated peach can be readily hybridized with native populations of all the other wild species of peach. Successful hybrids have also been produced between peach and almond, apricot, plum and sour cherry. In most cases, these wide hybrids are largely sterile, although F_1 of almond and peach can be highly fertile and can be employed as rootstocks for both peach and almond [4].

Different characters have been used to uniquely identify genotypes and cultivars in peach. The Community of Plant Variety Office (CPVO), an European Union organization, considers 69 traits to be used in DUS (distinctness, uniformity and stability) test (Table **1**). The International Union for the Protection of New Varieties of Plants (UPOV), an intergovernmental organization with headquarters in Geneva (Switzerland), considers 68 traits for peach, all included in the CPVO list. Three traits are related to the tree (size, vigor and habit), six to the flowering shoot (thickness, length of internodes, color *etc.*), 10 to the flower and its organs (petals, stamens, anthers, ovary *etc.*), 14 consider the leaf and petiole, 24 the fruit (size, shape, symmetry, ground color, over color, pubescence, sweetness, acidity *etc.*), seven the stone, and five traits regard some phenology (time of leaf bud burst, time of beginning of flowering, time of maturity for consumption and so on). All the traits are qualitative, this means that no measurements are needed but only a comparison with one or more test samples (other peach cultivars).

Table 1: Characteristics to be used in DUS test and description of peach and nectarin cultivars (from CPVO documentation)

CPVO Trait Number	Character	Type
1	Tree: size	Very small, small, medium, large, very large
2	Tree: vigor	Weak, medium, strong
3	Tree: habit	Upright, semi-upright, spreading, drooping, weeping
4	Flowering shoot: thickness (excluding brindilles)	Thin, medium, thick
5	Flowering shoot: length of internodes (excluding brindilles)	Very short, short, medium, long, very long
6	Flowering shoot: anthocyanin coloration (excluding brindilles, side away from sun)	Absent, present
7	Flowering shoot: intensity of anthocyanin coloration (excluding brindilles, side away from sun)	Weak, medium, strong
8	Flowering shoot: density of flower buds (excluding brindilles)	Sparse, medium, dense
9	Flowering shoot: general distribution of flower buds (excluding brindilles)	Isolated, in groups of two or more
10	Flower type	Non showy, showy
11	Calyx: color of inner side (opened flower, before falling of petals)	Greenish yellow, orange
12	Corolla: predominant color (inner side)	White, very light pink, light ping, medium pink, dark pink, violet pink, red
13	Petal: shape	Narrow elliptic, broad elliptic, round
14	Petal: size	Very small, small, medium, large, very large
15	Petals: number	Five, more than five
16	Stamens: position compared to petals	Below, same level, above
17	Stigma: position compared to anthers	Below, same level, above
18	Anthers: pollen	Absent, present
19	Ovary: pubescence	Absent, present
20	Young shoot: length of stipule	Short, medium, long
21	Leaf blade: length	Short, medium, long
22	Leaf blade: width	Narrow, medium, broad
23	Leaf blade: ratio length/width	Small, medium, large
24	Leaf blade: shape in cross section	Concave, flat, convex
25	Leaf blade: recurvature of apex	Absent, present

Table 1: contd….

26	Leaf blade: angle at base	Acute, approximately right angle
27	Leaf blade: angle at apex	Small, medium, large
28	Leaf blade: color	Greenish yellow, green, purplish red
29	Leaf: red mid-vein on the lower side	Absent, present
30	Petiole: length	Short, medium, long
31	Petiole: nectarines	Absent, present
32	Petiole: shape of nectarines	Round, reniform
33	Petiole: predominant number of nectarines	Two, more than two
34	Fruit: size	Very small, small, medium, large, very large
35	Fruit: shape (in ventral view)	Broad oblate, oblate, round, ovate, elliptic
36	Fruit: shape of pistil end	Prominently pointed, weakly pointed, flat, weakly depressed, strongly depressed
37	Fruit: symmetry (viewed from pistil end)	Asymmetric, symmetric
38	Fruit: prominence of suture	Weak, medium, strong
39	Fruit: depth of stalk cavity	Shallow, medium, deep
40	Fruit: width of stalk cavity	Narrow, medium, broad
41	Fruit: ground color	Green, cream green, greenish white, cream white, cream, pink white, greenish yellow, cream yellow, yellow, orange yellow
42	Fruit: over color	Absent, present
43	Fruit: hue of over color	Orange red, pink, pink red, light red, medium red, dark red, blackish red
44	Fruit: pattern of over color	Solid flush, striped, mottled, marbled
45	Fruit: extent of over color	Very small, small, medium, large, very large
46	Fruit: pubescence	Absent, present
47	Fruit: density of pubescence	Very sparse, sparse, medium, dense, very dense
48	Fruit: thickness of skin	Thin, medium, thick
49	Fruit: adherence of skin to flesh	Absent or very weak, weak, medium, strong, very strong
50	Fruit: firmness of flesh	Very soft, soft, medium, firm, very firm
51	Fruit: ground color of flesh	Greenish white, white, cream white, light yellow, yellow, orange yellow, orange, red
52	Fruit: anthocyanin coloration directly under skin	Absent or very weakly expressed, weakly expressed, strongly expressed
53	Fruit: anthocyanin coloration of flesh	Absent or very weakly expressed, weakly expressed, strongly expressed
54	Fruit: anthocyanin coloration around stone	Absent or very weakly expressed, weakly expressed, strongly expressed

Table 1: contd....

55	Fruit: texture of the flesh	Not fibrous, fibrous
56	Fruit: sweetness	Low, medium, high
57	Fruit: acidity	Low, medium, high
58	Stone: size compared to fruit	Small, medium, large
59	Stone: shape (in lateral view)	Oblate, round, elliptic, obovate
60	Stone: intensity of brown color	Light, medium, dark
61	Stone: relief of surface	Small pits, large pits, grooves, pits and grooves
62	Stone: tendency of splitting (at peak harvest)	Absent or very low, low, medium, high, very high
63	Stone: adherence to flesh	Absent, present
64	Stone: degree of adherence to flesh	Weak, medium, strong
65	Time of leaf bud burst	Very early, early, medium, late, very late
66	Time of beginning of flowering	Very early, early, medium, late, very late
67	Duration of flowering	Short, medium, long
68	Time of maturity for consumption	Very early, early, medium, late, very late
69	Tendency to pre-harvest drop	Absent or very weak, weak, medium, strong, very strong

1.3.1. Peach and Nectarine Cultivars

Peach and Nectarines are characterized by a large numbers of varieties (2768 licensed cultivars from 1980 to 2008) [5]. The classification of all the cultivars present in the market is linked to some pomological aspects of the fruit, in particular: the presence of pubescence typical of peaches [*Prunus persica* (L) Batsch var. *persica*] or its absence typical of nectarines [*Prunus persica* (L) Batsch var. *nucipersica* (L.) C.K. Schneid].

The flesh colour, mainly white or yellow, gives to the peach or nectarine a peculiar taste. White flesh is sweeter than the yellow one and very much appreciated by consumers, but some defects in the flesh hardness limit handling and marketability of the fruits. A short and scalar ripening period are other negative aspects.

High anthocyanin accumulation in the mesocarp, giving it a red-violet colour, is commonly called blood-flesh. Accessions with this trait have been known for centuries; they were mainly grown in France even if in Italy some ancient clones such as 'Pesco delle Vigna' can also be found. In Tables **2** to **5** the main peach and nectarine varieties are listed based on their pomological characteristics.

All over the world, most peaches and nectarines commercialised for fresh market are characterised by a melting flesh. During ripening, this fruit typology undergoes a process of flesh softening, gradual in the early phase but increasing more and more. At the end, the fruit firmness is so compromised that the fruit can be easily damaged resulting in a very short life span. However, there are some peach cultivars as 'Rich Lady', 'Vistarich' (Table **5**) and the nectarine 'Diamond Ray' (Table **3**) that show a very slow melting phase and a greater resistance of the fruit on the tree, which makes it edible when still relatively immature. The nectarine 'Big Top' (Table **3**) has also other peculiar fruit traits such as early and extensive anthocyanins accumulation in the skin, crunchy flesh, and low acid flavor [6].

Non-melting flesh is the requested character for the canning peach category. Endopolygalacturonases, the enzymes responsible for the cell wall hydrolysis, are not synthesized during the ripening phase and the flesh maintains its hardiness also after the canning process at high temperatures. Canning peach cultivars (Table **6**) have also no blush and no anthocyanin in the flesh, not too high acidity levels and do not form unpleasant compounds during canning process. There are non-melting cultivars licensed for the fresh market such as 'Crown Princess' and 'Crimson Lady' (harvest date 24 days before Rome Star) that have an attractive extended over colour, yellow flesh and medium organoleptic characters.

The stony hard trait (SH), controlled by a single recessive gene (hd/hd) [7] induces peculiar crispy flesh texture in the fruit. There is a very scarce, quite none, production of ethylene during ripening [8] and there no changes in pectin profiles during shelf life are observed [9]. The deanthocyanic series 'Ghiaccio' (Fig. **1a**), derived from an open pollination of the Korean cv Yumyeong, has stone hard flesh, white-cream flesh colour, a complete absence of over color, the flesh adherent to the stone and sub-acid taste [10, 11].

Flat shape peach and nectarine varieties are increasing in number in the last years, mainly in USA and France. The Italian series 'UFO' is also well known and is widespreadly cultivated in Spain and Italy.

UFO series, from UFO 1 to UFO 9 (Table **7**), covers a wide ripening calendar of the peach fruit presence on the market from the mid of May till the end of September, and exhibits a high performance in terms of cropping, fruit size and aroma. These flat peaches succeeded in overcoming the typical flat shape defects such as cracking in apex while maintaining positive characteristics such as the sweet taste, ease of eating and the smooth skin [12, 13].

All flat shape nectarines (Table **8**) are characterized by early flowering time, moderate vigor and small fruit size [14]. The fruits have a pronounced and attractive skin blush and rank high in terms of fruit appearance (Fig. **1b**).

Table 2: List of the main white flesh nectarine cultivars: their country of origin, harvest date respect to the reference cv Rome Star, flowering date at 10 % of open flowers, productivity and some of their pomological characteristics [12, 13]

Cultivar	Country of origin	Ripening time (dd ± Rome Star)[1]	Flowering period [66]	Yield[3]	Fruit size[34]	Flesh Firmness[50]	Overcolor[43]	Over color extent (%)	Taste type[8]
Extra-early and early ripening time									
Silver King®Prita*	FRA	-41	M	M	M	M	DR	80	A
Early Silver*	ITA	-37	E	E	L	M	DR	70	E
Neve*	ITA	-30	M	E	VL	F	DR	80	A
Jade®Momée*	FRA	-25	M	M	VL	F	DR	70	E
Caldesi 2000*	ITA	-23	M	M	VL	M	MR	80	E
Intermediate ripening time									
Magique® Maillarmagie*	FRA	-18	E	E	VL	F	MR	90	S
Emeraude®Monnude*	FRA	-14	M	E	VL	M	MR	80	S
Caldesi 2010*	ITA	2	M	M	VL	F	DR	70	E
Silver Giant*	ITA	14	M	M	VL	F	DR	60	A
Late and very late ripening time									
Silver Moon*	ITA	27	VE	ME	L	F	DR	70	E
Caldesi 2020*	ITA	39	L	ME	L	F	MR	60	A

1 Rome Star ripens in the North of Italy from 1st to 5th August, in the Center Italy from 25th to 30th July, in the South Italy from 20th to 25th July
66cpvo DUS test trait number = Flowering period: (VE= very early; E= early; M= medium; L= late; VL= very late)
3 Yield: (S= scarse; MS= medium-scarse; M= medium; E= high; ME= very high)
34 cpvo DUS test trait number = Fruit size: (VS=very small; S= small; M= medium; L= large; VL= very large)
50 cpvo DUS test trait number = Flesh firmness: (VS= very soft; S= soft; M= medium; F=firm; VF= very firm)
43 cpvo DUS test trait number = Hue of over color: (OR= orange red; P= pink; PR= pink red; LG= light red; MR= medium red; DR= dark red; BR= blackish red)
8 Taste type: (A= acid (<130meq/L); S= Subacid (< 69 meq/L); E= equilibrate (70-129 meq/L))

Table 3: List of the main yellow flesh nectarine cultivars: their country of origin, harvest date respect to the reference cv Rome Star, flowering date at 10 % of open flowers, productivity and some of their pomological characteristics [15, 16]

Cultivar	Country of origin	Ripening time (dd ± Rome Star)[1]	Flowering period[66]	Yield[3]	Fruit size[34]	Flesh Firmness[50]	Over color[43]	Over color extent (%)	Taste type[8]
Extra-early and early ripening time									
Gran Sun*	USA	-51	VE	M	S	F	DR	80	A
Big Bang® Maillara*	FRA	-41	E	E	M	F	DR	100	S
Rose Diamond*	USA	-40	E	M	S	F	OR	70	E
Rita Star® Bratastar*	USA	-40	E	M	M	F	DR	90	E
Ambra*	ITA	-31	E	E	L	M	DR	90	A
Laura*	USA	-31	E	E	S	F	MR	70	A
Big Top® Zaitabo*	USA	-24	M	ME	M	VF	MR	80	S
Intermediate ripening time									
Alitop*	ITA	-18	E	E	VL	VF	MR	90	S
Maria Laura	ITA	-18	M	E	M	F	DR	80	A
Diamond Ray*	USA	-16	M	M	L	F	DR	90	A
Amiga*	ITA	-15	E	M	L	F	DR	90	A
Stark Redgold®	USA	-1	M	E	VL	F	DR	60	A
Nectaross	ITA	6	M	E	L	F	MR	70	A
Venus	ITA	7	M	M	L	F	MR	50	A
Maria Aurelia	ITA	8	M	M	L	M	DR	80	A
Orion*	ITA	12	M	E	L	M	MR	60	A
Late and very late ripening time									
Sweet Red®	ITA	16	M	M	VL	M	DR	60	A
Maria Dolce*	ITA	17	M	M	L	M	LR	60	S
Sweet Lady*	ITA	23	M	M	VL	M	DR	60	E
Lady Erica®	ITA	30	M	M	L	F	DR	60	S
Morsiani 90®	ITA	36	M	M	VL	M	DR	60	A
Fairlane	USA	42	M	M	VL	M	DR	60	A
August Red	USA	45	M	M	L	F	MR	50	A
Max 7®	ITA	45	M	M	VL	F	DR	60	A
California	USA	49	L	ME	VL	F	LR	40	A

See legend in Table **2**.

Table 4: List of the main white flesh peach cultivars: their country of origin, harvest date respect to the reference cv Rome Star, flowering date at 10 % of open flowers, productivity and some of their pomological characteristics [15, 16]

Cultivar	Country of origin	Ripening time (dd ± Rome Star)[1]	Flowering period[66]	Yield[3]	Fruit size[34]	Flesh Firmness[50]	Over color[43]	Over color extent (%)	Taste type[8]
Extra-early and early ripening time									
White Crest®	ITA	-54	M	MS	M	S	MR	60	E
Primerose®	USA	-51	M	MS	S	M	LR	60	E
Hermione®Zailice*	USA	-39	M	E	L	F	DR	80	E
Alexandra®	USA	-36	M	E	M	M	MR	70	E
Crizia*	ITA	-35	M	MS	L	F	DR	70	E
Iris Rosso	ITA	-24	M	MS	VL	F	MR	80	E
Alipersiè*	ITA	-19	M	E	L	F	DR	80	E
Intermediate ripening time									
Greta*	ITA	-15	E	M	VL	F	LR	50	E
Alirosada*	ITA	-9	M	ME	VL	F	DR	70	E
Maria Bianca	ITA	-9	L	E	VL	M	LR	50	E
Benedicte® Meydicte*	FRA	2	M	E	L	F	LR	50	E
Late and very late ripening time									
Tendresse® Julie*	FRA	11	M	M	L	M	MR	60	E
Aliblanca*	ITA	15	M	ME	VL	F	DR	60	E
Regina Bianca	ITA	16	M	E	VL	M	DR	50	E
Maria Angela	ITA	17	L	ME	VL	F	LR	50	E
Maria Delizia	ITA	26	L	M	VL	M	LR	50	E
Douceur*	FRA	33	E	E	L	F	PR	40	S
Gladys®Zailati*	USA	39	M	M	VL	F	LR	50	E
Regina di Londa	ITA	40	M	E	L	M	DR	30	E

See legend in Table **2**.

Table 5: List of the main yellow flesh peach cultivars: their country of origin, harvest date respect to the the reference cv Rome Star, flowering date at 10 % of open flowers, productivity and some of their pomological characteristics [15, 16]

Cultivar	Country of origin	Ripening time (dd ± Rome Star)[1]	Flowering period[66]	Yield[3]	Fruit size[34]	Flesh Firmness[50]	Over color[43]	Over color extent (%)	Taste type[8]
Extra-early and early ripening time									
Françoise	FRA	-56	E	M	S	M	DR	80	E
Rich May	USA	-51	E	M	M	F	DR	90	A
Maycrest	USA	-46	E	M	M	M	DR	90	E
Lolita	ITA	-44	M	E	M	F	ML	70	E
Springcrest	USA	-44	E	M	M	M	DR	80	E
Spring Lady® Merspri*	USA	-37	M	M	M	F	DR	90	E
Rubyrich® Zainoar*	USA	-37	E	ME	L	VF	DR	80	E
Alix®	ITA	-31	L	M	M	F	DR	70	E
Royal Glory® Zaifer*	USA	-28	M	ME	M	F	DR	100	S
Intermediate ripening time									
Vistarich® Zainove*	USA	-19	E	M	VL	F	DR	90	E
Rich Lady*	USA	-17	E	M	L	F	OR	80	E
Summer Rich*	USA	-15	M	M	L	F	DR	90	E
Grenat® Monafi*	FRA	-8	E	M	L	F	MR	80	S
Doris	FRA	-7	M	M	M	F	DR	80	E
Diamond Princess*	USA	-2	M	ME	L	F	DR	80	E
Rome Star* (Referring CV)	ITA	0	E	ME	VL	F	LR	90	E
Elegant Lady	USA	4	M	ME	L	F	DR	80	E
Symphonie*	FRA	4	L	ME	L	F	MR	90	E
Zee Lady®	USA	4	M	M	L	F	DR	80	E
Late and very late ripening time									
Kaweah®	USA	15	L	M	L	F	DR	80	E
Summer Lady®	USA	19	L	M	L	F	MR	80	E
Gilda Rossa*	ITA	23	L	M	L	F	DR	90	E
O'Henry®	USA	28	L	E	VL	M	DR	70	E
Tardibelle	FRA	34	L	ME	VL	F	MR	50	E

Table 5: contd….

Red Star*	USA	35	L	E	VL	VF	MR	80	E
Guglielmina	ITA	35	L	E	VL	F	DR	50	E
Red Late	USA	46	M	M	VL	F	LR	40	E
Fairtime	USA	56	M	M	VL	F	LR	40	E

See legend in Table **2**.

Table 6: List of the main canning peach cultivars: their country of origin, harvest date respect to the reference cv Rome Star, flowering date at 10 % of open flowers, productivity and some of their pomological characteristics [15, 16]

Cultivar	Country of origin	Ripening time (dd ± Rome Star)[1]	Flowering period [66]	Yield[3]	Fruit size[34]	Flesh Firmness[50]	Over color[43]	Over color extent (%)	Taste type[8]
Extra-early and early ripening time									
Jonia *	ITA	-42	M	M	M	F	LR	20	E
Egea *	ITA	-35	M	M	S	VF	LR	20	E
Federica *	USA	-30	M	M	L	VF	LR	10	E
Tirrenia *	ITA	-23	L	ME	L	F	MR	20	E
Romea	ITA	-18	L	E	L	VF	LR	20	E
Intermediate ripening time									
Loadel *	USA	-12	L	M	L	VF	LR	5	E
Villa Doria *	ITA	-6	M	M	M	VF	LR	5	E
Carson	USA	-4	L	E	L	F	LR	30	E
Eolia *	ITA	2	L	M	M	F			E
Bowen	USA	10	L	M	M	VF	P	5	E
Andross	USA	12	L	E	M	F	LR	30	E
Late and very late ripening time									
Jungermann	USA	19	L	ME	L	F	LR	20	E
Babygold 9	USA	25	L	M	L	VF	LR	10	E

See legend in Table **2**.

Table 7: List of the main flat peach cultivars: their country of origin, harvest date respect to the reference cv Rome Star, flowering date at 10 % of open flowers, productivity and some of their pomological characteristics [15, 16]

Cultivar	Country of origin	Ripening time (dd ± Rome Star)[1]	Flowering period [66]	Yield[3]	Fruit size[34]	Flesh Firmness[50]	Flesh ground color[51]	Over color[43]	Over color extent (%)	Taste type[8]
UFO 1	ITA	-59	E	M	M	F	W	DR	70	A
UFO 2	ITA	-55	E	E	M	F	W	MR	85	E
Platiforone Sweet Ring	ITA	-46	E	E	M	F	Y	DR	80	S
UFO 3	ITA	-43	M	E	M	F	W	MR	70	S
UFO 4	ITA	-40	M	E	L	F	W	DR	70	S
UFO 5	ITA	-25	M	E	L	F	Y	DR	70	E
Platifortwo Pink Ring	ITA	-21	E	E	L	F	W	DR	90	S
UFO 6	ITA	-18	E	E	L	F	Y	DR	70	E
Sweet Cap®	FRA	0	L	E	VL	F	W	MR	50	E
UFO 7	ITA	23	M	E	VL	F	Y	DR	80	S
UFO 8	ITA	46	L	E	VL	F	Y	MR	60	S
UFO 9	ITA	58	L	E	VL	F	W	MR	60	S

See legend in Table **2**.

Table 8: List of the main flat nectarine cultivars: their country of origin, harvest date respect to the reference cv Rome Star, flowering date at 10 % of open flowers, productivity and some of their pomological characteristics [15, 16]

Cultivar	Country of origin	Ripening time (dd ± Rome Star)[1]	Flowering period [66]	Yield[3]	Fruit size[34]	Flesh Firmness[50]	Flesh ground color[51]	Over color[43]	Over color extent (%)	Taste type[8]
PLATINET 1	ITA	-38	M	M	L	F	W	MR	80	E
PLATINET 2	ITA	-34	M	M	L	F	Y	DR	70	A
CONCETTINA	ITA	-16	E	E	L	F	W	DR	90	S
PLATINET 3	ITA	-5	M	M	L	F	W	MR	75	A
PLATINET 4	ITA	10	E	M	VL	F	W	DR	80	E

See legend in Table **2**.

Figure 1: Peach and nectarine varieties (source: Bevilacqua D. & Di Cintio A. - CRA-FRU).

a) The deanthocyanic stony hard peach 'Ghiaccio 1' (ripening date: -8 days from the reference 'Rome Star'); **b)** The flat shape nectarine 'Platinet 3' (harvesting date: -5 days from RS).

1.4. Apricot

Prunus armeniaca is not native of Armenia, where it has been continuously cultivated since at least the first century AD. It was brought to Armenia from a more eastern center of origin and remains were found in pre-Christian archeological excavation sites.

Nowadays fruits are used in both fresh and dry form, canned or otherwise preserved as jam and marmalade or pulp [17].

Apricots have been placed definitely in the genus *Prunus* but, depending on the botanical authority, opinions have been mixed on whether apricot should be placed within the sub genus *Prunophora* or *Amygdalus* since it shares some morphological characteristics with both. The classification of Potter [2] distinguished plums from apricots on the basis of ovary pubescence, being absent or glabrous in the plums and present or pubescent in apricots.

China's immense size and varied topography, as well as its numerous geographical and climatic zones provided enormous genetic diversity in many plant families that were unknown in Western countries. Apricot classification in China parallels that of western taxonomists to the sub-family level of the *Prunoideae*: this is then divided into nine genera and *Armeniaca* is one of these. The genus *Armeniaca* is divided into 10 species with the mention of an 11th species that is not present in China [17].

Three centers of origin are recognized for apricot. One is located in the mountainous regions and adjacent lands of central and western China, known as the 'Chinese Center'. A second region, the 'Inner Asiatic Center', was defined by the approximate boundaries of northwestern India, Afghanistan, Tajikistan, Uzbekistan and western Tien-Shan mountains. The last center of origin for apricot described by Vavilov [18] was known as 'Asia Minor Center'. The region is defined as the lands of Transcaucasia, Iran and Turkmenistan [17, 19].

The cultivated apricot is diploid and has a chromosome number of 2n=2x=16.

Different characters have been used to uniquely identify genotypes and cultivars in apricot. The Community of Plant Variety Office (CPVO) and UPOV consider a total of 57 traits to be used in DUS (distinctness, uniformity and stability) test (see Table **9**). Four traits are related to the tree (vigour, habit, degree of branching, distribution of flower buds), three to the shoot (young shoot anthocyanin coloration, one year old shoot color, one year old shoot size of bud support), four consider the flower and its organs (petals, stamens, anthers, ovary *etc.*), 16 the leaf and petiole, 26 regard the fruit (size, shape, symmetry, ground colour, over colour, pubescence, sweetness, acidity *etc.*), two the stone and two traits take into account some phenology (time of beginning of flowering, time of beginning of ripening). As already mentioned for peaches, all these traits are qualitative, thus no measurements are needed and only a comparison with one or more test samples (other apricot cultivars) is carried out. In Table **10** traits related to apricot fruit characteristics and the main reference cultivars are reported.

Table 9: Characteristics to be used in DUS test and description of the apricot cultivars (from CPVO documentation). Only the fruit traits are reported

CPVO Trait Number	Character	Type
28	Size	Very small, small, medium, large, very large
29	Shape in lateral view	Triangular, ovate, oblong, elliptic, circular, oblate, obovate, oblique rhombic
30	Shape in ventral view	Triangular, ovate, oblong, elliptic, circular, oblate, obovate
31	Height	Short, medium, tall
32	Lateral width	Narrow, medium, broad

Table 9: contd....

33	Ventral width	Narrow, medium, broad
34	Ratio height/ventral width	Small, medium, large
35	Ratio lateral width/ventral width	Small, medium, large
36	Symmetry in ventral view	Symmetric, slightly asymmetric, clearly asymmetric
37	Suture	Raised, slightly sunken, moderately sunken, deeply sunken
38	Depth of stalk cavity	Shallow, medium, deep
39	Shape of apex	Acute, rounded, truncate, retuse
40	Presence of mucron	Absent, present
41	Surface	Smooth, bumpy
42	Skin pubescence	Absent, present
43	Glossiness of skin	Absent, medium, strong
44	Ground color of skin	Not visible, white, yellowish, yellow green, light orange, medium orange, dark orange
45	Relative area of over color	Absent, small, medium, large, very large
46	Hue of over color	Orange red, red, pink, purple
47	Intensity of over color	Light, medium, large
48	Pattern of over color	Isolated flecks (spots), solid flush, covered all over with very small dots
49	Color of flesh	Whitish green, white, cream, light orange, medium orange, dark orange
50	Texture of flesh	Fine, medium, coarse
51	Firmness of flesh	Very soft, soft, medium, firm, very firm
52	Ratio weight of fruit/weight of stone	Small, medium, large
53	Adherence of stone to flesh	Absent, weak, medium, strong

1.4.1. Apricot Cultivars

In these last years, the market supply of apricot fruits is meeting the many requests of consumers. Main distinctive characters of the new apricot cultivars are a product of quality, sweeter and juicy, firm, big in size, very attractive thanks to an extensive blush (Fig. **2a**) and available for an ample temporal arc.

Main critical aspects of the apricot cultivation are the self-incompatibility, the high chilling requirement and the susceptibility to Sharka virus.

There are some old autochthon cultivars very well adapted to the local climatic conditions of the Centre-South Italy as 'Boccuccia Spinosa' (Fig. **2b**), 'Portici', 'Pellechiella' (Fig. **2c**) and 'Vitillo', which are characterized by a yellow aromatic fruit. These varieties have a very good productivity and guarantee good incomes in the local markets. Consumers recognize these products and appreciate their flavour and their historical and cultural ties with the territory.

On the other hand, the French self-fertile 'Carmingo Series' (Fig. **2d**), with red over coloured fruits and very late ripening represents a fresh novelty in the varietal panorama.

Table 10: List of the main apricot cultivars: their country of origin, ripening time expressed in days from the reference cv S. Castrese, flowering date at 10% of open flowers, average yield and some characteristics of the fruits [20].

Cultivar	Country of origin	Ripening time (dd ± S. Castrese)[1]	Flowering period[56]	Yield[3]	Fruit size[28]	Flesh Firmness[50]	Hue of over color[46]	Over color extent (%)	Cracking %
Extra-early and early ripening time									
Carmingo® Primarel*	FRA	-28	L	M	S	S	R	20	0
Spring Blush® EA 3126 TH*	FRA	-28	L	M	M	F	R	45	10
Carmingo® Primando*	FRA	-26	L	H	S	M	O	60	
Luna*	ITA	-26	VL	S	VS	M	R	20	0
Carmingo® Primaya*	FRA	-24	M	M	L	M	R	25	0
Wonder Cot®	USA	-24	M	M	L	F	R	35	25
Carmingo® Primaris*	FRA	-20	L	M	L	F	R	20	10
Carmen Top® Carmen*	USA	-18	M	M	L	M	O	15	4
Aurora*	ITA	-18	E	M	M	M	R	30	5
Solaria	USA	-18	M	S	L	F	R	30	5
Sweet Red® Red Silver*	FRA	-17	L	H	L	F	R	50	20
Bora	ITA	-16	M	H	L	S	R	20	0
Magic Cot® RM22	USA	-15	M	M	VL	F	R	40	40

Table 10: contd…

Sweet Cot® Toyuda*	USA	-14	M	M	L	M	O	10	15
Perle Cot®* APR38-4	USA	-13	L	M	L	F	R	60	0
Tom Cot® Toyaco*	USA	-12	M	H	S	M	R	30	0
Pinkcot® Cotpy*	FRA	-12	M	M	VL	F	R	40	5
Antonio Errani	ITA	-12	M	M	M	M	R	20	5
Carmingo® Priabel*	FRA	-12	L	S	M	M	R	25	0
Carmingo® Primarina*	FRA	-12	L	M	M	M	R	5	
Intermediate ripening time									
Robada*	USA	-11	M	M	L	F	R	50	0
Orange Rubis® Couloumine*	FRA	-11	L	H	L	F	R	25	5
Big Red® EA 4006*	FRA	-10	L	M	M	F	R	50	5
Bella d'Imola	ITA	-7	M	M	L	M	R	10	5
Carmingo® Medaga*	FRA	-7	L	S	L	F	O	40	0
Goldrich	USA	-6	M	M	L	F	O	5	5
Laura	ITA	-4	M	M	VL	M	O	35	5
Vitillo	ITA	-4	L	M	VL	M	R	15	5
Boreale	ITA	-4	M	M	VL	F	R	35	0
San Castrese	ITA	0	M	H	M	M	O	5	5
Laycot®*	CAN	0	M	M	M	M	R	45	10
Kioto*	FRA	1	L	M	L	F	R	55	10
Flavorcot® Bayoto*	USA	1	M	H	S	F	R	25	5
Portici	ITA	1	L	M	L	M	P	15	15
Pellecchiella	ITA	3	L	M	L	F	R	30	15
Boccuccia spinosa	ITA	4	M	M	S	F	R	5	0
Zebra® Priboto*	FRA	9	M	M	VL	F	O	15	5
Pisana	ITA	10	L	M	L	M	R	30	10

Table 10: contd…

Late and very late ripening time									
Pieve Tardiva	ITA	11	L	S	M	M	R	40	0
Carmingo® Faralia*	FRA	16	L	M	L	F	R	40	5
Carmingo® Farbaly*	FRA	38	L	M	L	F	P	45	15
Augusta 3*	ITA	45	VL	M	M	F	R	15	15
Carmingo® Farfia*	FRA	48	M	S	M	F	R	20	0
Carmingo® Farius*	FRA	48	L	S	M	F	R	35	0
Carmingo® Fardao*	FRA	52	L	S	M	F	R	15	0
Carmingo® Farclo*	FRA	54	L	S	M	F	O	30	5

[1]San Castrese ripens in the North of Italy around the 3rd of July, in the Center Italy around the 30th of June, in the South Italy the 24th of June
56 cpvo DUS test trait number = Time of beginning flowering period: VE= very early; E= early; M= medium; L= late; VL= very late
3 Yield: S= scarse; MS= medium-scarse; M= medium; E= high; ME= very high
28 cpvo DUS test trait number = Fruit size: VS=very small; S= small; M= medium; L= large; VL= very large
51 cpvo DUS test trait number = Firmness of Flesh: VS= very soft; S= soft; M= medium; F=firm; VF= very firm
46 cpvo DUS test trait number = Hue of over color: OR = orange red; P = pink; R= red; Pu= purple

Figure 2: Apricot cultivars showing the broad variability of the fruits (source: Bevilacqua D. & Di Cintio A. - CRA-FRU).

a) 'Orange Rubis'® (ripening date -11 days from the reference cv San Castrese); **b)** 'Boccuccia Spinosa' (+4 days from SC); **c)** Pellecchiella (+3 days from SC); **d)** Carmingo® Faralia* (+16 days from SC)

1.5. Cherries

The major cherry species commercially grown for their fruits are sweet and sour cherry (*Prunus avium* L. and *Prunus cerasus* L., respectively).

Cherry production is limited to temperate regions that experience moderately cold winter temperatures. Sweet cherries are best eaten fresh. Sour cherries are excellent for cooking and are used mainly in pies, jams, juices and pastry products.

The two main groups of cherries have been reported to be originated in the area that includes Asia Minor, Iran, Iraq and Syria. Sour cherry is believed to be arisen through natural hybridizations between ground cherry (*P. fruticosa* Pall.) and sweet cherry.

Sweet cherry is diploid (2n=16) and sour cherry and ground cherry are tetraploid (2n=32). Sweet cherry exhibits a classic gametophytic self-incompatibility (GSI) system that is a common genetic mechanism promoting outcrossing in flowering plants. In GSI, self-incompatibility (SI) is determined by a single multi-allelic locus, called the S-locus that contains a minimum of two genes, one controlling stylar specificity and the other controlling pollen specificity of the SI reaction [21]. Different characters have been used to uniquely identify genotypes and cultivars in sweet cherry. The Community of Plant Variety Office (CPVO) and UPOV consider 41 traits to be used in DUS (distinctness, uniformity and stability) test.

Three of these traits consider the tree (vigor, habit, branching), five the shoot (young shoot anthocyanin coloration, pubescence of apex, length of internode, number of lenticels, thickness), four are related to the flower and its organs (petals, stamens, anthers, ovary *etc.*), seven to the leaf and petiolc, 18 regard the fruit (size, shape, symmetry, ground color, over color, pubescence, sweetness, acidity *etc.*), two the stone and two traits consider some phenology (time of beginning of flowering, time of beginning of ripening). As also said for the two other *Prunus* species, the traits considered are all qualitative and so no measurements are needed. The trait scoring is carried out by comparison with one or more test samples (other cherry cultivars). In Table **11** some traits related to some fruit characteristics in the main reference cultivars are reported.

Table 11: Characteristics to be used in DUS test and description of the sweet cherry cultivars (from CPVO documentation). Only the fruit traits are reported

CPVO Trait Number	Character	Type
20	Size	Very small, small, medium, large, very large
21	Shape	Cordate, reniform, oblate, circular, elliptic
22	Pistil end	Pointed, flat, depressed
23	Suture	Absent, weakly conspicuous, strongly conspicuous
24	Length of stalk	Very short, short, medium, long, very long
25	Thickness of stalk	Thin, medium, thick
26	Abscission layer between stalk and fruit	Absent, present
27	Color of skin	Yellow, yellow with blush, orange red, light red, red, brown red, dark red, blackish
28	Size of lenticels on skin	Small, medium, large
29	Number of lenticels on skin	Few, medium, many
30	Thickness of skin	Thin, intermediate, thick
31	Color of flesh	Cream, yellow, pink, medium red, dark red
32	Color of juice	Colorless, light yellow, pink, red, purple
33	Firmness	Soft, medium, firm, very firm
34	Acidity	Low, medium, high
35	Sweetness	Low, medium, high
36	Juiciness	Weak, medium, strong

1.5.1. Sweet Cherry Cultivars

The most interesting characters of the new cultivars and varieties currently cultivated can be summarized in: big size (fruits weighting not less than 10g), a wide pedo-climatic adaptability, self-fertility, yield, good holding of the harvest fruits on the tree, rapid entry into production and possibility to pick up cherry without stalk. 'Margit', 'Germersdorfi Orias 3', 'Linda' and the self-fertile 'Enrica' (Fig. **3a**) are considered suitable for a mechanical harvest. One other great result obtained in varietal improvement was the extension of the ripening calendar that now covers 6-7 weeks (since half of May till end of June, Table **12**).

Cracking susceptibility and the scarce tolerance to some diseases like the brown rot (Fig. **3b**) are the main problems highly limiting the quality and the profit of sweet cheery cultivation. In fact, the precocious cultivars (Table **12**) while having excellent fruits with superior aesthetic characters and productivity respect to the reference and 'ancient' cv Burlat (Fig. **3c**), also exhibit the same scarce flesh firmness and over all, the high sensibility to cracking of the fruits. For this reason, rain protection systems are suggested. The high number of varieties that ripen in the early and intermediate season time allow an appropriate choice in different environments. The group of the late cultivars has only recently been enlarged by the presence of three Canadian cultivars (Fig. **3d**) extending the ripening calendar of about one week.

Table 12: List of the main sweet cherry cultivars: their country of origin, ripening time expressed in days from the reference cv Burlat, beginning of flowering period at 50% of open flowers; production, and some pomological characters [22]

Cultivar	Country of Origin	Ripening time (dd ± Burlat)[1]	Flowering period[40]	Yield[3]	Fruit size[20]	Flesh Firmness[50]	Taste[6]	Self-Fertile Y/N	Cracking Y/N
Extra-early ripening time									
Rita*	HUN	-6	VP	VS	M	H	S	N	Y
Earlise®-Early Lory*	FRA	-3	VP	S	M	M	M	N	Y
Early Bigi®-Bigi Sol*	FRA	-3	VP	S	L	M	M	N	Y
Sweet Early ®-Panaro 1*	ITA	-2	VP	M	L	S	M	N	Y
Burlat	FRA	0	P	S	S	S	S	N	Y
Burlat C1	ITA	0	VP	M	M	M	M	N	Y
Early Star®-Panaro 2*	ITA	5	P	S	VL	M	M	Y	Y
Early ripening time									
Chelan®	USA	7	P	M	M	M	M	N	N
Brooks*	USA	8	VP	M	VL	H	G	N	Y
Tieton®-PC 71446*	USA	9	P	M	VL	H	M	N	Y
Celeste®-Sumpaca*	CAN	9	P	H	L	H	M	Y	Y
Cashmere®	USA	10	I	M	VL	M	G	Y	N
Grace Star*	ITA	11	I	H	M	M	M	Y	N
Margit	HUN	13	I	VS	L	S	M	N	Y
Samba®-Sumste*	CAN	13	P	M	VL	M	M	N	Y
Giorgia	ITA	14	I	M	VL	H	M	N	N
Index*	USA	14		S	L	M	G	Y	N
New Star	CAN	15	P	M	L	H	G	Y	Y

Table 12: contd…

Vera* (III-15/6)	HUN	16	VP	VS	VL	M	M	Y	N
Carmen*	HUN	16	L	VS	VL	M	G	Y	N
Intermediate ripening time									
Sandra Rose	CAN	17	P	VS	VL	M	M	Y	Y
Enrica	ITA	17	VP	M	VL	H	S	Y	N
Van	CAN	17	P	M	VL	H	M	N	N
Giulietta	ITA	17	P	VS	L	M	S	Y	N
Canada Giant®-Sumgita*	CAN	18	I	VS	VL	M	G	N	N
Cristalina®-Sumnue*	CAN	18	I	S	VL	H	G	N	Y
Summit	CAN	19	I	VS	VL	S	M	N	N
Techlovan*	CZE	19	I	S	VL	M	G	N	N
Black Star*	ITA	20	L	M	VL	M	G	N	N
Germersdorfi Orias 3	HUN	20	I	M	VL	M	G	N	N
Gégé®	FRA	20	I	VS	VL	H	M	N	N
Medium-late ripening time									
Somerset	USA	20	VP	M	L	H	S	N	Y
Sylvia	CAN	20	L	M	L	M	G	N	N
Aida*	HUN	21	P	M	L	H	G	N	N
Kordia-Attika®	CZE	21	VP	VS	M	M	G	N	N
Ferrovia	ITA	21	I	VS	L	M	G	N	N
Ferrovia Spur	ITA	21	P	S	VL	M	G	N	N
Lapins	CAN	21	VP	S	L	H	G	Y	N
Lucrezia	ITA	24	P	H	L	M	M	Y	N
Big Star*	ITA	24	I	H	VL	H	G	Y	N
Linda	HUN	24	I	S	VL	H	M	N	Y
Late ripening time									
Patty-Skeena®*	CAN	27	I	M	VL	H	M	Y	Y
Regina	DEU	27	I	M	VL	H	G	N	N
Sweetheart®-Sumtare*	CAN	31	P	H	VL	M	G	Y	N
Late Lory®*	FRA	33	L	S	L	H	G	N	N
Alex®-Axel*	HUN	35	L	M	L	M	M	Y	N
Symphony-Selina®	CAN	38	I	H	L	M	M	Y	N
Staccato®-Summer Charm-13S2009*	CAN	39	P	H	L	M	G	Y	N

1 Burlat ripens in the North of Italy from the 5th to 10th of June, in the Center Italy from the 25th to the 30th of May, in the South Italy from the 15th to the 20th of May.
40 cpvo DUS test trait number = Time of beginning flowering period: VE= very early; E= early; M= medium; L= late; VL= very late.
3 Yield: S= scarse; MS= medium-scarse; M= medium; E= high; ME= very high.
20 cpvo DUS test trait number = Fruit size: VS=very small; S= small; M= medium; L= large; VL= very large.
33 cpvo DUS test trait number = Firmness of Flesh: VS= very soft; S= soft; M= medium; F=firm; VF= very firm.
6 Taste: S= scarse; M= medium; G= Good.

Figure 3: Sweet cherry varieties (source: Bevilacqua D. & Di Cintio A. - CRA-FRU).

a) the self-fertile 'Enrica' (ripening date: +17 days respect to the reference 'Burlat'); **b)** 'Grace Star' (+11 dd from B) showing susceptibility to brown rot; c) the reference 'Burlat'; d) Staccato®-Summer Charm-13S2009*.

1.6. Plums

Most of the plums commercially grown fall into one of the two groups: European (exaploid) types or Japanese (diploid) types. European plums (*Prunus domestica*) are generally better adapted to cooler regions than Japanese types [17]. The term 'Japanese plum' was originally applied to *Prunus salicina*, but now includes all the fresh-market plums developed by intercrossing various diploid species with the original species [23]. A wide array of skin and flesh colours is available, but in recent years the market has been dominated by black skin with light yellow or red flesh.

Plums are placed in the *Prunophora* sub-genus, section *Euprunus* and contains Asian and European species, distinguished by 1-2 flowers per bud, stone often sculptured, and leaves rolled in the bud. Section *Prunocerasus* contains the American plum species and also the Asian species appear to fit better both taxonomically and horticulturally. *P. domestica* may be descended from polyploidy forms of *P. cerasifera* (Myrabolan plum), which have a long history of local use and selection across continent, and have a range of fruit color and palatability. *Prunus simonii*, sometimes described by Western botanists, is probably a variant of *P. salicina*. It was used in developing California cultivars because of its firm flesh and strong flavor [24].

Different characters have been used to uniquely identify genotypes and cultivars in European and Japanese plum. The Community of Plant Variety Office (CPVO) and UPOV consider 62 traits to be used in DUS (distinctness, uniformity and stability) test for European plum.

Two traits consider the tree (vigor, density of crown), 11 the shoot (attitude, thickness length of internode, pubescence, number of lenticels, *etc.*), 13 are related to the flower and its organs (number of flowers, diameter, pedicel characteristics, petals, stamens, anthers, ovary *etc.*), 16 to the leaf and petiole, 12 consider the fruit (size, shape, symmetry, ground color, over color, pubescence, sweetness, acidity *etc.*), six the stone (shape, development of keel, size, *etc.*) and two traits regard some phenological traits (time of beginning of flowering, time of beginning of ripening). All the traits are qualitative, and no measurements are needed but only a comparison with one or more test samples (other plum cultivars). Some of the traits related to fruit in the main reference cultivars are reported in Table **13**.

For the Japanese plums the Community of Plant Variety Office (CPVO) and UPOV consider 61 traits to be used in DUS (distinctness, uniformity and stability) test: three traits are related to the tree (type of bearing, vigor, habit), five consider the shoot or the bud (shoot color, spur length, bud size, bud shape of apex, position of vegetative bud in relation to shoot), seven consider the flower and its organs (number of flowers, diameter, pedicel characteristics, petals, stamens, anthers, ovary *etc.*), 12 the leaf and petiole, 25 regard the fruit (size, shape, symmetry, ground color, over color, pubescence, sweetness, acidity *etc.*), seven the stone (size, shape, symmetry, texture, width of stalk end) and two traits some phenology (time of beginning of flowering, time of beginning of ripening). As for the European plums no measurements are needed. In Table **14** traits related to fruit characteristics in the main reference cultivars are reported.

Table 13: Characteristics to be used in DUS test and description of European plum cultivars (from CPVO documentation). Only fruit traits are reported

CPVO trait number	Character	Type
43	Size	Very small, small, medium, large, very large
44	Shape in lateral view	Oblong, elliptic, circular, oblate, ovate, obovate
45	Symmetry in ventral view	Symmetric, asymmetric

Table 13: contd...

46	Depth of suture towards stalk end	Shallow, medium, deep
47	Depression of apex	Absent, intermediate, strong
48	Pubescence at apex	Absent, present
49	Depth of stalk cavity	Shallow, medium, deep
50	Ground color of skin	Greenish white, green, yellowish green, yellow, orange yellow, red, light violet, purplish violet, dark violet, violet blue, dark blue
51	Color of flesh	Whitish, green, yellowish green, yellow, orange, red
52	Firmness of flesh	Soft, medium, firm
53	Juiciness	Low, medium, high
54	Degree of adherence of stone to flesh	Non-adherent, semi-adherent, adherent

Table 14: Characteristics to be used in DUS test and description of Japanese plum cultivars (from CPVO documentation). Only fruit traits are reported

CPVO trait number	Character	Type
28	Length of the stalk	Short, medium, long
29	Size	Very small, small, medium, large, very large
30	Height	Short, medium, tall
31	Width	Narrow, medium, broad
32	Shape in lateral view	Oblong, elliptic, circular, oblate, cordate, obovate, obcordate
33	Symmetry	Symmetric, moderately symmetric, strongly symmetric
34	Shape of base	Pointed, truncate, depressed
35	Shape of apex	Pointed, rounded, truncate, depressed
36	Depth of stalk cavity	Shallow, medium, broad
37	Width of stalk cavity	Narrow, medium, broad
38	Depth of suture	Absent, shallow, medium, deep
39	Bloom of skin	Absent, weak, medium, strong, very strong
40	Ground of color skin	Not visible, green, yellowish green, yellow
41	Relative area of over color	Absent, small, medium, large, very large
42	Over color of skin	Yellow, orange yellow, medium red, dark red, purple, dark blue, black

Table 14: contd…

43	Pattern of over color	Flecks only, mottled, solid flush only
44	Number of lenticels	Few, medium, many
45	Size of lenticels	Small, medium, large
46	Color of flesh	Whitish, green, yellowish green, yellow, orange, medium red, dark red, purplish
47	Firmness	Soft, medium, firm
48	Juiciness	Low, medium, high
49	Acidity	Low, medium, high
50	Sweetness	Low, medium, high
51	Adherence of stone to flesh	Non-adherent, semi-adherent, adherent
52	Amount of fiber	Low, medium, high

1.6.1. Japanese and European Plum Cultivars

In the last decade, there was a varietal renewal that allowed a better choice among the numerous and different cultivars but not an enlargement of the cultivated surface.

The new cultivars, mainly belonging to the Japanese group, are very attractive for their pomological aspects like the big size and the different skin colour: green, yellow ('Bragialla' in Fig. **4a**), red (Aphrodite* in Fig. **4b**) and black (Black Gold® in Fig. **4c**), with a yellow or red flesh and high organoleptic characters. Japanese cultivars, find only in warm and dry weather the best conditions to guarantee a suitable qualitative and productive standard.

The extreme susceptibility of the Japanese plums to the European Stone Fruit Yellows phytoplasma (ESFY) and the damages caused to all the stone fruits by the Plum Pox Virus (PPV) limit drastically the cultivation of plums.

European plum cultivars are destined to the fresh consumption with a high qualitative value: big size, high firmness and good organoleptic characters. 'D'Ente 707' (Fig. **5a**) is the only cultivar that can be processed and dried. Even if the European varieties are more tolerant to phytoplasma, more rustic and with high productivity, very few can satisfy market demands.

Table 15: List of the main Japanese plum cultivars: their country of origin, ripening time expressed in days from the reference cv Shiro, beginning of flowering period at 10% of open flowers; yield, a few pomological characters and the cultivars used as best pollinizers [25]

Cultivar	Country of origin	Ripening time (dd ± 'Shiro')[1]	Flowering period[60]	Yield[3]	Fruit size[29]	Firm-ness[47]	Taste[6]	Skin over color[42]	Flesh color[46]	Pollinizers
Early ripening time										
Sorriso di Primavera	ITA	-14	P	H	S	S	M	MR	Y	Shiro, Morettini 355, Santa Rosa
DOFI Sandra*	ITA	-10	I	VH	L	M	G	DB	Y	Sorriso di Primavera, Shiro, Black Gold, Angeleno, Blackamber
Carmen Blu®	ITA	-8	I	M	L	M	G	B	Y	Sorriso di Primavera, Shiro
Shiro	RUS	0	I	VH	M	M	G	Y	Y	Sorriso di Primavera, Angeleno®, Santa Rosa, Morettini 355
Intermediate ripening time										
Anne Gold®*	ITA	8	I	H	L	H	G	Y	YG	Sorriso di Primavera, Shiro
Black Gold®	USA	9	P	M	L	H	G	B	MR	Blackamber, Angeleno®, Laroda, Black Diamond®, Ozark Premier
Blackamber	Unknown	10	P	M	L	H	M	B	Y	Friar, Santa Rosa, Laroda, Black Star®
Aphrodite*	USA	15	I	M	L	H	G	MR	Y	Angeleno®, Friar, Shiro, Morettini 355, Santa Rosa, Sangue di Drago
Black Diamond®	USA	24	I	M	L	H	G	B	MR	Sorriso di Primavera, Friar, Angeleno®, Black Star®, Laroda

Table 15: contd…

Laroda	USA	26	I	M	M	H	G	DR	O	Ozark Premier, Friar, Sorriso di Primavera, Black Gold®, Santa Rosa
Fortune	USA	28	P	M	M	H	G	MR	Y	Friar, Laroda, Santa Rosa
Late and very late ripening time										
Golden Plumza*	ITA	29	I	S	L	H	G	Y	W	Green Sun®, Yellow Sun, July Sun, Friar, Angeleno®
Friar	USA	32	P	VH	M	H	G	B	Y	Blackamber, Laroda, Santa Rosa, Ozark Premier, Morettini 355
Green Sun®	USA	40	I	M	L	H	G	Y	YG	Friar, Obilnaja, Yellow Sun, July Sun
TC Sun*	USA	43	I	H	M	H	G	Y	Y	Tracy Sun®, Green Sun, Fortune
Bragialla*	ITA	50	I	H	M	H	G	Y	YG	Angeleno®, Sorriso di Primavera, Anne Gold®*
Angeleno®	Unknown	63	I	M	M	H	M	B	Y	Black Gold®, Black Diamond®, Sorriso di Primavera, Obilnaja, Friar
Autumn Giant®	USA	68	I	M	M	H	M	MR	Y	Simka, Santa Rosa

1 Shiro ripens in the North of Italy from the 10th to 15th of July, in the Center Italy from the 5th to the 10th of July, in the South Italy from the 1st to the 5th of July.
60 cpvo DUS test trait number = Time of beginning flowering period: VE= very early; E= early; M= medium; L= late; VL= very late.
3 Yield: S= scarse; MS= medium-scarse; M= medium; E= high; ME= very high.
29 cpvo DUS test trait number = Fruit size: VS=very small; S= small; M= medium; L= large; VL= very large.
6 Taste: S= scarse; M= medium; G= Good.
42 cpvo DUS test trait number = Over color of Skin: Y = yellow; MR= medium red; DR = Dark red; P= purple; DB = Dark blue; B= Black.
46 cpvo DUS test trait number = Color of flesh: W= whitish; YG= yellowish green; Y= yellow; O= orange; MR= medium red.

Table 16: List of the main European plum cultivars: their country of origin, ripening time expressed in days from the reference cv Stanley (Fig. **5b**), beginning of flowering period at 10% of open flowers; production, some pomological characters and the cultivars used as best pollinizers

Cultivar	Country of Origin	Ripening time (dd ± 'Stanley')[1]	Flowering period[60]	Yield[3]	Fruit size[29]	Firm-ness[47]	Taste[6]	Skin over color[42]	Flesh color[46]	Pollinizers
Intermediate ripening time										
Firenze 90*	ITA	-35	P	M	M	H	M	DB	Y	President, Sugar, Empress
Late ripening time										
Emperor	USA	-11	I	M	L	M	M	P	Y	President, Stanley
D'Ente 707	FRA	-3	L	S	S	M	G	MR	YG	Bluefre, President, Stanley
Stanley	USA	0	I	VH	M	H	G	P	Y	Bluefre, Giant, President
President	GBR	9	P	H	VL	H	G	P	Y	Bluefre, Grossa di Felisio, Stanley, Giant

1 Stanley ripens in the North of Italy from the 5th to 10th of September, in the Center Italy from the 1st to the 5th of September, in the South Italy from the 25th to the 31st of August.

60 cpvo DUS test trait number = Time of beginning flowering period: VE= very early; E= early; M= medium; L= late; VL= very late.

3 Yield: S= scarse; MS= medium-scarse; M= medium; E= high; ME= very high.

29 cpvo DUS test trait number = Fruit size: VS=very small; S= small; M= medium; L= large; VL= very large.

6 Taste: S= scarse; M= medium; G= Good.

42 cpvo DUS test trait number = Over color of Skin: Y = yellow; MR= medium red; DR = Dark red; P= purple; DB = Dark blue; B= Black.

46 cpvo DUS test trait number = Color of flesh: W= whitish; YG= yellowish green; Y= yellow; O= orange; MR= medium red.

Figure 4: Japanese plum varieties (source: Bevilacqua D. & Di Cintio A. - CRA-FRU).

a) 'Bragialla' (ripening date: +50days from 'Shiro'); **b)** 'Aphrodite' (+15 dd from S); **c)** 'Black Gold'® (+9 dd from S).

Figure 5: European plum varieties.

a) 'Stanley', the reference cultivar for the ripening date; **b)** 'D'Ente 707' (ripening date -3 dd from S).

2. PHYSIOLOGY RELATED TO PRODUCTION

2.1. From Flower to Fruit

Peach and sour cherry produce flower buds almost exclusively on one-year old shoots, while apricot, plum and sweet cherry differentiate their flower buds mostly on very short shoots (spurs) developed on two-year-old and/or older wood. The number and distribution of flower buds can vary with tree vigor, cultivar, and light exposure under which the shoot develops.

Flowers of stone fruit crops are hermaphrodite, *i.e.* featured with both male (stamens) and female (pistil) reproductive organs (Fig. **6**).

Figure 6: Peach flower (top view and longitudinal section; source: Liverani A. & Giovannini D.).

The scale below the flowers is in cm

2.2. Induction

The first and important event towards crop formation is flower initiation, by which the apical meristem of a number of vegetative buds becomes committed to form flowers. Bud initiation, preceding that of flowering and cropping, occurs in late spring or early summer (Table **17**) generally at the end of intense shoot growth phase. Leaves act as promoters of the differentiation, so a canopy not well exposed to sunlight or a defoliation process have a negative effect on flower differentiation.

Table 17: Time of flower initiation in stone fruit species (rearranged from Westwood [26])

Crop	Induction period	Type of wood bearing the flower buds
Almond	Mid August ÷ mid September	1-year-old shoots
Peach	Late June ÷ late July	1-year-old shoots
Apricot	Early August	1-year-old shoots & 2-year-old spurs
European Plum	Late June ÷ mid August	1-year-old shoots & 2-year-old spurs
Japanese Plum	Mid July ÷ early August	1-year-old shoots & 2-year-old spurs
Sweet Cherry	Early July	2-year-old spurs
Sour Cherry	Mid July	2-year-old spurs

Following the initiation *stimulus*, detectable morphological changes occur in the central part of the apical meristem of the bud, which shows an intense mitotic activity. Thereafter, histological differentiation of all floral verticils - sepals, petals, stamens and pistil – runs slowly also during winter dormancy. Micro and macro-sporogenesis ends shortly before blooming, with formation of male (pollen grains) and female gametes (egg apparatus).

In recent years, the identification and characterization of a number of homeotic genes acting as key-regulators in the flower bud formation of model angiosperm species, such as *Antirrhinum majus* L. and *Arabidopsis thaliana* (L.) Heynh., have allowed to seek their homologues in fruit species [27-29]. Given the similarity between *Arabidopsis* and peach (sharing the same taxonomic underclass and the presence of structurally similar genes), models created on the former can be transferred in the latter or, in general, in stone fruit species. A group of genes referring to the 'ABCDE' model (Table **18**), encoding transcription factors regulating the development of the flower structures, are involved: A-function genes control the calyx differentiation, ABE-function genes the corolla, BCE the stamens and CDE the pistil and the ovule [30].

Table 18: Homeotic genes present in the 'ABCDE' domains

Domain	***Arabidopsis thaliana***
A	*APETALA1 (AP1), APETALA2 (AP2)*
B	*APETALA3 (AP3), PISTILLATA (PI)*
C	*AGAMOUS (AG), SHATTERPROOF (SHP)*
D	*SEEDSTICK (STK)*
E	*SEPALLATA (SEP)*

2.3. Dormancy and Chilling Requirement

At the end of the summer, stone fruit trees enter a dormant period in which the buds are made insensitive to growth-promoting signals. This state is called endo-dormancy (or true dormancy) and prevents the trees from resuming growth in winter in response to transient warm temperatures, avoiding damage by subsequent frost(s) in late winter or early spring (Fig. **7**) [31]. The release from endo-dormancy requires the exposition of the tree to low temperatures for a certain period. The amount of cold needed to overcome this state is called chilling requirement (CR), and varies greatly according to the crop and among cultivars of the same crop (Table **19**). The practical interest in the real-time evaluation of the progress of the endo-dormancy has led to the elaboration of a set of predictive models based on temperature. The model of Weinberger [32], based on the linear accumulation of chilling hours below the critical temperature of 7.2°C (45°F), has evolved to more complex models such as the Utah model [33] and the dynamic model based on a Chilling Units (CU) response curve [34], which takes into consideration the nonlinear response to chilling temperatures over time. Impediments to accurate CR evaluation of a given variety includes the negative effect of high temperatures to CU accumulation [34, 35], as well as the identification of the precise moment where CU accumulation begins [34].

Figure 7: Prolonged flowering due to insufficient chilling accumulation.

(Courtesy of David Byrne, Texas A&M University, College Station, Texas, USA).

CR is considered as a stable characteristic of a cultivar, which can be estimated with a good approximation in experiments using constant thermal conditions or, more easily, deduced from the comparison in open field with a reference cultivar of known CR. It is a polygenic trait, with quantity expression; in peach, however, there is some evidence for one or a few genes with major effects [36], skewing the distribution curve of the progeny towards the lower-chilling parent. Several worldwide breeding programs have the objective to select cultivars with low chilling requirement, which would ensure yield also in environments where winter temperatures allow insufficient CU accumulation. In such areas, indeed, medium to high chilling requirement cultivars would experience a late, weak and extended blooming.

Table 19: Chilling requirement (expressed as amount of hours below 7.2°C) to break endo-dormancy for different stone fruit species. The cultivars listed were classified according to their CR level (data elaborated from different sources)

Species	Chilling requirement	Low Chilling Requirement	Medium Chilling Requirement	High Chilling Requirement
Peach	100 to 1,050	'August Pride', 'Babcock', 'Bonita', 'Desertgold', 'Early Amber', 'Earligrande', 'FlordaGrand', 'FlordaPrince', 'Midpride', 'Tropic-berta', 'TropicSweet'	'Bicentennial', 'Bounty', 'Carolina Belle', 'Goldcrest', 'Marigold', 'Topaz'	'Surecrop', 'Velvet', 'Summer Pearl', 'Reliance'

Table 19: contd…

Nectarine	400 to 1,050	'Desert Dawn', 'Desert Delight', 'Rose', 'Panamint', 'Pioneer', 'Silver Lode'		
Japanese Plum	400 to 750	'Gemade Ouro', 'Golden Talisma', 'Kelsey Paulista', 'Roxa de Itaquera', 'Sanguinea', 'Amerelinha', 'Carmesim', 'Pluma-7' Reubennel', 'Harry Pickstone 'Gulfblaze', 'Gulfbeauty', 'Gulfrose	'Beauty Burgundy', 'Delight', 'Howard', 'Miracle', 'Kelsey', 'Mariposa', 'Meredith', 'Methley', 'Santa Rosa', 'Satsuma', 'Sprite'	
European Plum	700-1100		'Stanley', 'President', 'D'Ente 707'	
Cherry	1000-1300			'Cristobalina', 'Brooks', 'Ruby', 'Somerset', 'Burlat', 'New Star', 'Marvin'
Apricot	300-1000	'Goldkist', 'Early Gold', 'Newcastle', 'Currot'		'Bulida', 'Orange Red', 'San Castrese', 'Goldrich'
Almond	100-500	'Desmayo', 'Marcona', 'Ferragnès'		

The endo-dormancy status can be overcome with spraying growth regulators (cytokinins and gibberellic acid) or caustic substances (hydrogen cyanamide, unsaturated fatty acids chain, urea) about 30-40 days before the expected bud opening. Hydrogen cyanamide, usually commercialised as Dormex®, has been tested in many crops, particularly in kiwifruit, peach, plum and apricot. The application of Dormex® in extra-low-chill peach cultivars was found to be recommendable only after exceptionally warm winter conditions not fulfilling cultivar chilling requirement [37]. Among growth regulators, gibberellic acid and cytokinins have also been positively tested [38, 39].

Once endo-dormancy is overcome, the tree enters the physiological phase of eco-dormancy, where growth resumption is prevented by the environmental conditions being a certain amount of time with temperatures above the thermal threshold (about 4.5 °C for most of species), called heat requirement (HR), needed. HR fulfillment determines the reactivation of a series of morphogenetic processes, such as bud growth resumption and the completion of flower organ differentiation. Heat requirement is often expressed as Growing Degree Hours (GDH) or Growing Degree Days (GDD). The GDH model, initially proposed by Richardson for peach, has been widely used to estimate the time of bud break and the subsequent phenological stages, including the date of fruit ripening. In its simplified version, the effectiveness increases linearly in the range between 4.5 °C and 25 °C, for maximum stay above 25 °C.

2.4. Pollination

In the pollination event, pollen grains shed by the anthers reach and adhere to the stigma. Cross pollination is usually needed to grant adequate yields in stone fruit crops, as most cultivars are self-incompatible. Insects, such as honeybees (*Apis mellifera* L.), are the most important pollen vectors for *Prunus* species, covering up to 700-900 m a distance with favorable weather conditions. However, to grant successful cross pollination, it is recommended to plant cross-fertile individuals at a maximum distance of 50-100 m. For adequate yields, a sufficient proportion of flowers must set fruits, which normally occurs after flowers have been pollinated and fertilized. At the stigma surface, pollen tube growth is autotrophic, depending on its own reserves; growing down the style towards the ovary, the pollen tubes become heterotrophic and find their nutritive source in the extra-cellular matrix within the pistil.

2.5. Self-Incompatibility

Self-incompatibility (SI), which prevents inbreeding, is a strategy developed by many hermaphrodite species to safeguard genetic variability, hence to maintain the ability to adapt to changing surrounding conditions. In stone fruit species, various degrees of SI are found: at the extremes are almond, where self-incompatibility is the norm, and peach, almost completely self-fertile; in plum, prune, cherry and apricot, there are incompatible, self-fertile and partial self-fertile cultivars.

Most *Prunus* species exhibit a gametophytic self-incompatibility (GSI) system, controlled by a single, multi-allelic locus called S-locus [40]. The male (S-Fbox) and female (S-RNase) determinant genes of the SI system have been recently identified, tightly linked in the S-locus, so as to form a genetic unit, called S haplotype [41]. In the pollination event, the SI reaction is triggered when the S-allele in the pollen is identical to one of the two S-alleles in the recipient style, which is always the case with self-pollination. S-RNases glycoproteins produced by the style, and able to degrade RNA, are absorbed by the growing pollen tubes. The F-box at the S-locus is able to recognize and bind specifically the S-RNase with the same 'S allele' specificity, preventing it from degrading. The active S-RNase inhibits protein synthesis in the growing pollen tube, leading to a halt in its growth and its death [42].

The pollination event between two individuals can lead to a total SI, when the S genotype is identical; semi-compatibility, when only one of the two S alleles is common; and full compatibility, when both the S alleles are different. Semi-compatible condition is generally compatible with regular yielding as the quantity of pollen produced is high and adequate to grant sufficient fruit set. In commercial orchards of crops where the SI phenomenon is common, adequate fruit set can be ensured only by inter-planting cultivars belonging to different SI groups and having well-overlapping flowering period. On the practical side, the identification of the pistil *S* and the pollen *S* determinants has led to the development of PCR-based *S* genotyping and the classification of cultivars to specific SI groups. In sweet cherry, for example, over 700 varieties cultivated in Europe have been assigned to one of the forty-seven incompatibility groups identified to date and this information is available for users [43].

Since decades, the development of self-fertile cultivars is an important goal of the breeding programs of stone fruit species, especially cherry and almond. In sweet cherry, the first two self-compatible selections (JI2420 and JI2434) were obtained in 1954 in the UK through X-ray irradiation [44]. Both selections carry pollen-part mutations, JI2420 has the S4 allele mutated (S4'), and JI2434 the S3 (S3') allele. To date, S4' has been most extensively used in breeding programs worldwide combining self-compatibility with the high fruit quality of the cultivar Stella. The first self-compatible cherry cultivar released carries this allele, as well as nearly all the other self-compatible cultivars available (Table **20**), deriving from Stella [45].

Table 20: List of cherry self-compatible cultivars (rearranged from Schuster [43])

Self-compatible cultivars	S-genotype	Parentage (if known)
Axel (Alex)	$S_3S_{3'}$	Van × Cherry Self Fertile 46
Blackgold	$S_{4'}S_6$	
Blaze Star	$S_{4'}S_6$	Lapins × Durone Compatto di Vignola
Celeste (Sumpaca)	$S_1S_{4'}$	Van × Newstar
Columbia (Benton)	$S_{4'}S_9$	Stella × Beaulieu
Compact Stella	$S_3S_{4'}$	Irradiated Stella
Cristobalina	S_3S_6	
Dame Roma	$S_{4'}S_{13}$	Black Douglas × Stella
Early Star	$S_{4'}S_9$	Burlat × Stella compact
Endiablà	S_3S_3	
Glacier	$S_{4'}S_9$	Stella × Early Burlat
Grace Star	$S_{4'}S_9$	Burlat × o.p.
Habunt†	$S_3S_{4'}$	Valeska × Sunburst
Index	$S_3S_{4'}$	Stella × o.p.
JI 2420	$S_4S_{4'}$	Emperor Francis × Napoleon X-ray pollen
JI 2434 (AHR)	$S_3S_{3'}$	Emperor Francis × Napoleon X-ray pollen
JI 2434 (EM)	$S_{3'}S_4$	Emperor Francis × Napoleon X-ray pollen
Kronio	$S_{5'}S_6$	
Lapins	$S_1S_{4'}$	Van × Stella
Newstar	$S_3S_{4'}$	Van × Stella
Pál	$S_{4'}S_9$	Burlat × Stella
Paulus	$S_{4'}S_9$	Burlat × Stella
Petrus (syn. Péter)	$S_3S_{4'}$	Burlat × Stella
Ripoll	S_3S_6	
PiSue192	$S_{4'}S_9$	Nabigos × Stella
Runar	S_3S_3	
Sandor	$S_{4'}S_9$	Burlat × Stella
Sandra Rose	$S_3S_{4'}$	(Star × Van) × Sunburst
Santina	$S_1S_{4'}$	Stella × Summit
Selah (Liberty Bell)	$S_3S_{4'}$	(Rainier × Bing) × Stella
Senyoreta Primerenca	S_3S_6	

Table 20: contd...

Sir Don	$S_{4'}S_{13}$	Black Douglas × Stella
Sir Hans	$S_2S_{4'}$	Stella × Stella
Skeena	$S_1S_{4'}$	(Bing × Stella) × (Van × Stella)
Sonata (Sumleta)	$S_3S_{4'}$	Lapins × (Van × Stella)*
Staccato	$S_3S_{4'}$	Sweetheart × o.p.*
Stardust	$S_1S_{4'}$	2N-63-20 × Stella**
Starkrimson	$S_3S_{4'}$	Garden Bing × Stella*
Stella	$S_3S_{4'}$	Lambert × JI 2420
Sumesi	$S_3S_{4'}$	Van × (Stella × o.p.)*
Sunburst	$S_3S_{4'}$	Van × Stella
Sweetheart (Sumtare)	$S_3S_{4'}$	Van × Newstar
Symphony	$S_1S_{4'}$	Lapins × Bing
Tehranivee	$S_3S_{4'}$	Van × Stella
Temprana de Sot	S_3S_6	
Vandalay	$S_3S_{4'}$	Van × Stella
Waiyin No.7	$S_3S_{4'}$	

2.6. Fruit Set and Growth

2.6.1. Fruit Set

Stone fruit crops generally produce a number of flowers and set a number of fruits higher than those that the tree is able to bear and bring to maturation. Hence, in the weeks following blooming completion, a great number of young fruits are usually shed in subsequent 'waves', through which the tree tries to self-regulate the crop load to the amount it is able to support. The first wave of abscission is made up of unfertilized flowers; later on, also fertilized fruits are shed, as the result of phenomena of correlative inhibition between vegetative (shoots) and reproductive (fruits) sinks, or between reproductive sinks. However, the tree self-regulation is often insufficient, and some crops need fruit thinning to bring to maturation fruits that meet market requirements.

2.6.2. Fruit Growth and Development

In stone fruits, fruit is a drupe, characterized by a fleshy part (mesocarp and exocarp), and a stony part (endocarp) enclosing the seed (Fig. **8a**). The ultimate fruit size is the result of the combined effect of cell division, cell enlargement and intercellular air space formation events. The curve obtained by measuring the fruit diameter from blooming to ripening follows a double-sigmoid pattern, where two periods of very active growth are spaced by a phase, also called lag period, where the size of the fruit remains quite steady, although its dry mass increases (Fig. **8b**).

In peach, the fruit developmental cycle can be parted into four stages. Cell division dominates Stage I, hence the fruit growth potential [46]. Cell enlargement also occurs at this stage, and the fruit size grows exponentially. Stage II (the lag period) coincides with the endocarp lignification and embryo growth: its duration varies greatly according to the genotype, from few days in the very early ripening (where the curve is almost single-sigmoidal) to 60-70 days in the late ripening cultivars. In Stage III, the fruit size resumes exponential growth, as a consequence of the rapid enlargement of the mesocarp cells due to the increase in size of the vacuoles and in volume of the intercellular spaces. Stage IV is characterized by the onset of ripening syndrome. The growth pattern of cherry fruits is identical to that of peach.

Figure 8: A tipical stone fruit and its developmental stages (source: Giovannini D. & Liverani A.).

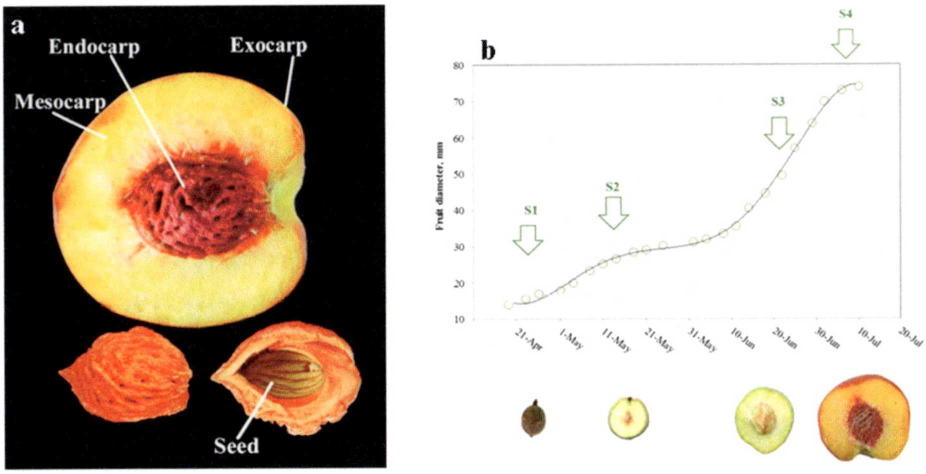

a) A peach fruit; **b)** Double-sigmoid growth curves for 'Redhaven' fruit.

Fruit growth cycle is regulated by hormones which, following different dynamics, promote cell division and enlargement, regulate metabolite accumulation in the fruit, as well as the functionality of the phloem sap transport system. Auxins, citokinins, gibberellins, but also abscissic acid and ethylene are the most important hormones [47]. The fruit final size is genotype dependent, although environmental conditions and tree management, in particular the crop load regulation trough thinning and irrigation, can significantly affect cell number and/or size and growth rates.

2.6.3. Fruit Ripening

Fruit ripening is a complex genetically-programmed physiological syndrome, regulated by a number of physiological and biochemical changes well-documented in literature, such as sugar content increase, acidity and flesh firmness decrease, chlorophyll degradation and biosynthesis of the aromatic compounds responsible of the flavor [48]. These events finally lead to a soft, edible ripe fruit with desirable quality traits [49]. Flesh softening is one of the ripening-related events that most affects fruit quality and shelf-life. Its extent varies according to the stone fruit species and genotype, and involves several processes, including disassembly of polysaccharide networks in the primary cell wall and middle lamella, and loss of turgor due to transpiratory water loss [50, 53].

Based on the occurrence or not of respiration and ethylene biosynthesis increase at the final stage of fruit growth, stone fruits can be either climacteric (apricot, peach and plum) or non-climacteric (cherry).

According to the flesh softening pattern during the ripening stage, peaches can be divided into melting or non-melting type. In the melting ones, the loss of firmness shows a typical biphasic pattern, where the initial slow rate (softening) is followed by a rapid firmness decrease (melting) in correspondence to the onset of the ethylene climacteric [54]. Due to the mutation in a polygalacturonase gene [55], the non-melting peaches do not experience the melting phase and maintain their firmness. This type of fruit has been typically directed to the processing sector although the interest for the fresh market has increased because of their higher resistance to handlings. A third type of peach, so-called stony hard, is characterized by a lack of ethylene production at ripening due to a single recessive gene [56, 57]. As a consequence, the firmness remains high at complete ripening and even at post-harvest stage.

2.6.4. Harvest Time

As discussed above, peaches and nectarines, apricots and plums are climacteric fruits, displaying a rapid increase in ethylene production at the onset of ripening associated with rapid changes in flesh color, texture, aroma and other biochemical characteristics; cherry, however, is a non-climacteric fruit. From the commercial point of view, while cherries are ready to eat at harvest time and senesce very quickly, the other stone fruit are usually picked in a pre-climacteric state (mature-firm) in order to elongate their market life.

The identification of the proper harvest time is of outmost importance, since the physiological maturity of fruit at picking time greatly affects both its potential storage and the quality perceived by the consumer.

In peach, the link between 'on-tree physiological maturity' and the evolution of the most important fruit quality features during the postharvest phase is particularly close [58]. A premature harvest increases the fruit handling resistance and, usually, extends its commercial life, but hampers the achievement of full fruit quality at consumption [59]; a harvest delayed to a more advanced ripening stage, on the other hand, while improving the organoleptic quality of the fruit, accelerates its decay. The good balance between these two conflicting instances is hard to find.

Criteria to evaluate the proper harvest time are generally based on visual (fruit skin color and size) and sense of touch evaluations and/or destructive analytical methods (flesh firmness, soluble solids content, titrable acidity, *etc*). Traditional analyses imply the destruction of the fruits, hence are performed on samples of a limited number of fruits. Extensive research has focused on the development of methods to assess non-destructively ripening-related changes in the fruit on the tree or to predict the internal quality of the fruit postharvest.

The visible/near infrared (vis/NIR) spectroscopy has been extensively studied, and many manufacturers of on-line grading lines have now implemented NIR systems to measure various quality attributes [60]. Predictions obtained so far on several crops are quite accurate, also at the single fruit level, for parameters such as the soluble solid content or the dry matter [61, 63], but not so satisfactory for other important parameters such as flesh firmness and titrable acidity [63, 65]. In addition, it was pointed out that commercial implementation of NIR technology requires large datasets (obtained through destructive measurements), spectral pre-processing techniques, and continuous updating of the calibration curves to increase the accuracy of forecasts [60, 66, 67]. The vis spectroscopy has been used to develop an easy-to-use method to monitor the evolution of fruit ripening [66]. A new index, called I_{AD} (Index of absorbance difference), was developed, which can be regarded as a reliable marker of fruit maturity, and a promising tool to establish the proper harvest time and to group harvested fruit into classes homogeneous for ripening stage. This index has already been used on stone fruits: peach [68, 69], plum [70] and apricot [67]. The electronic nose technology has been investigated to discriminate differences in fruit quality, and separate fruit according to their maturity stage. Experiences on apples, peaches and nectarines [47, 71, 72], and oranges showed promising results.

Manual picking is the standard method for the fresh market: peach, Japanese plum and apricot ripe gradually on the tree, and more pickings are required to harvest good-quality fruits; machines, instead, have been developed for harvesting stone fruit especially for processing purposes.

Once harvested, peach, nectarine, and plum fruits deteriorate quickly at ambient temperature. The rapid drop in temperature in the immediate post-harvest, as well as storage at temperatures near 0 °C, are useful practices to delay of a few days softening, flavor losses, and decay development [74].

2.7. Physiological Disorders in Stone Fruits

Split pit. Early ripening peaches, cherries and, to a certain extent, plums are more prone than medium-late ones to a phenomenon called ‘split pit’, where the endocarp separates in two halves, making the consumption of the fruit very inconvenient. Pit splitting has been also associated with high nitrogen fertilization, high rate of irrigation, large fruit size [75].

Fruit cracking. Stone fruit may suffer from fruit cracking or splitting, but these phenomena are particularly limiting factors in sweet cherry production (Fig. **9**). Cracking susceptibility varies from high to moderate according to the genotype, but cherry cultivars immune to the phenomenon are still unknown. Rainy weather during fruit maturation phase can cause important losses due to cracking. During the final fruit swelling phase, the cuticular membrane which protects and covers the epidermis cells is subjected to high and continuous strain. Since at this stage the new synthesis of cutin and wax - the principal components of the cuticle - is almost none, the membrane becomes thinner and thinner and is not able to expand without fracturing. Cracked fruits are easily infected by fungi, such as *Botrytis*, and can be a source of infection for nearby non-cracked fruits.

Figure 9: Cherries affected by cracking (source: Liverani A. & Giovannini D.).

The exclusion of the highly susceptible cultivars is among the strategies to adopt in order to reduce rain-induced cracking damages in sweet cherry industry. Lists of cultivars ranked for their susceptibility are available and should be consulted [76]. However, sometimes the same genotype can behave as susceptible or moderately tolerant in different environments, with no clear association with the amounts of rainfall [77]. A regular water supply, for example through trickle irrigation, can reduce the cuticular strain during fruit growth and alleviate cracking severity. Breeding strategies for less cracking susceptible fruit should focus on selecting for low strain and high thickness; also genotypes that deposit cutin and wax during Stage III would be most interesting because their cuticular membrane is likely to be less strained [78].

The use of plastic coverings over the rows has proven to be successful in reducing, although not eliminating, crop losses. Although very expensive, in Europe coverings are being increasingly used in sweet cherry industry, trying to reduce the costs of the protective structures limiting tree size by the use of dwarfing rootstocks or growth retardants.

3. AGRONOMIC TECHNIQUES AND ORCHARD MANAGEMENT

3.1. Orchard Planting Systems

The planting system is the combination of tree density and spacing in the orchard, inextricably linked to the size and the shape of the trees [78].

Planting orchards with higher numbers of tree per hectare is the trend in most fruit growing areas worldwide. Indeed, high density plantings (HDP) offer several advantages compared to the low (LDP) or medium (MDP) density ones: shorter unproductive period, higher yields, especially in the first years of planting, easier mechanization, economies on labor costs, better adaptation to the varietal turnover [80, 81].

Conversely, HDP require greater capital investments when setting up the orchard, availability of vigor-controlling rootstocks, technically skilled labor, increased labor requirements.

These needs have limited, at least for stone fruit crops, the spreading of HDP even in the most advanced fruit-growing areas, where plantings with low to medium (400÷700) number of trees/ha are still prevailing.

3.1.1. Modern Orchard Systems

A modern orchard should be designed to optimize the use of available resources (*e.g.* sunlight, water, soil fertility) and allow easy access of labor and equipment to the alleys and the trees. No matter the planting system used, all orchards should combine high levels of light interception with adequate light distribution, so to minimize shading and ensure high yields of good market quality [82, 83].

The practical implementation of the principles of minimal amount of pruning [84, 86] in the training phase of the trees is a milestone in the transition from traditional to modern orchards.

The strategy of minimal pruning is based on some rules which have proven to be powerful to speed up the onset of the full productive phase of the tree [87]:

1. reduce the number of pruning cuts in the first 2-3 years of planting;
2. remove entire shoots (thinning cuts) and avoid, whenever possible, heading cuts;
3. privilege summer pruning (and alternatives to cuttings such as bending, twisting, pinching growing shoots) to winter pruning;
4. exploit early cropping as a tool to control tree vigor.

The production of well-feathered trees in the nursery can further contribute to hasten the productive life of the tree, which can start in the very first year of planting [88, 89].

3.1.2. Main Training Systems

Training systems for stone fruit crops are based on a few basic tree shapes, but with many variants and adaptations to the different species and growing conditions: the open center (vase), the central leader (fusetto), the vertical hedgerow (Palmette) and the inclined hedgerow (Y-shaped) types (Table **21**). A description of some of the most frequently adopted in the different stone fruit crops is given below.

3.1.2.1. Peach

Among stone fruit crops, peach has probably experienced the earliest and greatest dynamism in the development of new training systems (or variants to the already developed ones) and in the innovation of the pruning techniques to achieve the new forms. Changes have been targeted to speed up returns of the invested capitals, by hastening the entry into production, increasing yields and labour efficiency, particularly for pruning, thinning and harvesting operations.

3.1.2.2. Open Centre Systems

The Open-centre or Open Vase is probably the most diffuse shape in all peach-growing countries. However, the traditional forms, with 3-4 primary scaffold branches, geometrically spaced and dressed with lower hierarchy branches are no longer popular, being too long and laborious to obtain and late to entry into production. In the last few decades, several variants have been developed, where pruning is minimal and/or simplified, the number of the main branches is flexible, and the trees are smaller and manageable almost entirely by the ground.

A successful variant, developed in Italy at the end of the '70s and still diffuse, is the so-called Delayed vase (Fig. **10**). With this shape, the natural basitony of the young peach tree is seconded [87]. Young trees are pruned minimally, mostly in summer, and the few thinning cuts are mainly targeted to prevent the weakening of the central leader (maintained for the first 3-4 years of planting) from the competition of too vigorous shoots. Since the first growing season, the tree assumes a cone-shape with an expanded base [88].

Figure 10: Peach trees trained to delayed vase in the second (left) and the third (right) growing season (courtesy of Mennone C., ALSIA, Italy).

At the end of the third or the fourth year of planting, the central leader is finally cut, and the tree attains the open vase final shape, hence the origin of the 'delayed' attribute. The delayed vase performs best when using not too vigorous cultivar/rootstock combinations and choosing low-vigour inducing soils. Its main advantages are the small size of the trees, whose height should not overcome 2.5 m, and the limited requirement of labour, almost entirely manageable from the ground (Table **21**). A recent variant is the Catalan vase or Spanish bush (Fig. **11**). This small vase, developed in Spain, is also raising some interest in other peach-growing areas, as it combines limited labour requirements (pruning is partly mechanizable and does not require highly skilled labour) with early and high yields [90]. In each of the first two years of planting, two summer pruning operations, consisting in topping the canopy, are performed to stimulate the emission of feathers from the growing shoots and allow the tree to fill the allotted space as fast as possible. In the late summer of the second year, a manual pruning operation is required to open the centre of the bushy canopy, by removing excessive vegetation, and select the 4-5 permanent branches [91]. At their final size, trees are about 2.5 m high, and can be managed from the ground or using small ladders [90].

Figure 11: The Catalan vase or Spanish bush.

During the training phase, two topping operations are performed to stimulate the emission of laterals. From the second year of planting, the topping can be made mechanically (courtesy of Guarino F., O.P. Sibari, Italy).

3.1.2.3. Hedgerow Systems

The hedgerow systems require the use of trellises, to which the permanent branches are tied and which provide support to the productive wall. They are suited to the use of mechanical platforms for pruning, thinning and harvesting, to the improvement of labour efficiency.

The traditional 'Palmetta', developed in Italy in the '50s, has now been replaced by modern variants, in which the rapid formation of a vertical productive and functional wall is privileged over the achievement of perfectly designed single trees. The entry into production of modern Palmetta-type systems can be anticipated by following the principles of minimal pruning. If well-feathered trees are made available from the nursery, no heading cuts are needed at planting: the trees can be allowed to grow freely with limited thinning of laterals badly positioned or in excess. In the first growing season (or the second, according to the amount of growth) the first tier of branches can be selected and tied on the trellis, at a height not less than 80-100 cm from the ground; the upper tiers are chosen the following seasons. Simplified variants have been recently developed in Italy, in which the permanent structure is made of a unique tier of two (U-Palmetta) or three ('Candelabrum', Fig. **12**) branches, which are trained vertically and parallel one to another. These permanent branches are dressed of short branches to close the space between the trees along the row.

The high capital investments needed for platforms and trellises have resized the interest towards the Palmetta, which persists in areas with very invigorating conditions, not allowing the adoption of training systems based on small-sized trees [87].

However, as they are based on tall trees (3.5÷4.0 m high), hedgerow systems are to be preferred in areas at a high risk of late frosts: because of the temperature inversion gradient phenomenon, the upper half of the tree is better preserved from freezing damages than the lower half [89]. Finally, the tall and narrow canopies are better designed to the application of anti-hail nets to protect the crop and the trees against hail damages.

Figure 12: Candelabrum-shaped trees.

At the second leaf (left) and at the fifth leaf (right). Trees are spaced 4.5 (between rows) x 3.5 m (635 trees/ha; source: Giovannini D. & Liverani A.).

3.1.2.4. Central Leader

This system, conceived for HDP, requires the use of trellises and wires to sustain the trees, and mechanical platforms to help labour to perform pruning, thinning and harvesting operations. Trained to the Central Leader (or Fusetto) system (Fig. **13**), the trees assume a conical shape, larger at the base and progressively narrower at the top, which rapidly attains the height of 3.5-4.0 m. During the training phase, the adoption of minimal pruning strategies combined with the early onset of competition among the neighbor trees on the row (planted at the distance of 1.5 ÷ 2.0 m) foster the early entry into production of the trees (Table **21**).

Figure 13: Fusetto-shaped trees.

Young trees at the second leaf (left), showing high fertility at the end of flowering. Trees at the 9th leaf (right), showing good vegetative growth. Spacing is 4.5 (between rows) x 1.5 m (1480 trees/ha). This training system is suited to the use of anti-hail nets (source: Giovannini D. & Liverani A.).

Vigor control is the major problem with this training system, and probably the main reasons why it has not spread widely, despite the very early and high cropping potential. Few (4-5) years after planting, indeed, excessive shading and loss of bearing wood in the lower part of the canopy can be exacerbated, especially at the highest densities. The Fusetto is more suited to fertile, medium-late ripening varieties, while it has proven to be unsuited to early varieties, where a long season after harvest with no fruit competition can create problems of excessive vegetative growth. Recent availability of medium to low-vigor rootstocks could mitigate the vigor control problems.

A variant to Fusetto is the Central axis, where trees along the row are planted at a closer distance, *i.e.* about 1.0 m, and the tree woody skeleton is made up of an axis on which short productive branches are inserted.

3.1.2.5. Y-Shape

Designed for densities ranging between 1500÷2000 trees, Y-shaped training systems are featured with continuous, inclined double walls (Table **21**). Trellises and wires are required to sustain and give the chosen inclination to the two 'wings' of each tree, selected in the first growing season and trained to grow perpendicularly to the row direction. This training system reaches very high leaf area per unit land area (LAI) values and very high interception levels of the available light (above 70%) compared to either open centre or hedgerow systems; moreover, such high levels can be maintained throughout the orchard life.

A variant of the Y-shape (Fig. **14**) has been developed and diffused in Southern Italy areas characterized by prolonged vegetative growing and high radiation levels: the inclined walls of neighboring rows are allowed to join in the middle of the alley, so to maximize light interception [92, 93]. Such a system reaches early and high yields of good marketable quality. The Y-shape proved to be well designed to the use of PVC tunnels to anticipate ripening. In those environments suited to early productions, advances in ripening can be obtained by combining early-season cultivars with the use of plastic coverings. Supporting arches can be positioned with their ridge over the zone where the two 'wings' of adjacent rows converge.

Figure 14: A variant to the Y-shaped training system.

The inclined walls of neighboring rows are allowed to join in the middle of the alley (courtesy of Mennone C., ALSIA, Italy).

Another variant is the V system, where the double inclined wall is obtained by planting inclined trees alternated along the row. Compared to the Y-system, however, the V is much less widespread, because it doubles the costs of the trees and, with the planting ageing, the vegetative growth is hard to manage.

3.1.2.6. Sweet Cherry

Breeding progress in the selection of early vigor-controlling rootstocks and new knowledge on the pruning and training strategies to manage vigor and crop load, have opened new and interesting ways for increasing planting density and economic benefits of sweet cherry orchards. In the recent past, good profits in cherry industry were indeed limited by traits typical of this stone fruit crop: the big size of adult trees, the strong upright habit with low inclination to form lateral shoots, and the long unproductive phase.

The main sweet cherry growing countries have shifted from the traditional LDP (about 300 trees/ha) to the MDP (500÷800 trees/ha), currently the most diffuse [94, 97]. However, also HDP has become a reality [98].

Tree shapes such as the modern variants of the open vase are particularly suited to MPD, and have been successfully adopted in a number of countries of the Mediterranean Basin. In the Spanish bush (or Catalan vase) system, the trees are spaced 4-5 x 3 m and maintained small-sized (2.5 m tall), so to manage most of the harvest and the pruning operations from the ground. Repeated pruning of the canopy, the use of controlled water stress and the application of growth regulators are the strategies used to keep the trees short-statured [95, 99]. The first yields are achieved already in the third year, and most cultivars become full productive from the fifth year of planting. Variants such as the Modified Spanish bush have been proposed more recently, where chemicals (*e.g.*, Promalin) or pruning modifications (bending, ringing and/or woodcut rather than cuts) have been introduced to stimulate branching, and to anticipate of one year the entry into production [95, 99].

The Central leader systems, of which several variants and denominations have been proposed [97, 100, 102], are suited to both MPD and HDP cherry orchards. Tree spacing can vary from 5.0 x 3.0 m (667 trees per ha) to 3.5 m x 1.5 m (1714 trees/ha). In Hungary, the semi-intensive system called Modified Brunner spindle [103], and the intensive Hungarian slender spindle have been developed [94, 104]. In the Hungarian slender spindle, shoot bending, light summer pinching and leader maintenance (no heading cuts) combined with appropriate cultivar/rootstocks combinations and the use of BA (benzylaminopurine) containing bio-regulators, allow early entry into production, high quantity and quality crop harvestable 70-100% from the ground. The Central leader tree form is ideally suited to dwarfing and semi-dwarfing rootstocks.

Hedgerow systems, like the Palmette-type or the Marchand systems, suited for the MDP orchards (500-800 trees/ha), have become more widespread in the Mediterranean growing areas because they better fit soil and climate conditions, grower skills and available farm equipment [96]. Trellises and wires are required to sustain the canopies, which are tall and narrow across the row, and platforms can help in the management of pruning and harvesting operations.

Tatura Trellis, the V-shaped system developed in Australia [105], is suited to cherry HDP (800÷1200 trees/h). Many variations in materials, lengths of arms and internal angles, number of wires, distance between rows, and distance between poles can occur. Minimal pruning during the first three years and only bending and tying induce earlier fruiting and contribute to reduce vigor. The best quality fruit is produced at the base of one year-old shoots and on young fruit spurs. When the trees attain the productive phase, annual renewal pruning should be adopted to ensure maintenance of good fruit quality.

Finally, very high planting density (up to 5,000 trees/ha) orchards, totally manageable from the ground (so-called ladderless orchards), are in an advanced phase of study, with some commercial attempts [96, 98]. Trees are trained to a central axis, on which short fruiting branches, renewed periodically, are inserted. These systems require the use of dwarfing rootstocks, such as 'Gisela 5', cultivars with an open, not too upright habit and a good feathering aptitude, such as 'Ferrovia', 'Kordia' and 'Regina' [96]. Production should be concentrated on the flower buds differentiated at the base of one-year old shoots more than on spurs.

These systems require very high start-up costs, due to the high number of trees and the need of protecting the crop (and the trees) from environmental injuries caused by hail, rain and birds; moreover, they require skilled labor to maintain a high level of efficiency of the orchard over time.

3.1.2.7. Apricot and Plum

Although the training systems developed for peach (Table **21**) might be considered valid also for apricot and plum, these stone fruit crops have not experienced the same dynamism in the planting systems as peach and cherry. However, also for apricot and plum the trend of modern plantings is to grow smaller-sized trees, to hasten the entry into production, to increase efficiency and reduce the costs of labor.

The open center form and its modern variants (Delayed vase, Spanish bush, Open vase) are the most diffuse for these species, and the varieties with spreading growth habit, like many Japanese plum varieties, are those which best adapt to this system.

Apricot is a crop difficult to train, as it is reluctant to be shaped into unnatural forms [106]. Winter pruning causes strong vegetative reactions, such as the production of water sprouts, or facilitates the onset of bacterial or fungal infections, conducing to the tree decline syndrome [107, 108]. Trees tend to have several growing waves over the same season, especially in some cultivars, cultivar/rootstocks combinations and growing sites. In addition, a large variability in fruiting habit exists among cultivars [109]. For all these reasons, when training this crop, it is of outmost importance to comply with some rules [110]: in the first 2-3 years of planting, the natural habit of the cultivar should be seconded as much as possible and the pruning should be very light. On adult trees, pruning should be performed in summertime, *i.e.* after harvest to avoid vegetative overreaction and to facilitate wound healing, and be completed at the end of the dormant season.

When designing the orchard, one should take into account that most of Japanese varieties, some European cultivars and recently released apricot cultivars are self-sterile, so the inclusion of a 10-12% of pollinators is needed to ensure an adequate cross-pollination.

Central leader (*i.e.* Fusetto) and Y-shaped planting systems are considered innovative for plum orchards. Y-shaped systems, although adopted for apricot in some growing areas, are not suited to all cultivars, due to different growth habits, and the high costs of investments and the need of skilled labor have limited their diffusion so far [111].

In apricot and plum, excessive crop load negatively influences the quality of the fruit, and could trigger biennial bearing and branch breakage. To prevent these phenomena, it is advisable to reduce the number of fruits by performing some pruning at the end of the winter, consisting in shortening the fruiting branches. This would also reduce the need of labor for the manual thinning of the fruit, to be performed after the natural fruit drop [112].

Table 21: Main training systems and planting densities for stone fruits (adapted from [115])

Crop	Training system	Rootstock vigor	Spacing	Density (tree/ha)	Tree height (m)	Notes
Peach	Open Vase	High Medium	6.0 x 5.5 5.0 x 5.0	300 400	3.5 2.5	Still the most common training system of commercial orchards in the world
	Delayed Vase	High Medium	5.5 x 4.0 4.5 x 2.5	450 890	2.5	The free growth in the first 3-4 years anticipates production relative to the open vase
	Catalan Vase/ Spanish Bush	Medium-high	5.0 x 3.0	670	2.5	Easy to train and partially mechanizable
	Free Palmette	Medium-high	4.5 x 3.0	740	3.5	The most diffuse among the hedgerow systems
	Candelabrum	Medium	4.0 x 3.0	840	3.5	A three-axes Palmette
	U-Palmette	Medium-high	4.0 x 2.0	1250	3.5	A two-axes Palmette
	Central leader	High Medium	5.0 x 3.5 4.5 x 2.5	570 890	3.5	A conical shape, seconding the natural tree growth;
	Fusetto	Medium	4.5 x 2.5 4.0 x 2.0	800 1250	3.0	Trees are columnar axes with short bearing branches periodically renewed
	Y-shaped	High Medium	5.0 x 1.5 5.0 x 1.0	1350 2000	3.0	For intensive orchards in well-lit environments
Cherry	Globe	High	7.0 x 5.0	285	4.0-7.0	The natural shape of a cherry tree
	Catalan Vase Spanish Bush	Medium	4.5 x 3.0 4.0 x 2.5	740 1000	2.5	Easy to train and partially mechanizable

Table 21: contd...

	Palmette	Medium Low	5.0 x 3.0 5.0 x 2.0	670 1000	4.0	Hedgerow system that needs platforms to help labor
	Marchand	Low	4.5 x 3.0	750	3.0	Inclination of trees and branches contains tree vigor
	Ladderless orchard	Dwarfing	4.0 x 1.5 3.5 x 1.0	1670 2850	2.5	A cherry orchard totally manageable from the ground
Apricot	Open Vase	High Medium	5.0 x 3.0 4.0 x 3.0	670 830	3.0	By far the most common training system of commercial orchards worldwide
	Free Palmette	High Medium	4.5 x 3.0 4.0 x 2.5	740 1000	4.0	Hedgerow system that needs platforms to help labor
	Y-shaped	Medium	5.0 x 1.5	1330	3.0	For intensive orchard in well-lit environments; suited cultivars are required
Plum	Open Vase	High Medium	5.0 x 5.0 4.0 x 4.0	400 625	4.0	The common version has 3 main branches and is also suited to mechanical harvest fruit for industrial use
	Delayed Vase	High Medium	4.5 x 3.0 4.0 x 2.5	740 1000	2.5	The whole tree is manageable from the ground
	Free Palmette	High Medium	4.5 x 3.0 4.0 x 2.5	740 1000	4.0	Hedgerow system that needs platforms to help labor

3.2. Pruning Mature Trees

Summer pruning operations such as bending, twisting and pinching the shoots are more effective than winter cuts in facilitating and speeding up the transition of the trees from the unproductive to the adult, full productive phase. On adult trees, selective removal of shoots and foliage in summer can prevent excessive shading in the inner and lower parts of the canopy. However, once the training phase is completed, the winter pruning increases its importance in maintaining the shape and the efficiency of the canopy, in preserving the shoot-root equilibrium, in contrasting the ageing of the fertile wood, in promoting the good quality of the crop. The amount of winter pruning varies with the species (*eg.* in peach, about a half of one-year old shoots are annually thinned, in cherry pruning is much less intense) and the age of the trees (increasing the pruning may stimulate re-growth in weakened trees). If postponed to early spring, it can also act as a partial fruit thinning operation, reducing the time of the manual thinning, which is labor expensive but absolutely necessary in some fruit crop species.

Figure 15: Five years old peach trees trained to delayed vase, before (left) and after (right) the winter pruning, delayed at blooming time (source: Giovannini D. & Liverani A.)

Compared to the past, pruning operations have become quicker and less accurate. Simplification has become increasingly necessary because of the limited availability of highly qualified labor.

Pruning in the bearing phase of the orchard should take into account also the differences among cultivars: in peach, for example, there are cultivars that bear most of their production on long (60-80 cm) one year-old shoots (*i.e.* those of the 'Rich' group); others, mainly producing on medium (30-60 cm) vigor shoots (*i.e.* 'Elegant Lady', 'Rome Star', 'Summer Lady'); other cultivars, among which the nectarines 'Big Top' and 'Diamond Ray', producing on medium or short (10-30 cm) one-year old shoots and on spurs; and a fourth group, to which many nectarines and most of the canning cultivars belong, producing best on short bearing wood, like spurs and short one year-old shoots.

3.3. Crop Load Management

The main objective of crop load regulation is to minimize competition for assimilates among fruits and between fruits and the other growing organs of the tree. Among the stone fruits, peach is the one where the adjustment of the crop load is a standard commercial practice, absolutely necessary to bring to maturation a marketable product. Indeed, peach trees usually set a number of fruits far higher than their bearing capacity. Over-cropping, on the other hand, penalizes the final quality of the fruit and might lead to branch breakage.

Pruning is the earliest and cheapest method to regulate crop load, though insufficient: fruit thinning is required to set the proper balance between the reproductive and the vegetative components of the tree. Final fruit size is related to the number of cells per fruit, and fruit cell division mainly takes place in Stage I of the fruit developmental period of stone fruits [113]. In the second half of Stage I, the growth of the fruit is resource-limited, due to the competition for assimilates with other growing organs of the tree [114]. Timely crop load adjustment would thus stimulate cell division of the remaining fruits, and to be more effective, thinning should be performed before the end of Stage I or soon after the Stage II onset.

Thinning is usually done manually. This practice reaches maximum effectiveness with skilled workers, being able to select the fruits to be removed first and establish the most appropriate distance among the remaining fruits (the number of leaves per fruit and the final fruit size are positively related), taking into account the type and the size of the bearing wood, the position of the fruit on the branch and the light environment in which the branch is located within the canopy. Manual thinning is a time consuming operation (150÷300 hours/ha according to the training system) [89]: labor costs increases and the high fertility of many of the current peach cultivars have fostered the research to focus on alternatives to hand thinning.

For the chemical thinning of peach, substances with caustic effect, such as hydrogen cyanide, soybean oil, calcium polysulfide or surfactants, like the fatty amine polymer called Armo Thin® have been tested on various stone fruit species. Flower buds at the 'pink button' stage or just opened flowers are the most susceptible to the caustic action of this compound. Therefore, the earlier the treatment, the more severe the thinning effect: application of Armo Thin® at a 2–3% concentration when 70–80% of the flowers are opened has yielded encouraging results in a number of growing sites and on different cultivars [116]. However, a winning strategy for chemical thinning of peach has not been found yet. The attempts to find a single chemical compound as an alternative to manual thinning have failed so far. An integrated approach in which different chemicals are applied on the same trees at specific and subsequent phenological stages might be possible: at flower differentiation with GA (Gibberellic Acid) containing chemicals or during bud dormancy (dormancy-breaking agents) and, if insufficient, at bloom (spraying surfactants) and even repeating the application with ethephon at the young fruit stage [116]. A tool recently developed to mechanically thin flowers is the Darwin machine (Darwin 300; Fruit-Tec, Deggenhausertal, Germany, Fig. **16**), equipped with a rotating part carrying whips that, while rotating, tear off and cut the flowers. The thinning severity can be regulated by adjusting the speed of the rotating tool or the forward speed of the machine; the hand thinning is later required to refine the spacing of the remaining fruits. The hedgerow planting systems are the most suited to this type of mechanical thinning [117, 118].

Figure 16: Mechanical peach thinning equipment.

Darwin 300 string thinner Fruit-Tec (Deggenhausertal, Germany) for bloom thinning peach and nectarine trees with narrow canopies (courtesy of Creso, Italy).

Overthinning is the greatest risk of using chemical and mechanical tools: although the final size of the fruits might be increased, this advantage would not be sufficient to economically compensate yield loss.

3.4. Rootstocks

Stone fruit trees are propagated worldwide onto rootstocks. The use of the appropriate rootstock provides the scion-cultivar with several benefits compared to its own roots, such as greater resistance against unfavorable climatic and soil conditions, as well as harmful soil-borne pests and diseases. In addition, the rootstock can influence other traits such as vigor (Fig. **17**), and positively affects precocity, crop efficiency, fruit size and quality [118, 120].

Figure 17: Differences in tree size and blooming time in a collection of stone fruit rootstocks (source: Liverani A. & Giovannini D.).

Although rootstocks raised from seeds are still prevailing worldwide, the tendency is towards the use of vegetatively propagated rootstocks, which can provide more uniform tree performance. Rootstock breeding is relatively recent. Until the middle of the XX century, indeed, rootstocks were generally selected from seedling populations, choosing the individuals mainly because of their good propagation abilities [121]. Seedlings rootstocks still dominate stone fruit industry all around the world, probably because of their lower cost and ease of propagation compared to the clonal ones. In the last decades, there has been a remarkable progress in the development of new rootstocks for stone fruits, with an increasing use of inter-specific hybridization in order to meet various agronomical requirements in a single rootstock [120, 122]. Many stone fruit species can be budded onto other *Prunus* species. As a result, peaches, plums, apricots and almonds can often benefit of rootstocks developed for each other through the breeding (Table **22**). Obtaining a new rootstock is an expensive and time consuming procedure: following the initial cross between the chosen parents, the selection requires, on average, 30 years. To reach conclusive information on the value of a new rootstock, long-lasting, multi-site trials are needed, where the new rootstocks are evaluated in combination with a number of cultivars and in comparison with well-known rootstocks as references.

3.5. Seedling Rootstocks

Peaches and nectarines are grown mainly onto seedling rootstocks, and almost every peach-growing area has developed its own selections. Wild peach trees (*P.vulgaris*, P. Mill.), commercial cultivars, and rootstock selections are the main seed sources for this crop. The cultivars most commonly used as rootstocks are 'Halford', 'Lovell', 'Bailey' in the USA and Canada; 'Jerez', Jalacingo' and 'Tetela' in Mexico; 'Cuaresmillo' in Argentina; 'Chucho Picudo' in Chile; 'Aldrighi' and 'Capdeboscq' in Brazil; 'Elberta' and 'Golden Queen' in Australia and New Zealand [5]. Rootstock selections commercially diffused are the Italian 'P.S.' series, the French 'Montclar', 'Rubirà' and 'Higama', or the US 'Nemaguard', 'Nemared' and 'Guardian'.

Apricot (*P. armeniaca*) cultivars are grown on apricot but also on peach and European plum (*Prunus domestica,* L.) seedlings. European plum is mostly grafted onto *P. domestica, P. insititia* (St. Julien), and *P. cerasifera* (Myrobalan), while *P. mahaleb* (Mahaleb) and *P. avium* seedlings are still widely used for sweet cherry.

3.6. Rootstock Resistance to Stresses

Stone fruit industries around the world are faced with challenges of manifold nature and importance. Suitability to cultivation and productivity can be limited by a several abiotic and biotic stresses, whose order of priority can change according to the different growing areas.

3.6.1. Abiotic Stresses

3.6.1.1. Calcareous Soil

Among stone fruit species, peach is particularly susceptible to lime-induced chlorosis, a nutritional disorder caused by low bioavailability of iron and the difficulty of Fe acquisition by the roots. In calcareous soils, characterized by high pH (7.4-8.5) and a high level of bicarbonate, peach trees become weak, unproductive and iron-deficient. Since tolerance against chlorosis is a priority issue for this species, most of the rootstocks currently available are calcareous soil-adapted, such as the hybrids 'GF677'and 'Garnem', with almond in their parentage, or the hybrids 'Cadaman' and 'Barrier 1', deriving their tolerance from a *P. davidiana* parent. Also many of the newly released rootstocks are characterized by various degrees of tolerance to chlorosis, like the 'Rootpack' series, 'Evrica' and 'Tetra' [124]. Good sources of tolerance for apricot and plum are Myrobalan, *P.insititia* and their interspecific hybrids and *P. mahaleb* and *P. cerasus* (and their hybrids) for sweet cherry.

3.6.1.2. Low Soil Temperatures

Roots are much more susceptible to freezing than aboveground parts of the trees. Root susceptibility to cold is influenced by the soil type, and damages occur more frequently on light as compared to heavier soils, because cold can penetrate deeper. Rootstocks are directly involved in tree susceptibility when they are themselves injured by freezes (Fig. **18**).

Mazzard rootstocks are considered more susceptible to cold than Mahaleb ones, which can tolerate -15 °C temperatures against the -l0 °C of the former. *P. cerasus* rootstocks are more tolerant than Mahaleb ones, and the 'Weiroot' selections show a good hardiness, partly maintained in the hybrids of *P. cerasus,* like the 'GiSelA' series [125]. In peach, the cold-hardy Canadian 'Siberian C' or the recently released Russian rootstocks 'Krymsk® 1', 'Krymsk® 2', 'Krymsk® 3', can be useful to prevent cold injuries to peach orchards. In plum, hardiness seems to be an issue of limited importance, probably because most rootstocks do not suffer significantly from this problem. Rootstocks might indirectly influence the scion susceptibility to winter cold damages. This is the case of rootstocks inducing vigorous growth until late in the season, which can increase the risks of scion damages by autumn- or early-winter freezes; also, late-winter and early-spring damages might be more frequent when a rootstock releases the tree relatively early from dormancy.

Figure 18: Frost damage symptoms in the cambium of a peach tree (source: Giovannini D. & Liverani A.).

3.6.1.3. Waterlogging

Sites with poor drainage, high water tables and high rates of precipitation should be avoided for stone fruit cultivation, because prolonged anaerobic soil conditions harm the root system. However, water logging is often a transient problem and adequate artificial drainage coupled with the use of appropriate rootstocks is usually sufficient to overcome major damages. Almond, apricot, cherry and peach rootstocks are generally very sensitive to water logging, whereas Myrobalan or P. *domestica* rootstocks are tolerant to resistant, and so are their hybrids [126, 127]. Surveys undertaken to determine the relative importance of the problems for stone fruit industries all around the world would indicate that the need for water logging tolerance has moved down in order of priority [118].

3.6.2. Biotic stresses

Root systems of fruit trees may be damaged or even killed by numerous pests and diseases. In sites where the soil pathogens concentration is particularly high, the use of tolerant or resistant rootstocks can be decisive for tree health and survival.

3.6.2.1. Nematodes

Most stone fruit growing areas around the world have significant problems with one or more species of nematodes, and the breeding to develop resistant rootstocks has been quite active in the last decades. Four kinds of nematodes are recognized as dangerous for stone fruits, which show various levels of resistance/susceptibility: root-knot nematodes (*Meloidogyne* spp.), ring nematodes (*Criconemella* spp.), root-lesion nematodes (*Pratylenchus* spp.) and dagger nematodes (*Xiphinema* spp.).

Meloidogyne genus, particularly *M. incognita* and *M. javanica,* are a threat to apricot, peach and plum production in some Mediterranean areas. Peach and plum seedlings, and the peach-almond hybrid GF677, are susceptible to highly susceptible rootstocks. In *Prunus*, two dominant genes have been identified: *Ma*, found in Myrobalan, conferring resistance against *M. arenaria, M. incognita, M. javanica,* and *M. floridiensis* and *RMia* (found in the peach 'Nemared'), inducing resistance against *M. javanica* [128]. Due to the relatively straightforward inheritance of resistance against this pest, many new rootstocks immune or resistant to one or more species of *Meloidogyne* have been developed and released in the last 20 years (Table **22**).

In resistant rootstocks, nematodes do not establish a feeding site nor develop, leading to the complete absence of galls in roots. Susceptible rootstocks have high percentages of galled roots with large nematode populations. Since the use of soil fumigants is increasingly banned, incorporation of root-knot nematode resistance into new rootstocks that also possess other valuable traits could be of great value.

The Ring nematode is a primary factor predisposing peach trees to the Peach Tree Short Life (PTSL) syndrome in the Southeast of US [129]. This nematode has a very broad host range, including species in diverse plant families. Attempts to identify *Prunus* germplasm unsuitable as host, hence resistant to this nematode, have been so far little successful [130]. Interestingly, peach trees on 'Guardian' rootstock display markedly greater resistance to PTLS than onto ‘Lovell’, even though ‘Guardian’ and ‘Lovell’ support similar populations of ring nematodes [129].

Among the lesion nematodes, *Pratylenchus vulnus* and *Pratylenchus penetrans* play the leading role in terms of dangerousness, so that the research of resistance sources to incorporate into rootstocks mainly focused on these two species. However, the development of commercial rootstocks with significant levels of resistance still lags behind, although screening methodologies have been developed for many stone fruits [131] and some sources of resistance identified [132, 135].

3.6.2.2. Soil Fungi

On fine-textured, high bulk density or poorly drained soils, stone fruit rootstocks are at risk of becoming infected by several species of *Phytophthora*, which causes crown rot, and several species of *Armillaria*, inducing the oak root rot (Fig. **19**). This latter one causes considerable damage to peach orchards as the hyphae of the fungus from an infected tree spread rapidly to neighboring trees. Both fungi are difficult to control or eradicate.

Apart from the use of tolerant rootstocks, the only control measure is to thoroughly remove roots and bootlaces of the fungal mycelium from the previous orchard. Various sources of resistance to *Armillaria* species have been identified [136], and some progress has been made in the development of rootstocks with resistance to *Armillaria mellea*. The inter-specific complex hybrids ‘Ishtara’ and ‘Myran’ have been reported as significantly more resistant to *A. mellea* than peach seedlings. Unfortunately, this resistance does not seem to extend to *A. tabescens* [137]. ‘Sharpe’, a putative plum hybrid rootstock of unknown origin, and ‘MP-29’ (peach x plum hybrid) have preliminarily shown good resistance/tolerance in trials carried out in Georgia [138]. In general, plum species native to the production region appear to provide the most usable resistance to this organism, and most programs attempting to develop resistant rootstocks for use with peach, plum, apricot and almond have utilized such germplasm.

Figure 19: White, fan-shaped mycelium between bark and wood (left) and mushrooms (right) of Armillaria root rot agent (*Armillaria* spp; source: Giovannini D. & Liverani A.)

3.6.2.3. Soil Sickness (Replanting)

Sites previously planted with the same or closely related fruit species can cause serious problems to the newly planted trees, generally indicated as 'replant diseases'. The most common effects of replanting are the dramatic reduction in tree growth and vigor, reduction in yield and a shortened productive life (Fig. **20**).

Figure 20: Replant effect on peach tree growth (source: Giovannini D. & Liverani A.).

The peach cv Redhaven grafted onto the peach rootstock PS B2 grown in a virgin soil (left); on a replanted soil (centre); and grafted onto GF 677 rootstock on a replanted soil (right). Pictures are in scale and were taken at the end of the third growing season.

Accumulated research results strongly indicate either abiotic (unbalanced soil nutrition, impaired soil structure, phytotoxic metabolites from roots of the previous crop, *etc.*) or biotic factors (a variety of soil-borne microorganisms including plant-parasitic nematodes and fungi, and others such as bacteria and cyanogenic microorganisms) as the main causes of the replant disease phenomenon. Conclusive evidence, however, is far from being ascertained. In a recent study, numerous bacteria belonging to the genus *Pseudomonas* (*P. pachastrellae, P. putida, P. fluorescens, P. straminea, P. fulva, P. taiwanensis, and P. monteilii*) were identified as associated with peach replant disease symptoms, exhibiting negative correlations with length and weight of the shoots [139]. Many fungi, in particular belonging to the *Fusarium, Penicillium, Aspergillus*, and *Trichoderm*a genera, were frequently isolated from the rhizosphere or roots of peach trees from the previous orchard, and found to be parasitic to the roots [140].

Whatever the cause, in the most important and traditional peach growing countries, such as Italy, France and Spain, replant disease is a priority issue and challenge. The continuation of a thriving and competitive peach industry in these countries has been allowed by the availability of tolerant rootstocks, particularly hybrids between peach and resistant related species, such as almond and *Prunus davidiana*, or also complex hybrids with other *Prunus* spp. in their pedigree. Invigorating rootstocks, finally, seem to be able to better cope with replanting disease than dwarfing ones.

3.6.3. Rootstock Controlling Tree Size

On high fertility soils, especially when using vigorous scion cultivars, a certain size control is desirable for facilitating orchard management and reducing pruning, thinning and harvest costs. As an indirect benefit due to reduced shading, less vigour is often accompanied by improved fruit qualitative traits such as red blush overcolor extension, size and sugar content [119]. Rootstocks with varying levels of dwarfing (some up to 50% and more than current industry standards), are available for peach, plum, apricot, cherry, and almond. Many of them have been bred or selected from *Prunus* species different than the species they were addressed to, or from inter-specific hybrids. (Table **22**). Highly vigorous species, such as sweet cherry, have benefitted more than others from the introduction of dwarfing and semi-dwarfing rootstocks (Fig. **21**) [141].

Figure 21: Influence of the rootstock on cherry tree size. From Lugli [141].

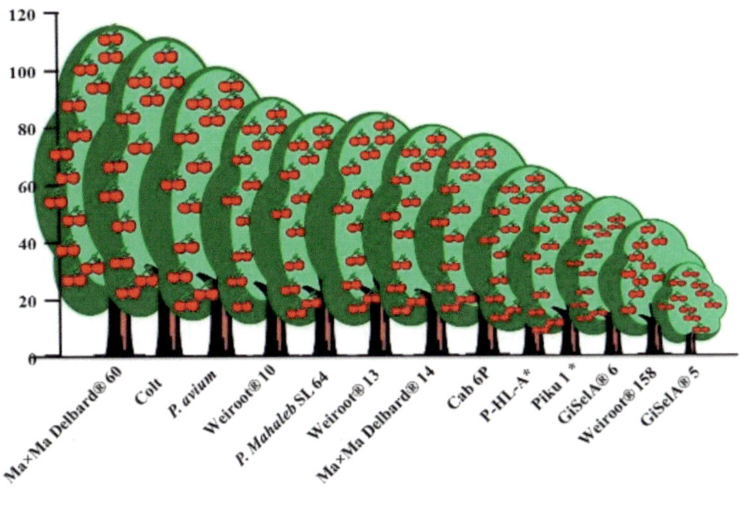

Among the most successful are the inter-specific 'GiSelA' series hybrids, with 'GiSelA® 5' being currently the most planted in Europe. It is important to point out that the potential benefits of the less invigorating rootstocks are fully exploited when both soil and climatic conditions are adequate. In evaluation trials on heavy and calcareous soils, trees grafted on dwarfing or very-dwarfing rootstocks such as 'GiSelA® 5', 'Damil' and 'Tabel-Edabriz' tended to dwarf excessively; the best agronomic performances (higher yield and fruit weight, and better skin overcolor) were obtained on intermediate or vigorous rootstocks such as 'Adarà', 'CAB 6P'and 'MaxMa 14' [142, 143].

In plum, many of the rootstocks developed to reduce tree size, like Pixy (*P. insititia*) or hybrids related to *P. besseyi* (like Plumina® Ferlenain), gave disappointing results because their dwarfing or semi-dwarfing effect is the result of their partial incompatibility with commercial scion cultivars [144].

In peach, several size-controlling rootstocks have been released in the last decades, and the most dwarfing ones have been obtained from other *Prunus* species or inter-specific hybrids (Table **22**). The degree of dwarfing of all of the above rootstocks varies with the variety, climate, soils, and site history. In most cases, however, tree performance onto size-controlling rootstocks has been not so satisfactory as hoped, due to incompatibility or excessively poor tree vigor having a depressing effect on yield, yield efficiency and/or fruit size. In a long-term multi-site comparative trial carried out in different Italian environments, 'Ishtarà-Ferciana' and 'Penta' showed good performances in terms of yield efficiency and fruit quality combined to a tree size reduction of 20-25% compared to the vigorous standard 'GF677', although these results were restricted only to those sites characterized by good fertile soils [145]. Size-controlling rootstocks for peach are being developed in California, and the recent releases 'Controller 5' (70% of the size of the commercial standard 'Nemaguard') and 'Controller 9' (90% of Nemaguard) maintained yield efficiency and fruit size despite significant tree dwarfing [146].

In studies on the growth pattern and physiology of trees onto various size-controlling rootstocks compared to the controls, cultivars grafted on the formers showed significantly lower shoot extension growth rates and consequent shorter internode length [147]. Also, the number and the diameter of xylem vessels, by influencing hydraulic conductance, seem to be key factors in determining the dwarfing potential of specific rootstock genotypes [148]. A reduced hydraulic conductance can cause stem water potential reduction (more negative values) during mid-day hours and, ultimately, decrease stomatal conductance and photosynthesis [149].

In some cases the dwarfing influence of rootstocks could be seen as a form of mild graft incompatibility. This could account for changes in the production and transport of minerals, assimilates and hormones. The shoot-root communication system through phloem and xylem translocation [150], where carbohydrates and auxins translocated in the phloem, and minerals, GA and cytokinins transported in xylem, have been intensively studied to understand the mechanism of rootstock growth control. Low auxin content in 'Weiroot' and 'GiSelA' rootstocks was found, but a direct rootstock effect on the auxin content of the scion has not yet been demonstrated.

Table 22: Main stone fruit rootstocks released in the last two decades

Rootstock	Year	Species	Origin[a]	Main objective[b]	Compatibility with	Reference
Adafuel*	1990	*P. dulcis* x *P. persica*	CSIC, Spain	Calcareous soil	Peach, almond	[151]
Adaptabil	2000	*P. besseyi* x *P. persica*	RIFG, Romania	Vigor control	Peach, european plum, apricot	[152]
Adarcias*	1994	*P. dulcis* x *P. persica*	CSIC, Spain	Calcareous soil	Peach, almond, japanese plum	[153]
Ademir*	1990	*P. cerasifera*	CSIC, Spain	Vigor control	Plum, apricot	[154]
Adesoto® 101*	1995	*P. insititia*	SPAIN	Vigor control, RKN	Peach, almond, plum, apricot	[155]
Cadaman® Avimag	1995	*P. persica* x *P. davidiana*	INRA, France and GYDV, Hungary	RKN	Peach	[156]
Castore	2003	*P. persica* x *P. dulcis*	UP, Italy	Vigor control, calcareus soil	Peach	[157]
Controller 5*	2004	*P. salicina* x *P. persica*	UC, USA	Vigor control	Peach	[158]
Controller 9*	2004	*P. salicina* x *P. persica*	UC, USA	Vigor control	Peach	[158]
Felinem*	1997	*P. dulcis* x *P. persica*	CITA, Spain	RKN, calcareus soil	Peach, almond, plum, apricot	[159]
Flordaguard	1991	*P. persica* x *P. davidiana*	UF, USA	Low chilling, RKN	Peach, japanese plum, almond	[160]
Garnem*	1997	*P. dulcis* x *P. persica*	CITA, Spain	RKN, calcareus soil	Peach, almond, plum, apricot	[159]
GiSelA® 5	1994	*P. cerasus* x *P. canescens*	GU, Germany	Vigor control	Cherry	[161]
GiSelA® 6	1994	*P. cerasus* x *P. canescens*	GU, Germany	Vigor control	Cherry	[162]
Guardian®	1997	*P. persica*	USDA &USC, USA	PTSL	Peach	[163]
Krymsk® 1	2002	*P. tomentosa* x *P. cerasifera*	KEBS, Russia	Vigor control, cold	Peach, plum, apricot	[164]
Krymsk® 86	2002	*P. cerasifera* x *P. persica*	KEBS, Russia	Vigor control, cold, waterlogging	Peach, plum, almond, apricot	[164]
Mayor®	2002	*P. dulcis* x *P.persica*	CIDA, Spain	Replant	Peach, almond	[165]
Miroper	2000	*P. cerasifera* × *P. persica*	RIFG, Romania	Vigor control	Peach, apricot	[166]
Monpol®	1997	*P. insititia*	CITA, Spain	RKN, calcareus soil, vigor control	Peach	[167]
Monegro*	1997	*P. dulcis* x *P. persica*	CITA, Spain	RKN, calcareus soil	Peach, almond, plum, apricot	[159]
Montizo®	1997	*P. insititia*	CITA, Spain	Calcareous soil, vigor control	Peach	[167]
Myrotop®	2012	*P. cerasifera*	INRA, France	RKN	Apricot, almond, Peach	[168]
Penta	1995	*P. domestica*	CRA-FRU, Italy	Vigor control	Peach, plum, Almond, Apricot	[169]

Table 22: contd…

PiKu 1*	2001	*P. avium* x (*P. canescens* x *P. tomentosa*)	JKI, Germany	Vigor control	Cherry	[170]
PiKu 3*	2001	(*P. canescens* x *P. tomentosa*) x *P. avium*	JKI, Germany	Vigor control	Cherry	[170]
PiKu 4*	2001	*P. cerasus* x (*P. canescens* x *P. incisa*)	JKI, Germany	Vigor control	Cherry	[170]
Polluce	2003	*P. persica* x *P. dulcis*	UP, Italy	Vigor control, calcareus soil	Peach	[157]
Pontaleb® Ferci*	1995	*P. mahaleb*	INRA, France	Vigor control	Cherry	[171]
Pumiselect® Rhenus 2*	1997	*P. pumila*	GRI, Germany	Vigor control, drought, cold	Peach, apricot	[171]
Rootpac® 40 Nanopac*	2008	(*P. dulcis* x *P. persica*) x (*P. dulcis* x *P. persica)*	AI, Spain	Vigor control	Peach, almond	[172]
Rootpac® 70 Purplepac*	2008	(*P. dulcis* x *P. persica*) x (*P. persica x P. davidiana*)	AI, Spain	Vigor control	Peach, almond	[173]
Rootpac® 90 Greenpac*	2008	(*P. persica* x *P. davidiana*) x (*P. dulcis* x *P. persica*)	AI, Spain	Vigor control	Peach, almond	[175]
Rootpac® R Replantpac*	2008	*P. cerasifera* x *P. dulcis*	AI, Spain	Replant, RKN	Peach, japanese plum, almond	[176]
Sharpe	2008	*P. angustifolia* x unknown Plum	USDA &UF, USA	RKN, PTSL, ARR, vigor control	Peach, plum	[177]
Sirio	1994	*P.persica* x *P. dulcis*	UP, Italy	Vigor control	Peach	[178]
Tabel® Edabriz*	1990	*P. cerasus*	INRA, France	Vigor control	Cherry	[179]
Tetra	1995	*P. domestica*	CRA-FRU, Italy	Vigor control	Peach, plum, apricot, almond	[169]
Toriplus®	2012	*P. domestica*	INRA, France	Vigor control	Apricot	[168]
Weiroot® 10	1990	*P. cerasus*	TUM, Germany	Vigor control	Cherry	[180]
Weiroot® 13	1990	*P. cerasus*	TUM, Germany	Vigor control	Cherry	[180]
Weiroot® 158	1990	*P. cerasus*	TUM, Germany	Vigor control	Cherry	[180]

[a] AI: Agromillora Iberia (Spain), CSIC: Consejo Superior de Investigaciónes Científicas (Spain), CIDA: Centro de Investigación y Desarrollo Agroalimentario (Spain), CITA: Centro de Investigaciónes y Tecnología Agroalimentaria de Aragón (Spain), CRA-FRU: Centro di Ricerca per la Frutticoltura (Italy), CSIC: Consejo Superior de Investigaciones Científicas (Spain), GRI: Geisenheim Research Institute (Germany), GU: University of Giessen (Germany), INRA: Institut National de la Recherche Agronomique (France), GYDV: Debrecen University (Hungary), JKI: Julius Kühn-Institut Federal Research Centre for Cultivated Plants (Germany), KEBS: Krymsk Experimental Breeding Station (Russia), RIFG: Research Institute for Fruit Growing Pitesti, Romania, TUM: Technical University Munich (Germany), UC: University of California (USA), UF: University of Florida (USA), USC: Clemson University (USA), UP: University of Pisa (Italy).
[b] ARR: *Armillaria* root rot, PTSL: Peach Tree Short Life, RKN: Root Knot Nematode.

ACKNOWLEDGEMENTS

Declared None.

CONFLICT OF INTEREST

The authors confirm that this chapter contents have no conflict of interest.

REFERENCES

[1] Potter, D (2012), Basic Information on the Stone Fruit Crops. In: Kole C, Abbott AG (eds) *Genetics, Genomics and Breeding of Stone fruits*. CRC Press Taylor & Francis Group, pp 1–21.

[2] Potter, D, Eriksson, T, Evans, R, Oh, S, *et al.* (2007), Phylogeny and classification of Rosaceae. *Plant Syst Evol* 266, 5–43.

[3] Rehder, A (1940), Manual of cultivated trees and shrubs hardy in North America exclusive of the subtropical and warmer temperate regions, Dioscoride. Portland, Oregon.

[4] Hancock, JF and Scorza, R and Lobos, G (2008), Peaches. *Temperate fruit crop breeding*. Springer, pp 265–298.

[5] Della Strada, G, Fideghelli, C (2010), The fruit varieties released in the world from 1980 through 2008. Rome, Italy.

[6] Bassi, D, Ghini, A, Mignani, I, Morgutti, S, *et al.* (2008), 'Big Top' nectarine, a case study for flesh texture. *VI Convegno Naz. sulla Peschicoltura Meridionale*. Caserta, Italy.

[7] Scorza, R, Sherman, W (1996), Peaches. In: Janick J, Moore JN (eds) *Fruit Breeding Tree and Tropical Fruits Vol. 1*. pp 325–440.

[8] Giovannini, D, Liverani, A, Merli, M, Brandi, F (2006), Breeding strategies to improve peach fruit quality. *Acta Hort* 713, 107–112.

[9] Fibiani, M, Lo Scalzo, R, Testoni, A (2004), Evoluzione post-raccolta del profilo pectico in accessioni di pesche a diverse caratteristiche di consistenza della polpa. *VII Giornate Scientifiche, S.O.I.* Napoli, Italy, p 343.

[10] Nicotra, A, Conte, L, Moser, L, Fantechi, P (2002), Serie Ghiaccio, 3 varietà di nuovo tipo di pesco. *L'Informatore Agrar* 8, 127–129.

[11] Conte, L, Bevilacqua, D, Di Cintio, A, Sartori, A, *et al.* (2011), Nuove tipologie di frutto nel programma di miglioramento genetico del pesco presso il CRA-FRU. *VII Convegno Peschicolo Meridionale*. Lamezia Terme - Sibari, Italy, p 42.

[12] Nicotra, A, Conte, L, Moser, L, Fantechi, P (2001), Serie Ufo, sei nuove cultivar di pesco a frutto piatto. *L'informatore Agrar* 50, 61–64.

[13] Nicotra, A, Conte, L, Moser, L, Fantechi, P (2003), Tre nuove cultivar di pesco a frutto piatto, tardive. *L'informatore Agrar* 18, 1–3.

[14] Conte, L, Nicotra, A (2006), L'esordio delle nettarine a frutto piatto. *L'Informatore Agrar* Suppl. 16, 1–4.

[15] Foschi, S, Liverani, A, Insero, O, Nencetti, V (2011), Pesche e nettarine - Liste Varietali, 2011. *Terra e Vita* Suppl., 49–50.

[16] Sartori, A, Terlizzi, M, Liverani, A, Insero, O, *et al.* (2012), Lista varietale del Pesco, molte le novità nel 2012. *L'Informatore Agrar* 31, 51–56.

[17] AA VV (2008), *The Future of Drylands: International Scientific Conference on Desertification and Drylands Research, Tunis, Tunisia, 19-21 June 2006*. Springer Science Jointly published with UNESCO, Paris, France.

[18] Vavilov, N (1992), The phyto geographical basis for plant breeding. In: Dorofeyev V (ed) *Origin and geography of cultivated plants*. Cambridge University Press, Cambridge, UK, pp 316–366.

[19] Ledbetter, CA (2008), Apricot. In: Hancock JF (ed) *Temperate fruit crop breeding: germplasm to genomics*. Springer, pp 39–82.

[20] Massai, R, Mennone, C (2012), Le varietà di Albicocco in Lista per il 2012. *L'informatore Agrar* 28, 51–54.

[21] Iezzoni, A (2010), Cherries. In: Hancock JF (ed) *Temperate fruit crop breeding*. Springer, pp 151–175.

[22] Palasciano, M, Godini, A, Bassi, G, Giongo, L, *et al.* (2012), Tre nuove varietà di ciliegio per gli impianti 2012. *L'Informatore Agrar* 22, 50–54.

[23] Okie, WR, Ramming, DW (1999), Plum breeding worldwide. *Hort Technol* 9, 162–176.

[24] Okie, WR, Hancock, JF (2008), Plums. In: Hancock JF (ed) *Temperate fruit crop breeding*. Springer, pp 337–358.

[25] Liverani, A, Sirri, S, Nenecetti, V (2012), Susino: due nuove entrate nella Lista 2012. *L'Informatore Agrar* 28, 55–58.

[26] Westwood, MN (1993), *Temperate Zone Pomology*. Timber Press, Portland, Oregon.

[27] Sung, SK, Yu, GH, Nam, J, Jeong, DH, *et al.* (2000), Developmentally regulated expression of two MADS-box genes, MdMADS3 and MdMADS4, in the morphogenesis of flower buds and fruits in apple. *Planta* 210, 519–528. doi: 10.1007/s004250050040.

[28] Wada, M, Cao, Q, Kotoda, N, Soejima, J, *et al.* (2002), Apple has two orthologues of FLORICAULA/LEAFY involved in flowering. *Plant Mol Biol* 49, 567–577. doi: 10.1023/A:1015544207121.

[29] Chen, Y, Jiang, P, Thammannagowda, S, Liang, H, *et al.* (2013), Characterization of Peach TFL1 and Comparison with FT/TFL1 Gene Families of the Rosaceae. *J Am Soc Hortic Sci* 138, 12–17.

[30] Soltis, DE, Chanderbali, AS, Kim, S, Buzgo, M, *et al.* (2007), The ABC model and its applicability to basal angiosperms. *Ann Bot* 100, 155–63. doi: 10.1093/aob/mcm117.

[31] Rohde, A, Bhalerao, RP (2007), Plant dormancy in the perennial context. *Trends Plant Sci* 12, 217–23. doi: 10.1016/j.tplants.2007.03.012.

[32] Weinberger, JH (1950), Chilling requirements of peach varieties. *Proc Am Soc Hortic Sci* 56, 122–8.

[33] Richardson, EA, Seeley, SD, Walker, DR (1974), A model for estimating the completion of rest for "Redhaven" and "Elberta" peach trees. *Hortic Sci* 9, 331–332.

[34] Seeley, S (1996), Modelling climatic regulation of bud dormancy ". In: Lang GA (ed) *Plant dormancy: physiology, biochemistry and molecular biology*. CAB INTERNATIONAL, Wallingford, pp 361–376.

[35] Erez, A, Lavee, S, Samish, RM (1971), Improved methods for breaking rest in the peach and other deciduous fruit species. *J Am Soc Hortic Sci* 96, 519–522.

[36] Scorza, R, Sherman, WB (1996), Peaches. In: Janick J, Moore JN (eds) *Fruit breeding Vol. I. Tree and Tropical Fruits*. John Wiley & Sons, Inc., New York, U.S.A., pp 325–440.

[37] Campoy, J, Ruiz, D, Egea, J (2011), Suitability of hydrogen cyanamide + oil application for dormancy overcoming in "Early Maycrest" peach in a warm-winter climate. *Eur J Hortic Sci* 76, 51–55.

[38] Lloyd, J, Firth, DJ (1993), Effect of hydrogen cyanamide and promalin on flowering, fruit set and harvest time of "Flordaprince" peach (Prunus persica (L.) Batsch) in subtropical Australia. *J Hortic Sci* 68, 177–183.

[39] Erez, A, Yablowitz, Z, Aronovitz, A, Hadar, A (2008), Dormancy breaking chemicals; efficiency with reduced phytotoxicity. *Acta Hortic* 772, 105–112.

[40] Tao, R, Iezzoni, AF (2010), The S-RNase-based gametophytic self-incompatibility system in Prunus exhibits distinct genetic and molecular features. *Sci Hortic (Amsterdam)* 124, 423–433. doi: 10.1016/j.scienta.2010.01.025.

[41] Sassa, H, Kakui, H, Minamikawa, M (2010), Pollen-expressed F-box gene family and mechanism of S-RNase-based gametophytic self-incompatibility (GSI) in Rosaceae. *Sexual plant reproduction* 23(1), 39–43.

[42] Yamane, H, Tao, R (2009), Molecular basis of Self-(in)compatibility and current status of S-genotyping in Rosaceous fruit trees. *J Japanese Soc Hortic Science*, 78(2), 137–157.

[43] Schuster, M (2012), Incompatible (S-) genotypes of sweet cherry cultivars (*Prunus avium* L.). *Sci Hortic* 148, 59–73.

[44] Lewis, D, Crowe, LK (1954), Structure of the incompatibility gene. IV. Types of mutations in *Prunus avium*. *Heredity* 8, 357-363.

[45] Sonneveld, T, Tobutt, KR, Vaughan, SP, Robbins, TP (2005), Loss of pollen-S function in two self-compatible selections of *Prunus avium* is associated with deletion/mutation of an S haplotype – specific *F-box* gene. *Plant Cell* 17(1), 37–51.

[46] Fishman, S, Génard, M (1998), A biophysical model of fruit growth: simulation of seasonal and diurnal dynamics of mass. *Plant Cell Envinron* 21, 739–752.

[47] Bonghi, C, Trainotti, L, Botton, A, Tadiello, A, Rasori, A, Ziliotto, F, Zaffalon, V, Casadoro, G, Ramina, A (2011), A microarray approach to identify genes involved in seed-pericarp cross-talk and development in peach. *BMC Plant Biol* 11, 107.

[48] Begheldo, M, Manganaris, GA, Bonghi, C, Tonutti, P (2008), Different postharvest conditions modulate ripening and ethylene biosynthetic and signal transduction pathways in Stony Hard peaches. *Postharvest Biol Tec* 48(1), 84–91.

[49] Prasanna, V, Prabha, TN, Tharanathan, RN (2007), Fruit ripening phenomena—an overview. *Crit Rev Food Sci* 47, 1–19.

[50] Rose, JKC, Bennett, AB (1999), Cooperative disassembly of the cellulose–xyloglucan network of plant cell walls: parallels between cell expansion and fruit ripening. *Trends Plant Sci* 4, 176–183.

[51] Brummell, DA, Harpster, MH (2001), Cell wall metabolism in fruit softening and quality and its manipulation in transgenic plants. *Plant Mol Biol* 47, 311–340.

[52] Saladie, M, Matas, AJ, Isaacson T (2007), A reevaluation of the key factors that influence tomato fruit softening and integrity. *Plant Physiol* 144, 1012–1028.

[53] Vicente, AR, Saladie, M, Rose, JK, Labavitch, JM (2007), The linkage between cell wall metabolism and fruit softening: looking to the future. *J Sci Food Agric* 87, 1435–1448.

[54] Tonutti, P, Bonghi, C, Ramina, A (1996), Fruit firmness and ethylene biosynthesis in three cultivars of peach (*Prunus persica* L. Batsch). *J Hortic Sci* 71, 141–147.

[55] Lester, DR, Sherman, WB, Atwell BJ (1996), Endopolygalacturonase and the Melting flesh (M) locus in peach. *J Am Soc Hort Sci* 121, 231–235.

[56] Haji, T, Yaegaki, H, Yamaguchi, M (2001), Changes in ethylene production and flesh firmness of melting, nonmelting and stony hard peaches after harvest. *J Japanese Soc Hortic Sci* 70, 458-459.

[57] Haji, T, Yaegaki, H, Yamaguchi, M (2005), Inheritance and expression of fruit texture melting, non-melting and stony hard in peach. *Sci Hortic* 105, 241-248.

[58] Crisosto, CH, Costa, G (2008), Preharvest factors affecting peach quality. In: *The Peach, botany, production and uses*. Wallingford, CABI, pp 536–549.

[59] Crisosto, CH, Mitchell, FG, Johnson, S (1995), Factors in fresh market stone fruit quality. *Postharvest News Inf* 6(2), 17–21.

[60] Nicolaï, BM, Beullens, K, Bobelyn, E, Peirs, A, *et al.* (2007), Nondestructive measurement of fruit and vegetable quality by means of NIR spectroscopy: A review. *Postharvest Biol Tec* 46(2), 99–118.

[61] Camps, C, Christen, D (2009), Non-destructive assessment of apricot fruit quality by portable visible-near infrared spectroscopy. *Food Sci Technol* 42, 1125-1131.

[62] Lurol, S, Mathieu-Hurtiger, V, Ciron, J, Tabariès, P (2010), L'outil Nir Case Sacmi. Mesures non destructives de la qualité des fruits et legumes. *Info Ctifl* 5, 261.

[63] Carvalho, ML, Almeida, D, Cardoso, J, Fialho, D (2010), Aplicação da espectroscopia NIR na determinação de alguns parâmetros de qualidade em ameixas. *2 ° Simpósio Nacional de Fruticultura*, Castelo Branco (Portogallo).

[64] Fu, XP, Ying, YB, Zhou, Y, Xie LJ, Xu HR (2008), Application of NIR spectroscopy for firmness evaluation of peaches. *J Zhejiang University Science* 9(7), 552-557.

[65] McGlone, VA, Jordan, RB, Martinsen, PJ (2002), Vis/NIR estimation at harvest of pre- and post-storage quality indices for 'Royal Gala' apple. *Postharvest Biol Technol* 25(2), 135–144.

[66] Ziosi, V, Noferini, M, Fiori, G, Tadiello, A, Trainotti, L, Casadoro, G, Costa, G (2008), A new index based on vis spectroscopy to characterize the progression of ripening in peach fruit. *Postharvest Biol Technol* 49(3), 319–329.

[67] Costa, G, Noferini, M, Fiori, G, Berthod, N, Besse, S, Rossier, J (2012), Methods to assess fruit quality focusing on non-destructive methods. *Acta Hort* 966, 275–284.

[68] Valero, C, Crisosto, CH, Slaughter, D (2007), Relationship between nondestructive firmness measurements and commercially important ripening fruit stages for peaches, nectarines and plums. *Postharvest Biol Technol* 44(3), 248–253.

[69] Lurie, S, Friedman, H, Weksler, A, Dagar, A, Eccher Zerbini, P (2013), Maturity assessment at harvest and prediction of softening in an early and late season melting peach. *Postharvest Biol Technol* 76, 10–16.

[70] Infante, R, Rubio, P, Contador, L, Noferini, M, Costa, G (2011), Determination of harvest maturity of D'Agen plums using the chlorophyll absorbance index. *Ciencia e investigación agrarian* 38, 199–203.

[71] Di Natale, C, Zude-Sasse, M, Macagnano, A, Paolesse, R, Herold, B, D'Amico, A (2002), Outer product analysis of electronic nose and visible spectra: application to the measurement of peach fruit characteristics. *Anal Chim Acta* 459(1), 107–117.

[72] Zhang, H, Wang, J, Ye, S, Chang, M (2009), Application of electronic nose and statistical analysis to predict quality indices of peach. *Food Bioproc Technol*, 5(1), 65–72.

[73] Rizzolo, A, Bianchi, G, Vanoli, M, Lurie, S, Spinelli, L, Torricelli, A (2012), Electronic nose to detect volatile compound profile and quality changes in 'Spring Belle' peach (*Prunus persica* L.) during cold storage in relation to fruit optical properties measured by time-resolved reflectance spectroscopy. *J Agr Food Chem* 61(8), 1671–1685.

[74] Crisosto, CH, Day, KR (2012), 10 Stone Fruit. In: Rees D (eds), *Crop Post-Harvest: Science and Technology*. Blackwell Publishing Ltd., pp 212–225.

[75] Woodbridge, CG (1978), Split-pit in peaches and nutrient levels. *J Amer Soc Hort Sci* 103(2), 278–280.

[76] Christensen, N (1996), Rain-induced cracking of sweet cherries: its causes and prevention. In: Webster AD, Looney NE (eds). *Cherries: Crop Physiology, Production and Uses.* CAB Inter- national, Wallingford, UK, pp 297-327.

[77] Cline, JA, Sekse, L, Meland, M, Webster, AD (1995), Rain-induced fruit cracking of sweet cherries: 1. Influence of cultivar and rootstock on fruit water absorption, cracking and quality. *Acta Agric Scand Sect B Soil Plant Sci* 45, 213-223.

[78] Peschel, S, Knoche, M (2012), Studies on water transport through the sweet cherry fruit surface : XII. Variation in cuticle properties among cultivars. *J Amer Soc Hort Sci* 137(6), 367–375.

[79] Wertheim, SJ (2005), 15 Planting system and tree shape. In: Tromp J, Webster AD, Wertheim SJ (eds) *Fundamentals of Temperate Zone Tree Fruit Production*, Backhuys Publishers, Leiden, The Netherlands, pp 190–203.

[80] Lakso AN, Robinson, TL (1997), Principles of orchard systems management optimizing supply, demand and partitioning in apple trees. *Acta Hort* 451, 405-415.

[81] Costes, E, Lauri, PÉ, Regnard, JL (2010), Analyzing fruit tree architecture: implications for tree management and fruit production. In: Janick J, (eds) *Horticultura Reviews, 32.* John Wiley & Sons, Inc., Oxford, pp 1–61.

[82] Jackson, JE, (1980), Light interception and utilization by orchard systems. *Hort Rev* 2, 208-267.

[83] Grossman, YL, Dejong, TM (1998), Training and pruning system effects on vegetative growth potential, light interception, and cropping efficiency in peach trees. *J Amer Soc Hort Sci* 123(6), 1058–1064.

[84] Baldassarri, T (1960), La vecchia e la nuova potatura. *Frutticoltura* 3, 18-25.

[85] Baldini, E (1976), *Arboricoltura.* Cooperativa Libraria Universitaria Editrice, Bologna, Italy.

[86] Mika, A (1986), Physiological responses of fruit trees to pruning. *Hortic Rev* 8, 337–378.

[87] Corelli-Grappadelli L, Marini, RP (2008), *Orchard planting systems. I*n: Layne DR, Bassi D (eds), *The Peach, botany, production and uses.* Wallingford, CAB International, pp 264 – 288.

[88] Sansavini, S, Neri, D (2005), Forme di allevamento e potatura del pesco. In: C. Fideghelli, S. Sansavini (eds). *Il pesco.* Edagricole, Bologna, pp 115-143.

[89] Giovannini, D, Neri, D, Di Vaio, C, Massai, R, *et al.* (2010), Efficienza gestionale degli impianti di pesco in un confronto Nord-Sud. *Frutticoltura* 7/8, 16-27.

[90] Iglesias, I (2013), Peach production in Spain: current situation and trends, from production to consumptio", *Proceedings of the 4th Conference Innovation in fruit growing.* Belgrade, pp 75–98.

[91] Montserrat, R, Iglesias, I (2011), I sistemi di allevamento adottati in Spagna: l'esempio del vaso catalano. *Frutticoltura* 7/8, 18–26.

[92] Caruso, T, Di Vaio, C, Guarino, F, Motisi, A, Nuzzo, V (2003), Le tipologie di impianto per la peschicoltura intensiva nell'Italia Meridionale. *Proceedings of the fourth South Italy Peach Conference.* Campobello di Licata, Agrigento, Italy, pp 44-51.

[93] Caruso, T, Motisi, A, Pernice, F, Di Vaio, C (2008), Peach planting systems in Southern Italy: ecophysiological aspects and technical developments. *Acta Hort* 772, 423–430.

[94] Hrotkó, K, Majar, L, Gyeviki, M, Simon, G (2008), Sistemi di allevamento e potatura per impianti ad alta densità: le esperienze dell'Ungheria. *Frutticoltura* 3, 10–19.

[95] Iglesias, I, Peris, M (2008), La produzione spagnola vince grazie a precocità, qualità, e organizzazione tecnico-commerciale. *Riv. di Frutticoltura*, 3, 28–32.

[96] Lugli, S, Musacchi, S, Grandi, M, Bassi, G, Franchini, S, Zago, M (2011), The sweet cherry production in Northern Italy: innovative rootstocks and emerging high-density plantings. *3rd Conference, Innovations in Fruit Growing*. Belgrad, pp 75–92.

[97] Lauri, PE, Claverie, J (2005), Training sweet cherries to improve fruit size and quality. An overview of some recent concepts and practical aspects developed at INRA France. *Acta Hort* 667, 361–366.

[98] Robinson, TL (2005), Developments in high density sweet cherry pruning and training systems around the world. *Acta Hort* 667, 269-272.

[99] Negueroles Pérez, J (2005), Cherry Cultivation in Spain. *Acta Hort* 667, 293- 302.

[100] Brunner, T (1975), Experiments on developing new intensive fruit training methods, with special reference to cherries. *Gyumolcstermesztes* 2,77–89.

[101] Zahn, FG (1990), Die Spindel beim Steinobst. *Erwerbsobstbau* 32(3), 60-66.

[102] Whiting, MD, Lang, G, Ophardt, D (2005), Rootstock and training system affect sweet cherry growth, yield, and fruit quality. *HortScience* 40(3), 582–586.

[103] Hrotkó, K, Simon, G, Magyar, L (1998a), Modified Brunner-spindle as a training system for semi-intensive sweet cherry orchards. *Acta Hort* 468, 459-464.

[104] Hrotkó, K, Simon, G, Magyar, L (1998b), Training of slender spindle trees for intensive sweet cherry orchards. *Acta Hort* 468, 465-470.

[105] Chalmers, DJ, Van den Ende, B (1975), The Tatura trellis: a new design in high yielding orchards. *J Agric Victoria* 73, 473-475.

[106] Bassi, D (1999), Apricot culture: present and future. *Acta Hort* 488, 35–40.

[107] Errani, A (1996), Nuovi orientamenti sulle forme di allevamento in frutticoltura. *Frutticoltura* 12, 71-73.

[108] Morvan, G (1995), Pathogenic organisms as causes of apricot decline. *Acta Hort* 384, 521-532.

[109] Neri, D, Morini, F, Massetani, F, Pirazzini, P (2010), Pruning: how to manage shoot growth. *Acta Hort* 862, 355-364.

[110] Jai, M, Lichou, J, Castes, E, Audubert, A (1995), L'abricotier. Etudes architecture et modelisation: consequences agronomiques. *Ctifl (F),* Serie Rapport n. 7/95.

[111] Sottile, F, Bellini, E, Nencetti, V, Peano, C, *et al.* (2010), Plum production in Italy: state of the art and perspectives. *Acta Hort* 874, 25-34.

[112] Christen, D, Devènes, G, Chassot, E, Jeltsch, P (2010), Chemical thinning of Apricot cultivars: promising preliminary results. *Acta Hort* 862, 373-380.

[113] Fishman, S, Génard, M (1998), A biophysical model of fruit growth: simulation of seasonal and diurnal dynamics of mass. *Plant, Cell Envinron* 21, 739–752.

[114] Grossman, Y, DeJong, T, (1995), Maximum fruit growth potential and seasonal patterns of resource dynamics during peach growth. *Ann Bot* 75, 553-560.

[115] Sansavini, S, Neri, D, Intrieri, C, Tombesi, A, *et al.* (2012), Impianti e forme di allevamento, potatura, controllo della fruttificazione e raccolta. In: S. Sansavini, G. Costa, R. Gucci, P. Inglese, A. Ramina, C. Xiloyannis (eds), *Arboricoltura Generale*, pp 333–398.

[116] Costa, G, Vizzotto, G (2000), Fruit thinning of peach trees. *Plant Growth Regulation* 31, 113–119.

[117] Schupp, JR, Baugher, TA, Miller, SS, Harsh, RM, Lesser, KM (2008), Mechanical thinning of peach and apple trees reduces labor input and increases fruit size. *Hort Technology* 18, 660–670.

[118] Foschi, S, Neri, D, Massetani, F (2012), Meccanizzare il pescheto per salvaguardare il reddito. *L'informatore Agrar* 24, 43–47.

[119] Beckman, TG, Lang, GA (2003), Rootstock Breeding for Stone Fruits. *Acta Hort* 622, 531–551.

[120] Wertheim, SJ, Webster, AD (2005), Rootstocks and interstems. In: Tromp J, Webster AD, Wertheim SJ (eds), *Fundamentals of Temperate Zone Tree Fruit Production*, pp 156–175.

[121] Schmidt, H, van de Weg, WE, (2005), Breeding. In: Tromp J, Webster AD, Wertheim SJ (eds), *Fundamentals of Temperate Zone Tree Fruit Production*, pp 136–155.

[122] Moreno, MA (2009), Rootstocks for stone and pome fruit tree species in Spain. *Proceedings International Conference on fruit tree rootstocks*. Pisa (Italy), pp 44-57.

[123] Reighard, GL, Loreti, F (2008), Rootstock Development. In: Layne DR, Bassi D (eds), *The Peach: Botany, production and Use*, CAB International, pp 193-220.

[124] Jiménez, S, Pinochet, J, Romero, J, Gogorcena, Y, *et al.* (2011), Performance of peach and plum based rootstocks of different vigour on a late peach cultivar in replant and calcareous conditions. *Sci Hortic* 129(1), 58-63.

[125] Webster, AD, Schmidt, H (1996), Rootstocks. In: Webster AD, Looney NE, (eds), *Cherries: Crop Physiology, Production and Uses*. CAB International, Wallingford, UK, pp 127-163.

[126] Crossa-Raynaud, P, Audergon, JM (1987), Apricot rootstocks. In: Rom RC, Carlson RF (eds), *Rootstocks for fruit crops*, pp 295–320.

[127] Okie, WR (1987), Plum rootstocks. In: Rom RC, Carlson RF (eds), *Rootstocks for fruit crops*, pp 321–360.

[128] Esmenjaud, D, Voisin, R, Ghelder, C, Bosselut, N, *et al.* (2008), Genetic dissection of resistance to root-knot nematodes *Meloidogyne* spp. in plum, peach, almond, and apricot from various segregating interspecific *Prunus* progenies. *Tree Genet Genomes* 5(2), 279–289.

[129] Nyczepir, AP, Okie, WR (1996), Occurrence of peach tree short life on a field site with no history of peach production. *HortScience* 31(1), 163.

[130] Reighard, GL, Newall, WC, Zehr, EI, Beckman, TG, etal. (1997), Field performance of *Prunus* rootstock cultivars and selections on replant soils in South Carolina. *Acta Hort* (ISHS) 451, 243-250.

[131] Culver, DJ, Ramming, DW, McKenry, MV (1989), Procedures for field and greenhouse screening of *Prunus* genotypes for resistance and tolerance to root lesion nematode. *J Amer Soc Hortic Sci* 114(1), 30–35.

[132] Alcañiz, E, Pinochet, J, Fernández, C, Esmenjaud, D, Felipe, A (1996), Evaluation of *Prunus* rootstocks for root-lesion nematode resistance. *HortScience* 31(6), 1013–1016.

[133] Melakeberhan, H, Bird, GW, Gore, R (1997), Impact of plant nutrition on *Pratylenchus penetrans* infection of *Prunus avium* rootstocks. *J Nematol* 29(3), 381–388.

[134] McFadden-Smith, W, Miles, NW, Potter, JW (1998), Greenhouse evaluation of *Prunus* rootstocks for resistance or tolerance to the root lesion nematode (*Pratylenchus penetrans*). *Acta Hort* 465(2), 723–729.

[135] Pinochet, J, Fernández, C, Calvet, C, Hernàndez-Dorrego, A, Felipe, A (2000), Selection against *Pratylenchus vulnus* populations attacking *Prunus* rootstocks. *Acta Hort* 35(7), 1333-1337.

[136] Beckman, TG, Okie, WR, Nyczepir, AP, Pusey, PL, Reilly, CC (1998), Relative susceptibility of peach and plum germplasm to Armillaria Root Rot. *HortScience* 33(6), 1062-1065.

[137] Beckman, TG, Pusey, PL (2001), Field testing peach rootstocks for resistance to Armillaria Root Rot. *HortScience* 36(1), 101-103.

[138] Beckman, TG, Chaparro, JX, Sherman, WB (2012), 'MP-29', a clonal interspecific hybrid rootstock for peach. *HortScience* 47(1), 128-131.

[139] Yang, JI, Ruegger, PM, McKenry, MV, Becker, JO, Borneman, J (2012), Correlations between root-associated microorganisms and peach replant disease symptoms in a California soil. *PloS one* 7(10), e46420.

[140] Manici, LM, Caputo, F (2010), Soil fungal communities as indicators for replanting new peach orchards in intensively cultivated areas. *Eur J Agron* 33(3), 188-196.

[141] Lugli, S, (2011), I portinnesti per i nuovi impianti in Italia. *Terra e Vita* 38, 22-25.

[142] Godini, A, Palasciano, M, Camposeo, S, Pacifico, A (2008), A nine-year study on the performance of twelve cherry rootstocks under non-irrigated conditions in Apulia (Southern Italy). *Acta Hort* 795, 191–198.

[143] Cantín, CM, Pinochet, J, Gogorcena, Y, Moreno, MÁ (2010), Growth, yield and fruit quality of 'Van' and 'Stark Hardy Giant' sweet cherry cultivars as influenced by grafting on different rootstocks. *Sci Hortic* 123(3), 329–335.

[144] Webster, AD, Wertheim, SJ (1993), Comparisons of species and hybrid rootstocks for European plum cultivars. *J Hortic Sci* 68(6), 861-869.

[145] Massai R, Loreti F (2010), Pesco. *Frutticoltura,* 7/8(Speciale portinnesti), 16-23.

[146] Dejong, TM, Johnson, RS, Day, KR (2011), Controller 5, Controller 9 and Hiawatha peach rootstocks: their performance and physiology. *Acta Hort* 903, 221-228.

[147] Weibel, A, Johnson, RS, Dejong, TM (2003), Comparative vegetative growth responses of two peach cultivars grown on size-controlling *versus* standard rootstocks. *J Amer Soc Hort Sci* 128(4), 463-471.

[148] Tombesi, S, Johnson, RS, Day, KR, DeJong, TM (2010), Relationships between xylem vessel characteristics, calculated axial hydraulic conductance and size-controlling capacity of peach rootstocks. *Ann bot* 105(2), 327-331.

[149] Solari, LI, Johnson, S, DeJong, TM (2006), Hydraulic conductance characteristics of peach (*Prunus persica*) trees on different rootstocks are related to biomass production and distribution. *Tree physiol* 26(10), 1343–50.

[150] Webster, AD, (1998), Strategies for controlling the size of sweet cherry trees. *Acta Hort* (ISHS) 658, 229–240.

[151] Cambra, R, (1990), Adafuel, an almond × peach hybrid rootstock. *HortScience* 25(5), 584.

[152] Duţu, I, Mazilu, C, Sergiu, A (2011), Romanian plum rootstocks newly released at the Research Institute for Fruit Growing (RIFG) Pitesti. Scientific Papers of the Research Institute for Fruit Growing. Pitesti, Romania, vol 27.

[153] Moreno, MA, Cambra, R, (1994), Adarcias: an almond x peach hybrid rootstock. *HortScience* 29(8), 925.

[154] Moreno, MA, Tabuenca, MC, Cambra, R (1995a), Ademir, a *Myrobalan* rootstock for plum. *HortScience* 30(7), 1475–1476.

[155] Moreno, MA, Tabuenca, MC, Cambra, R (1995b), Adesoto 101, a plum rootstock for peaches and other stone fruit. *HortScience* 30(6), 1314–1315.

[156] Edin, M, Garcin, A (1994), Un nouveau porte-greffe du pêcher CADAMAN® AVIMAG. *L'arboriculture Fruitière* 475, 20–23.
[157] Loreti, F, Massai, R (2006), 'Castore' and 'Polluce': two new hybrid rootstocks for peach and nectarine. *Acta Hort* 713, 275–278.
[158] Dejong, TM, Johnson, RS, Doyle, JF, Weibel, A, *et al.* (2004), Growth, yield and physiological behavior of size-controlling peach rootstocks developed in California. *Acta Hort* 658, 449–455.
[159] Felipe, AJ (2009), 'Felinem', 'Garnem', and 'Monegro' almond x peach hybrid rootstocks. *HortScience* 44(1), 196–197.
[160] Sherman, WB, Lyrene, PM, Sharpe, RH (1991), 'Flordaguard' peach rootstock. *HortScience* 26(4), 427–428.
[161] Franken-Bembenek, S, (2005), 'Gisela® 5' rootstock in Germany. *Acta Hort* (ISHS) 667, 167-172.
[162] Fideghelli, C, Loreti, F, (2009), Stone fruit rootstock. In: C, Fideghelli, F, Loreti (eds.) *Monografia dei portinnesti dei fruttiferi*, Italian Ministry of Agriculture.
[163] Beckman, TG, Reighard, GL, Okie, WR, Nyczepir, AP, *et al.* (1997), History, current status and future potential of Guardian™ (BY520-9) peach rootstock. *Acta Hort* 451, 251–258.
[164] Jimenez, S, Pinochet, J, Abadía, A, Moreno, MA, Gogorcena, Y (2008), Tolerance response to iron chlorosis of *Prunus* selections as rootstocks. *HortScience* 43/2, 304–309.
[165] Cos, J, Frutos, D, García, R, Rodríguez, J, Carrillo, A, (2004), *In Vitro* rooting study of the peach-almond hybrid Mayor®. *Acta Hort* (ISHS) 658, 623-627.
[166] Indreiaş, A, Stefan, I, Dutu, I, (2004), Apricot rootstocks created and used in Romania. *Acta Hort* (ISHS) 658, 509-511.
[167] Felipe, A, Carrera, M, Gòmez-Aparisi, J, Socias i Company, R (1997), 'Montizo' and 'Monpol', two new plum rootstocks for peaches. *Acta Hort* 451, 273–276.
[168] Duval, H, Masse, M, Jay, M (2012), Results of French apricot rootstock trials. *Acta Hort* 966, 37–42.
[169] Nicotra, A, Moser, L (1997), Two new plum rootstocks for peach and nectarines: Penta and Tetra. *Acta Hort* 451, pp. 269–271.
[170] Wolfram, B (1996), Advantages and problems of some selected cherry rootstocks in Dresden-Pillnitz. *Acta Hort* 410, 233–237.
[171] Claverie, J, Edin, M (1996), A new cherry rootstock. Ferci-Pontaleb®, INRA strain SL405. *Arboriculture Fruitière* 500, 41–47.
[172] Pinochet (2010a) http://www.freepatentsonline.com/PP21566.html.
[173] Pinochet (2010b) http://www.freepatentsonline.com/PP21432.html.
[174] Jacob, H (1992), *Prunus pumila* L., Eine geeignete schwachwachsende Pfirsichunteretlage. *Erwerbsobstbau* 34, 144-146.
[175] Pinochet, J (2009), 'Greenpac', a new peach hybrid rootstock adapted to Mediterranean conditions. *HortScience* 44(5), 1456–1457.
[176] Pinochet, J, (2010), 'Replantpac' (Rootpac® R) a Plum – almond hybrid rootstock for replant situations. *HortScience* 45(2), 299–301.
[177] Beckman, TG, Chaparro, JX, Sherman, WB (2008), 'Sharpe', a clonal plum rootstock for peach. *HortScience* 43(7), 2236–2237.
[178] Loreti, F, Massai, R (1994), 'Sirio', nuovo portinnesto ibrido pesco x mandorlo. *L'Informatore Agrar* 28, 47-49.

[179] Edin, M (1989), Tabel® Edabriz, porte-greffe nanisant du cerisier. *INFOS Ctifl* 55, 41- 45.
[180] Callesen, O (1998), Recent Developments in Cherry Rootstock Research. *Acta Hort* 468, 219–228.

CHAPTER 5

Genetics and Genomics of Stone Fruits

Sabrina Micali*, Elisa Vendramin, Maria Teresa Dettori and Ignazio Verde

Consiglio per la Ricerca e la sperimentazione in Agricoltura (Agricultural Research Council, Fruit Tree Research Centre, CRA-FRU), 00134 Rome, Italy

Abstract: Plant breeding has made remarkable progress in recent years and many agronomically important characters such as yield, quality and resistance to biotic and abiotic threats have been improved in many crop species. The use of molecular markers and, more recently, the availability of sequenced genomes are fostering the characterization and isolation of genes involved in key processes related to these traits in many species. This is in turn giving the opportunity to apply marker assisted selection (MAS) enabling the production of improved varieties in less time and at reduced costs compared to traditional breeding strategies. In *Prunus* spp, many linkage maps have been produced and several traits have been genetically localized leading, in a few cases, to the cloning of the genes underlying important traits. More recently, peach (*Prunus persica*) was chosen as the model species of the *Prunus* genus and its genomic sequence has been obtained (Peach v1.0). This will offer new opportunities to genomic-assisted breeding in all stone fruits. In the present chapter, the latest achievements in the genetics and genomics of the *Prunus* species are discussed together with the main relevant breeding outcomes.

Keywords: *Prunus*, germplasm conservation, genetic resources, Quantitative Trait Loci (QTL), Mendelian trait, marker assisted selection (MAS), RNAseq, transcriptomics, Next Generation Sequencing (NGS), candidate gene, Peach v1.0, Expressed Sequence Tag (EST).

1. GENETIC RESOURCES: GERMPLASM CONSERVATION, MOLECULAR CHARACTERIZATION, IDENTIFICATION AND MANAGEMENT OF CORE COLLECTIONS

1.1. Germplasm Conservation

During the last century, the notion that the world's agro biodiversity was rapidly dwindling led to a global effort to collect and preserve genetic resources.

***Corresponding author Sabrina Micali:** CRA - Centro di Ricerca per la Frutticoltura (CRA-FRU), Rome, Italy; Tel: +39 0679348193; Fax: +39 0679341630; E-mail: sabrina.micali@entecra.it

The first genebanks were established at the beginning of the seventies to conserve the long-term threatened crop landraces and, more recently their wild relatives. Most crops were, and are, conserved as seeds, but vegetatively propagated plants, mostly woody plants, and species that do not easily produce seeds, or those whose seeds cannot easily be conserved in a dried form, are conserved as clone collections in fields. In a few decades, many collections were assembled and in 1996, the Food and Agriculture Organization (FAO) already counted 1,300 genebanks maintaining over 6 million accessions all over the world [1]. Ever since, the total number of accessions held worldwide in genebanks has increased by approximately 20%, reaching 7.4 million in 2010; the number and size of genebanks have also increased [2]. It is estimated that only about 25-30% of the total worldwide conserved accessions are unique, the remainder being duplicates [2].

According to the second FAO State of the World report (SoW) published in 2010 [2], the four largest national genebanks are the Institute of Crop Germplasm Resources - Chinese Academy of Agricultural Sciences (ICGR-CAAS) in China, the National Center for Genetic Resources Preservation in the United States of America, the National Bureau of Plant Genetic Resources (NBPGR) in India and the N.I. Vavilov All-Russian Scientific Research Institute of Plant Industry. National genebanks housing more than 100,000 accessions are also found in Brazil, Canada, Germany, Japan and Korea.

1.2. Prunus Collections in the World

Several economically important species belong to the genus *Prunus*; for this reason germplasm collections are both numerous and distributed all over the world. According to the latest FAO SoW [2] 69,497 accessions conserved worldwide belong to *Prunus*. Table **1** summarizes data, regarding this genus, reported in the last FAO SoW. The thirteen largest National Genebanks are listed (FAO does not report single figures for *Prunus* species). The 69,497 *Prunus* accessions reported are divided into 223 genebanks. The biggest collections, containing over 6,000 accessions each (which amount to about 9% of the total *Prunus* accessions), are held in the Russian Federation and in the USA. In Europe the biggest collection, the third in the world, but only accounting for 3% of the total number, is housed in Italy at the CRA Fruit Tree Research Centre of Rome, closely followed by Hungary and Turkey.

Table 1: *Prunus* germplasm collections

Nation	Genebank	Accessions		Type of accession (%)	
		No.	%	Wild Species	Landraces
Russian Federation	VIR - N.I. Vavilov All-Russian Scientific Research Institute of Plant Industry	6579	9	18	13
USA	Horticulture Department, Michigan State University	6100	9	-	-
Italy	CRA-FRU Consiglio per la Ricerca e la Sperimentazione in Agricoltura - Centro di Ricerca per la Frutticoltura (CRA-FRU)	2421	3	<1	18
Hungary	Enterprise for Extension and Research in Fruit Growing and Ornamentals	2259	3	-	-
Turkey	Aegean Agricultural Research Institute	1874	3	<1	81
Ukraine	Crimean Pomological Station	1458	2	1	11
Switzerland	FRUCTUS - Association Suisse pour la Sauvegarde du Patrimoine Fruitier	1450	2	-	39
Japan	National Institute of Agrobiological Sciences	1423	2	1	13
France	INRA-Bordeaux Unité de Recherches sur Espèces Fruitières et Vigne	1220	2	<1	<1
Mexico	Instituto Nacional de Investigaciones Forestales, Agrícolas y Pecuarias	1116	2	3	-
Romania	ICPP Pitesti-Fruit Growing Research Institute Maracineni-Arges	1093	2	2	30
Iran	NPGBI-SPII - National Plant Gene Bank of Iran - Seed and Plant Improvement Institute	1006	1	-	-
Brazil	CPACT/EMBRAP-Embrapa Clima Temperado	1006	1	-	-
	Others (211)	40492	58	4	10
	Total	**69497**	**100**	**4**	**12**

Adapted from FAO [2].

The absence of the Far Eastern countries, except from Japan, and in particular the absence of China (which is represented in the FAO SoW with 5 institutions), in the list of the nations housing the first thirteen biggest *Prunus* genetic resources, is quite striking. Several *Prunus* species have their center of origin in China. In 1995 Faust and Timon [3] reported a body of evidence indicating that the peach is native to China and placed domestication, by means of historical and archeological evidence, before 3300 BC [3]. The peach sequencing and the extensive work of re-sequencing [4] confirmed Faust and Timon's hypothesis.

However, the center of origin of the apricot is still unclear, but most theories place it in China (NE, Central, Western). Alternatively, Central Asia could be the primary center and China the secondary one [5]. The center of origin of *Prunus salicina* (Japanese plum) is also placed in China, even if it is extensively cultivated in Japan [6] while that of *Prunus domestica,* a species unknown in the wild state, is in Europe. *P. domestica* is probably a hybrid involving the European tetraploid species *Prunus spinosa* [6, 7]. The cultivated cherries, mainly *P. avium* (sweet cherry) and *P. cerasus* (sour cherry), originated in Southwestern Asia; yet it is not clear whether Europe is a primary or secondary center [6, 8].

Li [9] reports that China is the world's largest conservatory of peach genetic diversity in the world, with about 800 accessions collected and conserved in China's three National Peach Germplasm Resources Repositories (at Beijing, Nanjing and Zhengzhou). The Chinese CIGRIS (Chinese Crop Information System (http://icgr.caas.net.cn/cgrisintroduction.html) reports 1,356 *Prunus* accessions, distributed as follows: 840 peaches, 283 apricots, 233 plums. Wang [10] states that China, being the center of origin of the apricot, has the richest germplasm resources and preserves 2,000 apricots, including seven species and 11 variations; Wang and his coworkers identified a core collection out of 1,501 apricot accessions belonging to Chinese collections.

In the USA, the national *Prunus* collection, belonging to the National Plant Germplasm System (NPGS), is held by the National Clonal Germplasm Repository for Tree Fruit/Nut Crops and Grapes (NCGR) at Davis (CA). One hundred and eight tetraploid cherries (among them 87 *P. cerasus* and 12 *P. fruticosa)* are housed at The Plant Genetic Resources Unit in Geneva (New York).

1.3. Databases: Web Resources

Providing a comprehensive list of the web-based catalogues of genetic resources for stone fruits is a complex task and we have not attempted to achieve this. Few international databanks are available; the information is spread over several countries, many of which lack a central database. Databases are mostly held by public institutions, but many small private databases are also available. They vary greatly in terms of the level of complexity, ranging from a simple list of names, which lack the possibility to perform any search, to highly refined databases providing complete information, which allow advanced searches. Here, information will be given relating to the largest internet-based databases available for stone fruits, favoring those presenting centralized data.

The USA has a central facility, the National Genetic Resources Program (NGRP), which "acquires, characterizes, preserves, documents, and distributes to scientists, germplasm of all life forms important for food and agricultural production". The Germplasm Resources Information Network (GRIN) provides this information at http://www.ars-grin.gov/npgs/searchgrin.html.

In China, the Chinese Crop Germplasm Information System (CGRIS) is the central repository for all types of plant genetic resource information. It is available at http://icgr.caas.net.cn/cgris_english.html but only partially in English (20/05/2013). CGRIS houses the management system of the National Crop Gene Bank (NCGB), of long and medium term storage facilities, of the National germplasm Resources Nursery and the databases of crop characterization and evaluation and also of germplasm exchange.

EURISCO (http://eurisco.ecpgr.org/-20/05/2013) is a web-based catalogue that provides information about *ex situ* plant collections maintained in Europe. It is based on a European network of *ex situ* National Inventories (NIs) and contains passport data on almost 1.1 million samples of crop diversity (updated May 2012). The database is very well organized and user-friendly. A search with "Prunus" as the keyword returns 18,313 hints distributed in 24 countries; however, many of these records may be redundant. The European *Prunus* Database (EPDB), maintained by the French National Institute for Agricultural Research (INRA) at Bordeaux under the initiative of the European Crop Programme for Genetic Resources (ECPGR), contains data of several European collections including all *Prunus* species and their related wild relatives. The portal (http://www.bordeaux.inra.fr/eupeachdb/) contains databases providing passport, characterization and evaluation data. It currently lists 1,273 accessions housed by 15 institutes from 10 countries. In Italy, there are several databases; the one belonging to the CRA-FRU, the third Institution in the world according to FAO SoW [2] for the number of *Prunus* accessions, is accessible at http://fru.entecra.it. It is currently being updated and will soon be moved to http://planta_res.entecra.it. Eastern European countries, such as Russia, Hungary,

Turkey and perhaps also the Ukraine, being a door towards the East, play an important role in the stone fruit germplasm collection but, to our knowledge, none maintains public internationally-oriented portals. In Spain, the Plant Genetic Resource Center (Centro Nacional de Recursos Fitogenéticos - CRF) has the responsibility of developing a national inventory of *ex-situ* collections of Spanish germplasm. The passport data are accessible at: http://wwwx.inia.es/crf/WWWCRF/CRFesp/Paginaprincipal.asp.

Regarding the UK, the National Fruit Collection includes over 3,500 accessions. The University of Reading maintains the collection and is currently developing the National Fruit Collection public database, accessible at http://www.nationalfruitcollection.org.uk/search.php. Stone fruits are represented by 304 plums, 328 cherries and 10 apricots. A Portal on Genetic Resources for Food and Agriculture (GRFA), with the aim of providing a single access point to all known information and technical data on the UK's farmed species, plants and animals, is available at http://grfa.org.uk/.

1.4. Core Collections

Collecting and preserving crops is mostly done by preserving seeds (in 1998 they accounted for approximately 90% of all the *ex situ* accessions - http://www.fao.org/sd/EPdirect/EPre0041.htm) but for woody species, which are vegetatively propagated as clones and for species which are difficult or impossible to store as seed, field genebanks are mainly used. Clonal collections are the most important and diffused means of preserving stone fruits diversity; they are very space consuming, labor-intensive and costly. In the USA the Plant Genetic Resources Unit (PGRU, Cornell University at Geneva) belonging to the National Plant Germplasm System (NPGS), calculated that the cost to maintain each tree in the field is $75 to $100 each year (http://www.ars.usda.gov/Aboutus/docs.htm?docid=6112; accessed 20/05/2013). Such a cost needs to be doubled or even tripled as, normally, two or three samples for each accession are maintained for security reasons, given that field collections are vulnerable to biotic and abiotic risks.

In a recent session devoted to the preparation of the "Draft Genebank Standards for plant genetic resources for food and agriculture" the Commission on Genetic Resources for Food and Agriculture (CGRFA) suggested maintaining a duplicate collection at another location for security reasons (http://www.fao.org/docrep/meeting/027/mf804e.pdf-p52) or, if this were impossible, by *in vitro* methods. *In vitro* conservation is used for maintenance of plant organs or plantlets in a medium-term period (some months up to some years) and generally, it is not the technique of choice for long-term conservation [11]. Cryogenic preservation, permitting cells or tissues to be stored for an indefinite period in liquid Nitrogen (-196°C) is considered to be an economic alternative to field collections. The PGRU calculated about $1 each sample/year, (http://www.ars.usda.gov/Aboutus/docs.htm?docid=6112; accessed 20/05/2013), though preservation requirements can be highly crop specific and specific techniques (and facilities) must be developed. In 1998, the FAO reckoned that only 1% of all maintained accessions were preserved by *in vitro* methods (http://www.fao.org/sd/EPdirect/EPre0041.htm).

An established collection requires managing but also phenotyping and, hopefully, molecular characterizing. Phenotypical and molecular characterization of collections is fundamental in supporting exploitation of the diversity contained within. Nowadays, faster molecular techniques have greatly improved the speed of the latter but phenotyping remains a slow and laborious task.

In 1964 Frankel and Brown [12, 13], foreseeing the problems posed by huge collections, proposed choosing a smaller collection ("core collection") to represent the complete collection. "Core" because the goal was representing as much as possible of the genetic diversity by means of a reduced number of individuals, eliminating as many redundancies as possible. The core collection can also be chosen to represent only a particular part of the diversity in the collection. In 1996, the Global Plan of Action for the Conservation and Sustainable Utilization of Plant Genetic Resources for Food and Agriculture [14] recommended establishing core collections as a means to improve the use of plant genetic resources. At the turn of the millennium, it was already generally recognized that the huge number of accessions collected in the effort to increase the diversity represented, hampered the ease of managing and using the collections themselves and unbearably increased maintenance costs. According to the International Plant Genetic Resources Institute (IPGRI), many collections were already facing major problems of size and organization [15]. The number of accessions to be chosen as "core" depends on many factors (the resources available for their maintenance, the dimensions of the initial collections, the genetic structure and diversity of the crop) but must be substantially smaller. In the early days of core collection establishment there were suggestions of limiting them to no more than 5-10% of the whole collection (representing at least 70% of the alleles) [16]. If the collection is very large, 10 % could still be impractical for careful evaluation. Mini-core collections, including 10% of the core collection (1% of the complete collection) have been also proposed [17] and nested core collections are often used because they allow greater flexibility, making it possible to choose both the accession to analyze and the size of the working collection [18]. Frankel's first proposal in the early days [12, 13], was followed by great debate about core collections and the methods employed to choose them. Van Hintum [15] presented a review: the first step is always the division of the collection into genetically meaningful groups and the decision concerning how many accessions to be taken from each group in order to obtain the core collection. Grouping (stratification) is crucial to represent effectively the whole, and should be done with the aim of maximizing variation between groups while keeping the variation within groups at a minimum. Many approaches can be chosen in both cases, depending on the available information and on the crop genetic structure. The collection curators could use a subjective approach directly using their experience and/or any available information (molecular if available, passports, geographic, genetic, phenotypic or agronomic) to divide the collection at hand into groups and decide the number, or statistical methods to exploit the information (multivariate analysis, dendrogram construction, comparison of marker diversity, *etc.*) could be used.

Molecular markers, mostly simple sequence repeats (SSRs, but single nucleotide polymorphisms - SNPs - are gaining popularity in the species in which they have been developed) are increasingly used to assess the genetic diversity of crops, and consequently, of collections. Nevertheless, there are not many examples of their use to create core collections of woody perennial fruit species. In 2008, Escribano [19] still reported only apple. A few new examples will be discussed later in this paragraph. Unlike phenotypic traits molecular markers are fast and not hampered by environmental effects. Once the collection diversity has been evaluated, the best allocation method to arrive at the core collection must be chosen. Several methods are frequently used, and the diversity of the core collections obtained are compared to choose the one which seems to offer the best results. A popular method is the Maximization strategy (M strategy), which focuses on the alleles. It individuates the combination of accessions that, while minimizing the number of accessions represented, maximizes the number of alleles and ensures the representation of all groups [15]. The M strategy performs particularly well in species that use self-fertilization as the primary means of reproduction or when the gene flow is low [20], which is often the case of clonal collections of woody perennials obtained sampling several geographically separated areas [19]. It is one of the most powerful methods to select the accessions with the most diverse alleles and to eliminate the redundancy coming from non-informative alleles [21]. It was used in herbaceous species like *Arabidopsis thaliana* [18], in perennials like the woody species *Annona cherimola* [19] and *Vitis vinifera* [22]. In *Arabidopsis thaliana* the first core collection, a nested core collection, was published in 2004 [18]. It maximized diversity, assessed by using SNPs, found in a worldwide set of 265 accessions. The M strategy was used to create 6 nested cores, increasing them by 8 accessions each (from 8 to 48). This is another example of the flexibility of nested cores: the core collection of 24 accessions captured the majority of molecular and morphological diversity, but the one of 48 accessions contains the rare alleles and was suggested as the best choice for SNP detection. In *Vitis vinifera*, the M-core collection was chosen in 2006 to represent a subset of 1,771 *V. vinifera* subsp. *sativa* cultivars, part of the cultivated compartment of the Vassal collection held in France. Fifty morphological traits were used to analyze variability [23]. The M-core collection comprises 141 individuals and, according to the authors, represents 86% of the diversity in the cultivated compartment of the Vassal collection [23]. Two years later Le Cunff [22] constructed four nested genetic core collections (G-cores: 12, 24,48 and 92 varieties each) representing respectively 58, 73, 83 and 100% of the total variability and increasing the unique genotypes of the cultivated compartment of the Vassal collection to 2,262. He used 20 SSR markers. Both G-cores and M-core were chosen by using the M strategy. Le Cunff [22] analyzed and compared the diversity retained in the G-cores and in the M-core by using several descriptors and concluded that even the G-12 core was more efficient in capturing diversity than the M-core. In both cases, the M-strategy was superior to random choice. The *Vitis* G-core collections are another example of the flexibility of nested cores: the G-12 core represents all the alleles with a frequency over 5%, the G-24 core over 3.5%. The G-48 core represents all the 271 alleles that show a frequency above 0.05% in the 2,262 genotypes, but if a researcher is interested in rare alleles, under 0.05%, they are all contained in the G-92. Escribano and coworkers [19] compared six methods to allocate a core collection in the woody perennial *Annona cherimola* exploring the diversity in the primary collection (279 accessions) by using SSRs. Two methods were based upon similarity dendrograms and four based on alternative methods, including the M strategy. They found that the best core collection was obtained using the M strategy. The chosen core included 40 accessions, even if all the alleles, including the rare ones, were already retained with 30. The 30-core had the same allele frequencies of the collection in at least 95% of the loci but was discarded because at one locus allele frequencies were significantly different. Consequently, the 40- final core is somewhat redundant. Escribano and co-workers [19] reported that the methods based on dendrograms failed probably due to the lack of a clear geographic pattern in the dendrogram groups.

Li and colleagues [24] analyzed the genetic diversity of a Chinese core collection of peach by SSRs. This core contained 56 cultivars (five were rootstocks) selected from 558 accessions listed in the National Peach Germplasm Resources Repository (Beijing, China) and represented 97% of the phenotypic traits in the original collection. National cores (France, Italy, Spain and China) are presently being prepared in the frame of the international EU "Fruit Breedomics" project. About 1300 peach cultivars, chosen to be representative of the genetic diversity present in germplasm collections in Europe and in China, were genotyped with the 9K peach SNP chip [25]. Four thousand three hundred SNPs, out of 8,144 in the chip, were used to characterize the collections and analyze the population structure [26].

Krichen and coworkers [27] obtained an apricot core collection exploring the diversity by morphological characters and molecular markers and using the M strategy. The Tunisian apricot germplasm (110 accessions) was represented by a core of 34 individuals. In China [10] 39 different core collections, each representing 1,501 apricots were obtained by developing 13 sampling schemes based on combining different grouping and sampling proportions and three total sample percentages of the whole collection. Grouping was carried out using the origin, or the geographical distribution in areas of China (Northeast, North, Northwest, Southwest), or the type of each accession (fresh, canning, ornamental, *etc.*); sampling proportions were calculated using square root, logarithm, fixed proportion or genetic diversity (assessed by phenotypical data). The size of the core subset was tried at 5, 10 and 15% of the whole collection. 10% was chosen as the best sampling percentage, so the final core contained 150 accessions. It was judged that the best method to obtain the apricot core was grouping by geographical distribution and sampling the groups logarithmically at random. The core chosen by using these parameters was analyzed by 22 SSRs.

2. MOLECULAR BREEDING IN STONE FRUITS

2.1. Molecular Markers

Genomic studies in stone fruits started in the early 1990s when several labs around the world started to develop the first available molecular markers, the Restriction Fragment Length Polymorphism (RFLP), extensively used in human genetics [28]. Initiatives started at Clemson University [29] and in several institutions across Europe where a *Prunus* network was initiated under the sponsorship of the European Union [30]. RFLP markers were developed in peach [31, 32], almond [33, 34] and cherry [35] and used to evaluate germplasm genetic diversity [36, 37] and to construct linkage maps [32, 33, 35, 36, 38, 39].

The advent of PCR technologies gave extra input to this field allowing the introduction of the first PCR based markers: Random Amplified Polymorphic DNA (RAPD) and Amplified Fragment Length Polymorphism (AFLP). These markers had many advantages (many loci are revealed in a single reaction, low cost per reaction, high level of polymorphism) but were rapidly abandoned due to several drawbacks such as dominant behavior, low reproducibility and transferability, which made them unsuitable for comparative analyses.

2.2. Simple Sequence Repeat (SSR)

In the meanwhile, a new class of molecular markers became available also for plants: the microsatellites or Simple Sequence Reapeats (SSRs) [40]. With regards to stone fruits, SSRs were first developed in peach at the end of the '90s [41-43] and in the following years several *Prunus* species followed. SSRs have many advantages compared to other available molecular markers: they are codominant, highly polymorphic, well spread across the genome, mainly single locus and multiallelic.

The first *Prunus* SSRs were developed and used in peach (*Prunus persica*) from genomic libraries [41-43] and cDNA clones [43]. SSRs were developed later in peach by using an enrichment approach [44-46], while a targeted approach using the peach Bacterial Artificial Chromosome library [47] allowed development of SSRs in specific genomic regions [48, 49]. The spread of expressed sequence tag (EST) sequencing allowed the individuation of SSRs in the peach transcriptome [50, 51]. The recent release of the peach genome sequence by The International Peach Genome Initiative [4] allowed the identification of the entire catalogue of SSRs spanning the genome. The complete set of SSRs > 19 bp can be viewed in a dedicated track of the peach genome browser at the following URL: http://services.appliedgenomics.org/fgb2/iga/prunus_public/gbrowse/prunus_public/.

A number of SSRs laying in unmapped and/or un-oriented whole genome shotgun (WGS) scaffolds of the Peach v1.0 assembly, were individuated and mapped in order to refine the assembly [4]. SSRs in peach have been widely used for germplasm characterization [52, 53] and map construction in several *Prunus* species. Microsatellites are widely conserved across the genus *Prunus* and can be easily transferred from one species to the other with a rate of about 80-100% [50, 54, 55]. The high rate of transferability makes the whole complete set of SSRs individuated in the peach genome sequence a valuable resource for the entire genus.

In apricot (*Prunus armeniaca*), SSRs were first developed from fruit cDNA libraries [56, 57] and from enriched genomic libraries [57-59]. A targeted approach was used to isolate SSRs in order to saturate genomic regions involved in the control of Plum Pox Virus (PPV) resistance [60]. Li and colleagues [61] developed a set of SSRs in the Chinese plum *Prunus mume* from cDNA libraries. These markers were extensively used for germplasm characterization [62] and linkage maps construction in apricot [63] and related *Prunus* species [51].

Microsatellites in plum were individuated in the diploid cultivated species *Prunus salicina.* To date only two reports of SSRs in this species have been published, one from enriched genomic library [54], and one from chloroplast DNA [64]. Nuclear markers were isolated from enriched genomic libraries and used to construct linkage maps in several *Prunus* projects [51, 65].

Several SSRs have been developed in cherries (*P. avium, P. cerasus* and *P. pseudocerasus*) from genomic libraries [66-68] and enriched genomic ones [66-70].

All the SSRs developed in *Prunus* spp are reported in Table **2**.

Table 2: SSRs available in *Prunus*

Marker	Species	Library	Reference
AMPA	*P. armeniaca*	Genomic and cDNA libraries	[57]
aprigms1	*P. armeniaca*	Targeted BAC clones	[60]
BPPCT	*P. persica*	Genomic library	[44]
CPDCT	*P. dulcis*	Enriched genomic library	[54]
CPSCT	*P. salicina*	Enriched genomic library	
CPPCT	*P. persica*	Enriched genomic library	[45]
EPDCU	*P. dulcis*	EST collection in GeneBank (from BU572502 to BU575295)	[51]
EPPCU (syn ECUPP)	*P. persica*	EST collection in GeneBank (from BU039022 to BU049005)	
EPPB	*P. persica*	EST collection in GeneBank	[71]
EPPISF	*P. persica*	cDNA library	[50]
EMPA	*P. avium*	Enriched genomic library	[69]
EMPaS	*P. avium*	Enriched genomic library	[70]
ES	*P. mume*	cDNA library	[61]
M	*P. persica*	cDNA library	[46]
MA	*P. persica*	Enriched genomic library	
MD	*P. persica*	EST collection	[72]
PAC	*P. armeniaca*	cDNA library	[56]
PaCITA	*P. armeniaca*	Genomic library	[58]
PceGA	*P. cerasus*	Genomic library	[66, 67]
Pchcms	*P. persica*	cDNA library	[43]
Pchgms	*P. persica*	Genomic library	[43, 47-49]

Table 2: contd...

Pgra	*P. grayana*	Genomic library	[73]
Pver	*P. verecunda*	DNA library	
PMS	*P. avium*	Genomic library	[66]
PS	*P. avium*	Enriched genomic library	
PTCR	*P. pseudocerasus*	Enriched genomic library	[74]
RPPG	*P. persica*	Peach v1.0 whole genome sequence	[4]
UCD-CH	*P. avium*	Genomic library	[68]
UDA	*P. dulcis*	Enriched genomic library	[75]
UDAp	*P. armeniaca*	Enriched genomic library	[59]
UDP	*P. persica*	Enriched genomic library	[41, 42]
TPScp	*P. salicina*	Chloroplast DNA	[64]

2.3. Single Nucleotide Polymorphism (SNP)

SNPs represent the most abundant class of molecular markers. They are widespread across eukaryotic genomes targeting both genic and intergenic regions providing more complete genome coverage than other marker types. The tremendous development of sequencing efforts in recent years has provided an important impulse to individuate SNPs in *Prunus* crops. The development of inexpensive high throughput technologies [76-79] for SNPs detection has resulted in the replacement of other DNA based marker systems as the genetic markers of choice [80].

In the last ten years, many EST libraries were constructed and sequenced in *Prunus*, providing a valuable resource for SNP identification [81-84]. SNPs can be developed readily from ESTs using bioinformatic tools due to their abundance in all regions of the genome. SNPs in candidate genes for flowering time were mapped in the *Prunus* reference map (T x E) [35, 85] while SNPs in candidate genes involved in fruit quality traits were mapped in peach maps [84, 86]. Ten *Knox* gene SNPs individuated in peach were also mapped in TxE [87, 88]. A set of 273 SNPs in candidate genes involved in metabolic pathways affecting fruit growth and maturity, texture, sugar and organic acid content, aroma and color, as well as putative allergen genes were bin mapped in T x E [89, 90]. Gene-anchored SNPs were also mapped in the almond (*Prunus dulcis*) linkage map constructed by using 'Nonpareil' x 'Lauranne' F_1 progeny [91]. A set of Rosaceae unigenes allowed the development of 1,039 Conserved Orthologous Set (RosCOS) [92]. For 857 RosCOS, primers flanking introns were designed and used to amplify the corresponding region from the T x E peach parent 'Earligold', the F_1 hybrid and six F_2 individuals selected for bin mapping. For 784 couples of primers (91%) an amplification product was obtained allowing the bin mapping of 607 RosCOS in T x E. From these, 126 were located in defined mapping bins on the *Fragaria* FV x FN map [93] and 545 showed a high level of DNA sequence conservation with apple, allowing their localization to precise positions on the *Malus* chromosomes. They were used to reveal colinearity, translocation and inversion events between *Prunus*, *Fragaria* and *Malus* genomes [94]. RosCOS markers, that are conserved gene-based molecular markers, appear extremely useful for the analysis of genome evolution among closely and distantly related species of the Rosaceae family [94]. However, as they are isolated from conserved regions, their polymorphism is lower than that obtained for SNPs identified in intergenic regions. The availability of the peach genome sequence by The International Peach Genome Initiative [4] coupled with the ability to acquire massive sequence datasets from next generation sequencers has rapidly changed the paradigm of variant identification in *Prunus* such as SNPs and small Deletion/Insertion Polymorphisms (DIPs). Several independent experiments using Next Generation Sequencing (NGS) of a great number of *Prunus* accessions allowed the identification of millions of SNPs in this important crop [4, 25, 95, 96]. Based on the large amount of SNPs identified in these species, two important genomic tools were developed in peach and cherry. The International Peach SNP Consortium (IPSC) released an Infinium array with 8,144 SNPs covering the whole peach genome: the IPSC 9 K peach SNP array v1 [25]. In cherry [96], the RosBREED 6K cherry SNP array v1 was released with 5,960 SNPs covering the sweet and sour cherry genomes. These tools are being used worldwide to foster cost effective marker-assisted selection strategies, whole genome fingerprinting, genome-wide association studies (GWAS), map-based gene cloning and population-based analyses.

2.4. Linkage Maps

Among the *Prunus* species and, in general, among fruit crops, the peach was traditionally held to be that with the higher level of genetic information. It has approximately 26 morphological markers which have been described [97]; however, co-segregation studies of major morphological genes were long hampered by the scarcity of information on the linkage between morphological markers, and the first maps were not available until after 1994 [98]. The first linkage mapping work in *Prunus persica* was carried out by Chaparro and colleagues [99] using, for the first time in *Prunus,* PCR based markers, namely Random Amplified Polymorphic DNAs (RAPDs). Though these markers allow to construct expeditiously genetic maps, they have several intrinsic disadvantages including a dominant behavior and a scarce reproducibility and have been progressively discarded for this application. Rajapakse and colleagues [36] in peach, and Viruel and colleagues [33] in almond, were the first to obtain *Prunus* maps by using Restriction Fragment Length Polymorphisms (RFLPs), the markers of choice for that time, in terms of reliability and reproducibility. Viruel and coworkers [33], in particular, detected eight linkage groups, providing the terminology and orientation of linkage groups in *Prunus*. The genetic variability in peach is quite low [100, 101] due to the restricted genetic base of the actual peach germplasm [53, 102]. This aspect has hampered the construction of genetic maps because a high proportion of the markers assayed in a particular progeny results monomorphic.

To solve this problem, the highly polymorphic F_2 progenies from interspecific almond × peach crosses were used to construct linkage maps. The higher saturation achieved enabled the detection of the expected eight linkage groups of the genus [35, 103, 104]. The map obtained from an F_2 progeny between the almond 'Texas' and the peach' Earligold' (T × E), mainly by using RFLPs, was adopted as the reference map for *Prunus*, following the terminology and orientation of linkage groups as in [33] and providing a large set of transferable markers that were used as anchors for constructing other maps of the genus [105]. The first T × E map was later saturated by adding 562 markers (361 RFLPs, 185 SSRs, 11 isozymes, and 5 sequence tagged site markers) [65, 106], for a total length of 519 cM and an average interval of 0.92 cM/marker achieving a high-density coverage of the *Prunus* genome. Several *Prunus* mapping populations have been used to construct over 40 linkage maps which have been extensively reviewed in [105, 107-109]. Most of these maps were constructed in peach and its hybrids with closely related species such as *Prunus davidiana*, allowing the identification of several important traits [32, 39, 71, 86, 110-117]. Other maps have also been established in almond [33, 91, 118, 119], apricot [63, 120-123], cherry [124-126] and in myrabolan [127]. Virtually, all contain a set of loci in common with the *Prunus* reference map thus allowing the recognition of each one of the eight linkage groups while providing good coverage as well as marker spacing of the *Prunus* genome. In 2005 Howad and colleagues [51] introduced a new approach of mapping in T x E using very few plants. They proposed selecting the most informative individuals from the T × E population for mapping purposes and selected six plants that, showing a large number of recombination events distributed along the genome, were sufficient to map almost every marker in the *Prunus* genome to a single "bin" of an average of 7.8 cM.

This approach had the two-fold advantage of reducing mapping time and costs whilst giving a high probability of finding polymorphisms due to the divergence between the almond and peach genomes. A bin mapping kit consisting of the six selected F_2 individuals and the population parents (the peach 'Earligold' and the F_1 hybrid) was also made accessible in the form of DNA preps to all interested parties.

Linkage maps of *Prunus* have generally been developed to study the genetic basis of agronomic traits for breeding purposes. The availability of anchor loci from the T x E map has also enabled comparison studies of major genes and or quantitative trait loci (QTLs) localized across different mapping populations. This network of maps from different populations, interconnected by common markers and their related major genes and QTLs, constitute a consensus map of the entire *Prunus* genus that is a major resource for geneticists and breeders working with peach and its related species [105].

The bin mapping method has been extensively applied to many relevant DNA sequences, such as those containing SSRs [51], candidate genes, particularly for fruit quality characters [86, 90], and to a set of Rosaceae conserved ortholog sequence (RosCOS) markers [92]. The T x E map was recently used, in an updated version that includes bin mapped markers (827 markers in total), to anchor, order and orientate the whole genome shotgun (WGS) scaffolds of the peach genome sequence [4]. To date, more than 2000 loci have been mapped or bin mapped in the reference map, with an average density of approximately one marker every 0.2 cM in the T x E genome. The recent availability of millions of SNPs and related arrays has speeded up the rapid construction of highly saturated linkage maps in peach [128-130] and cherry [131]. The future use of the SNPs arrays will greatly increase the number of loci mapped, providing a targeted tool for the saturation of specific regions of interest.

2.5. Mapped Agricultural Traits

Many important agronomic traits such as yield, quality and pest resistance have greatly benefited in the last century from the application of plant breeding programs; with the advent of the genomic era more and more successful utilization of this area of plant sciences is to be expected.

Genomic knowledge may help in obtaining superior cultivars when applied to breeding or genetic engineering programs. One area of genomics, DNA-marker technology, offers new possibilities for plant breeding. The genomic localization of important agronomic traits (both monogenic and polygenic) and their association with useful molecular markers can help during the selection process in shortening both the time and labor for obtaining superior genotypes by enabling the molecular screening of very large collections of plants and at a very early stage of their life. The use of DNA markers in plant breeding is called Marker Assisted Selection (MAS). Several agronomic traits have already been placed on the linkage maps developed in peach and in all the other *Prunus*, thus enabling the application of MAS for the species of this genus. In a few cases, important traits have also been fine mapped and their map-based cloning is being carried out. The recent availability of the genome sequence of the *Prunus* model species (peach), is expected to give a further boost to the application of genomics to plant breeding.

2.5.1. Mendelian Traits

Even prior to the genomic era, peach was the best genetically characterized species within the *Prunus* genus. A total of 42 morphological Mendelian traits had been described [132] and a few linkage relationships among them had also been determined [98]. The development of molecular marker maps has speeded up the localization of major genes involved in the control of important agronomical characters.

Twenty-eight major morphological, quality and agronomic peach traits showing simple Mendelian inheritance were localized on the *Prunus* consensus map [65, 133]. Three supplementary major genes have recently been mapped. One regards floral structure (Non-showy/showy, *Sh/sh*) and was localized on linkage group 8 (LG8) in a peach population [117]. Another one, affects fruit development and causes their premature abortion (Non-aborting fruit/aborting fruit, *Af/af*); it was first individuated and localized on LG6 [71] and was found to co-segregate with the flat fruit locus (*S/s*). The last gene was first identified by Pascal and colleagues [134]. It is responsible for resistance to powdery mildew (*Podosphaera pannosa* var. persicae; *Vr2/vr2*) in 'Pamirskij 5' and is closely associated to that for leaf color (red *vs.* green; *Gr/gr*). Both these two genes map in the vicinity of the breakpoint of a G6/G8 reciprocal translocation found in some red leaf cultivars [72, 104]. Following Pascal and colleagues [134], *Vr2* was tentatively placed in G6, where a major QTL for powdery mildew resistance was positioned in *P. davidiana*. Its position on G8 could not be excluded though, because *Vr2* had been tested in a 'Rubira' × 'Pamirskij 5' F_2, 'Rubira' being a red leaf peach. Almost all the major genes that have been mapped in peach are related to relevant traits in the peach industry. Some traits regard fruit characters (peach *vs.* nectarine; flat *vs.* round fruit; clingstone *vs.* freestone pit; melting *vs.* non-melting flesh; subacid *vs.* acid fruit flavor; yellow *vs.* white flesh), others are related to tree architecture (columnar *vs.* normal tree) and some others are related to disease resistance (root-knot nematode, powdery mildew). Due to the high saturation level of the *Prunus* syntenic maps, these traits can often be associated to molecular markers (generally microsatellites) that can be used for MAS. The recent peach genomic sequence is expected to promote the isolation of simple sequence repeats (SSRs) and single nucleotide polymorphisms (SNPs) as novel useful markers.

However, only in a few cases is marker assisted selection routinely used in breeding programs. An example is given by flesh softening and flesh adhesion (Melting/non-melting, *M/m* and Freestone/clingstone, *F/f,* respectively) traits. These two traits were initially thought to be under the control of two associated loci [135, 136]. Monet [137] and Peace and colleagues [138], starting from different approaches, came to another conclusion. In these authors' hypotheses, the two traits are controlled by a unique locus (*F*) with three different alleles (*F/f/f1*), the *F* allele being dominant and the *f1* recessive to the others. Peace and colleagues [138] proposed the gene coding for the endopolygalacturonase (EndoPG) enzyme, responsible for the degradation of the cell wall, as candidate for this trait, and identified microsatellite markers in its sequence. A PCR-based diagnostic test was also developed for the traits by designing four sets of primers on the sequence. Both the traits and the endo-PG markers were found to be co-localized in the distal end of LG4 [71, 86, 113, 139]. Further investigations suggested that the M and F traits are controlled by two concatenated copies of the endopolygalacturonase gene [140, 141]. Consistent with this evidence, two endoPG sequences were manually annotated in the Peach v1.0 assembly, in the distal region of pseudomolecule 4, between nt 22,649,519 and nt 22,687,159 [4], providing further evidence of the involvement of the *EndoPG* genes in the control of both traits.

Another important quality trait under the control of a single gene is fruit acidity. The low acidity of the peach fruit juice is determined by the dominant allele (*D*). The trait was mapped in the proximal end of LG5 [32, 71]. Recently, a high resolution map was developed in the vicinity of the *D* locus [142]. Sequence characterized region (SCAR) and Cleaved Amplified Polymorphisms (CAPS) markers were identified as closely associated to the trait and the interval locus was reduced to 0.4 cM. The contribution of the peach genomic sequence will give a definite aid in the completion of the map-based cloning of the *D* locus.

Most of the flat peach cultivars meet very positive feedback from consumers due to their taste and flavor. The flat fruit trait is controlled by a dominant gene (*S*). The locus was mapped on LG6 [32, 71] and co-segregated with several markers. One of them (MA014a) was proposed as a diagnostic marker for marker assisted selection.

The peach Mendelian traits are listed in Table **3**.

Some other traits have been described in other *Prunus* species. In almond, the following Mendelian traits have been individuated: anther color, self-incompatibility, kernel bitterness, shell hardness, leaf color [143] and blooming time [144]. In apricot, several authors reported the following traits: male sterility [145], self-compatibility [121, 145] and plum pox virus (PPV) resistance [120, 121, 146-149]; in cherry the self-compatibility locus, syntenic to other *Prunus* species, has been described [125, 126, 150].

More recently, candidate genes have been individuated for three additional Mendelian traits, playing an important role in the peach industry. The *carotenoid cleavage dioxygenase 4* (*PpCCD4*) was recently recognized as the gene responsible for the flesh color (white/yellow) in peach [151, 152]. A *Myb* transcription factor (*PpeMyb25*) was proposed as responsible for the nectarine trait and a diagnostic marker (indelG) was designed for the early screening of the trait for marker assisted breeding (MAB) [153]. Finally, Dardick and co-workers [154] individuated a candidate gene, *tiller angle control 1* (*TAC 1*), involved in the vertically oriented growth of branches, referred to as 'pillar' or 'broomy'.

2.5.2. Quantitative Traits

Fruit quality, yield, pest resistance and many other important agronomic traits, display a continuous phenotypic variation. They are determined by the segregation of multiple loci and are usually referred to as quantitative and their inheritance as polygenic. The individual loci controlling a quantitative trait are referred to as "Quantitative Trait Loci" (QTLs). Prior to the genomic era, these traits were analyzed in a statistical descriptive manner in order to estimate the approximate number of loci, the average gene action and the interaction among polygenes and with the environment [155]. The availability of genomic tools (*i.e.* molecular markers and linkage maps) has caused a tremendous impact on quantitative genetics. In peach, it allowed for the elucidation of the genetic architecture of some complex traits: the genomic regions involved were identified and their genetic effects were quantified.

Twenty-eight quantitative traits were localized on the consensus map of the *Prunus* genus [127, 156]. These QTLs, controlling important agronomic traits such as pest and disease resistance, fruit quality, tree architecture and phenological characters, were analyzed and mapped in different peach progenies and were localized on a consensus map obtained through the interconnected network of *Prunus* maps sharing common sets of anchor loci [35, 157-161].

Several other quantitative traits have recently been analyzed in intra and inter-specific crosses of peach and the genomic regions responsible of their control have been individuated.

A few works have dealt with resistance to Plum Pox Virus (PPV) carried by an accession of the peach related species, *P. davidiana*. Decroocq and colleagues [112] using an F_1 progeny (*P. persica* cv Summergrand x *P. davidiana*) reported six genomic regions involved in resistance to the Sharka potyvirus PPV, localized on LG1 (two regions), LG2, LG4, LG6 and LG7. Marandel and colleagues [115], using an F_2 progeny obtained from the same genetic background, detected six QTLs localized on LG1 (2), LG4, LG5, LG6 and LG7. The alignment of the two *P. davidiana* maps (F_1 and F_2) highlighted a good conservation of the QTLs over successive generations: of the six total QTLs identified in the F_1 map, three were found to be co-localized with the F_2 ones. In both studies QTLs were found to co-localize with translation initiation factors, a class of plant factors implicated in the process of resistance/susceptibility to the potyvirus. More recently, Martínez-Gómez and colleagues [116] using an F_1 offspring obtained from a cross with the same PPV resistance source of *P. davidiana* with the peach cv Rubira, recognized nine genomic regions responsible for the differential expression of symptoms to PPV infection, six of which had already been found in the aforementioned works. However, they observed a very lower level of resistance in this background compared to the other two progenies, thus indicating the presence of strong genotypic interactions.

A number of quality traits were localized in an advanced backcross population of *P. persica* obtained from a cross with the same *P. davidiana* male parent [111]. The authors individuated a number of regions spread throughout the genome involved in the following quantitative traits: sugar and organic acid concentration, fruit and stone size, skin and flesh color. They also found several alleles, coming from the male parent, positively contributing to the fruit quality, in clear contrast with the phenotype displayed by those of the *P. davidiana* accession.

In another work, peach genomic regions involved in chilling and heat requirement of the floral buds and in the blooming date traits, were studied using an F_2 progeny [117]. Several genetic factors controlling chilling requirement in peach were identified throughout the genome with the exception of LG2 and LG3. This is the first genetic information on this trait that has played a major role in the adaptability of the species to warm environments and could result of greater importance in the future, given the current global warming trend. Two QTLs, controlling bud heat requirement, were also mapped on LG1 and LG8 while QTLs affecting blooming time were identified in all linkage groups except LG3 and LG8. One genomic region of 2 cM, pleiotropic for the three traits, overlapped with the region of the evergrowing (EVG) trait sequenced in peach [162].

Recently, different peach progenies were used to study a series of disorders occurring in the peach fruit following prolonged storage in cold environments, collectively termed as chilling injury (CI) [139, 163, 164]. Significant quantitative trait loci (QTLs) were identified in LG4 for some of the major symptoms of CI, namely mealiness, graininess and bleeding. In this same linkage group, other QTLs involved in agronomic and fruit quality traits were also mapped [139].

Recently, the availability of SNP genomic tools allowed the fine mapping of QTLs for important traits. Yang and collaborators [129] individuated 14 QTLs controlling resistance to bacterial spot in peach using an F_2 segregant progeny while Eduardo and colleagues [130] identified about 30 regions involved in the control of volatile organic compounds (VOCs) in peach fruit in an F_1 peach progeny. Frett and colleagues [165] individuated four genomic regions involved in the determination of red in the skin coloration of peach fruits and proposed a candidate gene *(PprMyb10)* for the major QTL in LG3. In another recent work, a map-based approach coupled with NGS resequencing analysis, allowed the identification of a candidate gene, the *NAC transcription factor*, for a major QTL for maturity date (MD) on chromosome 4 [166].

In the other *Prunus* species, few other complex traits have been reported and mapped. In cherries (both sweet and sour), QTLs have been reported for skin and flesh color, fruit weight and size, blooming and ripening time [167-169]. In apricot, QTLs have been mapped for kernel bitterness [170], chilling requirement

[123] and PPV resistance [120, 148, 171]. In almond Sanchez-Perez and colleagues [172] mapped QTLs for kernel weight, blooming and ripening time.

Although a large number of QTLs have been identified for many important agronomic traits, no marker assisted breeding has been reported to date for any of them. The gap between QTLs identification and their practical application is probably due to the complex architecture underlying quantitative traits which requires greater knowledge of factors such as genetic loci and environment interactions, before marker assisted selection could be routinely applied in breeding programs. All the traits mapped in peach (major genes and QTLs) are reported in Table **3**.

Table 3: Description of major genes and QTLs reported in peach for morphological or agronomic traits, their localization in LGs, mapping populations used and reference

Characters	Symbol[b]	LG[a]	Populations	References
Architectural and Morphological Traits				
Evergrowing	*Evg*	G1	'Empress op op dwarf' x PI442380	[173]
Internode length	QTL	G1	(*P. ferganensis* x IF7310828) BC_1	[160]
Flower color	*B*	G1	'Garfi' x 'Nemared'	[174]
Broomy (or pillar) growth habit	*Br*	G2	Various progenies	[154, 175]
Double flower	*Dl*	G2	NC174RL x PI	[99]
Anther color (yellow/anthocyanic)	*Ag*	G3	'Texas' x 'Earligold'	[176]
Plant height	QTL	G4	'Venus' x 'BigTop'	[139]
Polycarpel	*Pcp*	G3	'Padre' x 54P455	[103]
Flower color	*Fc*	G3	'Akame' x 'Jusetou'	[177]
Plant height (normal/dwarf)	*Dw*	G6	'Akame' x 'Jusetou'	[177]
Leaf shape (narrow/wide)	*Nl*	G6	'Akame' x 'Jusetou'	[177]
Male sterility	*Ps*	G6	'Ferjalou Jalousia®' x 'Fantasia'	[32, 71]
Leaf color (red/yellow)	*Gr*	G6-G8	'Garfi' x 'Nemared'; 'Akame' x 'Jusetou'	[174, 177]
Leaf gland (reniform /globose/eglandular)	*E*	G7	(*P. ferganensis* x IF310828) BC_1	[39]
Flower morphology	*Sh*	G8	'Contender' x Fla.92-2C; Pop-DG	[117]
Phenological traits				
Chilling and Heat requirement, blooming date	QTLs	G1	'Contender' x Fla.92-2C	[117]

Table 3: contd...

Ripening time	QTLs	G2	(*P. ferganensis* x IF310828) BC_1	[160, 178]
Blooming date	QTL	G2	'Contender' x Fla.92-2C; 'Summergrand' x P1908	[111, 117]
Blooming time, ripening time, fruit development period	QTLs	G4	'Ferjalou Jalousia®' x 'Fantasia'; (*P. ferganensis* x IF7310828) BC_1; 'Venus' x 'BigTop'; 'Summergrand' x P1908; 'Contender' x 'Ambra'; PI91459 x 'Bounty'	[111, 139, 159, 160, 166, 178]
Chilling requirement, blooming date	QTL	G4	'Contender' x Fla.92-2C	[117]
Chilling requirement, blooming date	QTLs	G5	'Contender' x Fla.92-2C	[117]
Ripening time	QTLs	G6	(*P. ferganensis* x IF310828) BC_1	[160, 178]
Chilling requirement, blooming date	QTLs	G6	'Contender' x Fla.92-2C	[117]
Chilling requirement, blooming date	QTLs	G7	'Contender' x Fla.92-2C	[117]
Chilling requirement, heat requirement	QTLs	G8	'Contender' x Fla.92-2C	[117]
Fruit Quality Traits				
Flesh color (white/yellow)	*Y*	G1	'Padre' x 54P455; Pop-DG; 'Royal Prince' x 'Yoshihime'	[103, 128, 151, 152, 179]
Fruit skin color, soluble-solids content	QTLs	G2	(*P. ferganensis* x IF7310828) BC_1	[160, 178]
Fruit weight, fruit diameter, glucose content	QTLs	G3	'Suncrest' x 'Bailey'	[157]
Flesh color around the stone	*Cs*	G3	'Akame' x 'Jusetou'	[177]
Flesh adhesion (clingstone/freestone)	*F*	G4	(*P. ferganensis* x IF7310828) BC_1; 'Akame' x 'Jusetou'	[39, 160, 177]
Soluble-solids content, fructose, glucose	QTLs	G4	'Ferjalou Jalousia®' x 'Fantasia'; 'Venus' x 'BigTop'; 'Summergrand' x P1908	[111, 139, 159]
Chilling injury traits	QTL	G4	'Venus' x 'BigTop'	[139]
Fruit size	QTLs	G4	'Venus' x 'BigTop'; 'Summergrand' x P1908; 'Contender' x 'Ambra'	[90, 111, 139]
pH, titratable acidity	QTLs	G4	'Venus' x 'BigTop'	[139]

Table 3: contd…

Non-acid fruit	*D*	G5	'Ferjalou Jalousia®' x 'Fantasia'	[32, 158, 159]
Sucrose, malate, titratable acidity, pH	QTLs	G5	'Ferjalou Jalousia®' x 'Fantasia'; 'Summergrand' x P1908	[111, 159]
Skin hairiness (nectarine/peach)	*G*	G5	'Ferjalou Jalousia®' x 'Fantasia'; 'Padre' x 54P455	[32, 103, 158]
Fruit size	QTLs	G5	'Summergrand' x P1908	[111]
Kernel taste (bitter/sweet)	*Sk*	G5	'Padre' x 54P455	[103]
Fruit skin color, soluble-solids content	QTLs	G6	(*P. ferganensis* x IF310828) BC_1	[160]
Fruit shape (flat /round)	*S*	G6	'Ferjalou Jalousia®' x 'Fantasia'	[32, 71, 158]
Aborting fruit	*Af*	G6	'Ferjalou Jalousia®' x 'Fantasia'	[71]
Fruit skin color	*Sc*	G6-G8	'Akame' x 'Jusetou'	[177]
Quinase	QTL	G8	'Ferjalou Jalousia®' x 'Fantasia'	[159]
Volatile compounds	QTL	G2 G3 G4 G5 G6 G7	'Bolero' x 'Oro A'	[130]
Blush	QTL	G3 G4 G7	'Zin Dai Jiu Bao' x 'Crimson Lady' F_2	[165]
Resistance to biotic stresses				
Powdery mildew resistance	QTL	G1	'Summergrand' x P1908	[161]
PPV resistance	QTLs	G1	'Summergrand' x P1908; 'Summergrand' x P1908 F_2; 'Rubira' x P1908	[112, 115, 180]
Root-knot nematode resistance	*Mi*[(c)]	G2	P.2175 x GN22; 'Akame' x 'Jusetou', 'Lovell' x 'Nemared';'Garfi' x 'Nemared'; 'Padre' x 54P455	[103, 174, 177, 181, 182]
PPV resistance	QTLs	G2	'Summergrand' x P1908 ; 'Rubira' x P1908	[112, 180]
Leaf curl resistance	QTL	G3	'Summergrand' x P1908	[183]
PPV resistance	QTLs	G4	'Summergrand' x P1908; 'Summergrand' x P1908 F_2; 'Rubira' x P1908	[112, 115, 180]
PPV resistance	QTLs	G5	'Summergrand' x P1908 F_2; 'Rubira' x P1908	[115, 180]

Table 3: contd...

Powdery mildew resistance	*Vr2*	G6	'Rubira' x 'Pamirskij 5'	[134]
Powdery mildew resistance	QTL	G6	'Summergrand' x P1908	[161]
Leaf curl resistance	QTL	G6	'Summergrand' x P1908	[183]
PPV resistance	QTLs	G6	'Summergrand' x P1908; 'Summergrand' x P1908 F_2; 'Rubira' x P1908	[112, 115, 180]
Powdery mildew resistance	QTL	G7	(*P. ferganensis* x IF7310828) BC_1	[160, 178]
PPV resistance	QTLs	G7	'Summergrand' x P1908; 'Summergrand' x P1908 F_2; 'Rubira' x P1908	[112, 115, 180]
Powdery mildew resistance	QTL	G8	'Summergrand' x P1908 (*P. ferganensis* x IF7310828) BC_1	[160, 161]
Bacterial spot resistance	QTL	All LGs	'O'Henry' x 'Clayton'	[129]

Adapted and integrated from Arús and colleagues [105] and Verde and colleagues [4].
[a]LG=Linkage group; G6-G8 genes located close to the translocation break point between the two linkage groups.
[b]QTLs are included if they have been consistently found (at least in two independent measurements) in the indicated populations.
[c]One or two genes of nematode resistance with different notations and one QTL have been described in linkage group.

3. 'OMICS' BASED APPROACHES

In recent years, technical progresses made in our capability of acquiring large amounts of biological data are providing researchers with an unprecedented opportunity to analyze crop species from a holistic perspective. High-throughput sequencing technologies, methods for quantitative gene expression analyses and metabolic platforms are emerging as novel tools for picturing the genome, transcriptome, proteome, and metabolome of an organism, and to gain an insight of the mechanisms underlying plant life cycle.

Genomics analyzes the structure and function of the complete set of genes and DNA regulatory sequences of an organism. Transcriptomics studies the expression and structure of the complete set of RNAs encoded by the genome, whereas proteomics focuses on the structure and function of proteins transcribed from mRNAs. Finally, metabolomics involves the characterization and quantitation of small organic molecules found in an organism and, being the metabolome closely linked to the genome, it also offers the opportunity to investigate genome-phenotype relationships.

More recently, a systemic approach has been applied to the collection of detailed phenotypic data through the development and adoption of high-throughput and high-dimensional platforms for phenotypic characterization. 'Phenomics', which studies the full set of phenotypes of an individual (phenome), is being recognized as an independent discipline [184]. Copying with such a body of data requires the implementation of more powerful analysis tools. Novel disciplines such as bioinformatics are emerging, while new statistical methods are being developed. The integration of this body of information is the latest challenge for the so-called Systems Biology, a novel interdisciplinary approach trying to capture the utility of different -omics technologies, to analyze the complex and non-linear relations between different biological systems controlling plant growth and development.

In the *Prunus* genus, the "-omics revolution" can be traced back to the 90's with the early genetic works leading to the development of the first genetic linkage maps. These were mainly restricted to peach and almond, which, in contrast to other stone fruits such as cultivated plums and sour cherries, are diploid species with a low number of chromosomes (x = n = 8) and a small genome (nuclear DNA content assessed at ~230 Mbp) [4], less than twice that of *Arabidopsis thaliana* [185]. Since then, the *Prunus* scientific community has been very active in all fields of plant sciences and many -omics based platforms and tools have been developed. The goal of all these efforts has been to identify the key genetic components and the biological mechanisms responsible for important agronomic traits, and to provide breeders with powerful molecular instruments able to address environmental, quality, and healthy issues coming from both producers and consumers. In the following section, the latest achievements in transcriptomics and genomics of the *Prunus* species will be discussed together with their main relevant agricultural outcomes.

3.1. Transcriptomics

The transcriptome is the complete collection of all species of RNA molecules present in one or in a population of cells. It includes not only the coding fraction of messenger RNAs (mRNAs), but also the noncoding one represented by ribosomal RNAs (rRNAs), transfer RNAs (tRNAs) and the broad family of post-transcriptional small RNA regulatory molecules, such as small inhibitory RNAs (siRNAs) or interference RNAs, and microRNAs (miRNAs).

The analysis of protein-coding transcripts allows the simultaneous discovery of from a few up to thousands of genes and can be used both characterizing mRNA species expressed in a cell or a tissue in response to an environmental *stimulus*, and to study transcriptionally regulated changes in gene expression. Traditionally, different approaches have been developed for accessing the expressed gene catalogue and for functional genomics studies of a species. These can be divided into those not requiring any previous sequence knowledge, such as in the case of the cDNA Amplified Fragment Length Polymorphism (cDNA-AFLP) and Suppression Subtractive Hybridization (SSH) techniques, and those based on the availability of prior information, such as in microarray and candidate gene analyses, where a set of relevant transcripts must be initially identified in the same or in a species other than the one studied [186]. A major contribution to transcriptome exploration in both animal and plant species was historically given by the generation of Expressed Sequence Tags (ESTs) from cDNA libraries. Recent advancements in sequencing technology, together with the development of high-throughput and low cost next generation sequencing platforms, have opened the way to the massive paralleled sequencing of the transcriptome, allowing the sampling of almost all RNA molecules present in a cell or tissue, an approach known as RNA-seq.

3.1.1. Expressed Sequence Tags in the Prunus Genus

Expressed sequence tags (ESTs) are short (100-800 nucleotide bases) sequence reads, obtained by Sanger single-pass sequencing of either one of the 5' or 3' end of a given transcript. The procedure implies the reverse trascription of isolated mRNAs for the creation of a double stranded cDNA library, and the random selection of the clones to be sequenced with Sanger chemistry. Though characterized by sequencing error rates as high as 5% and requiring multi-step bioinformatic pre-processing to yield transcriptome information, ESTs are a key tool for a number of purposes including direct gene discovery, gene structure identification, alternative splicing, SNP characterization and proteome analysis [187].

Since the early 2000s [188], thousands of ESTs have been produced in *Prunus* spp from different tissues and conditions, mainly from peach but also almond and apricot, addressing various aspects such as fruit quality, tree development [82, 189], and postharvest physiological disorders [83].

With the aim of identifying genes related to fruit quality (mainly aroma-related volatile compounds), whose expression is correlated with critical developmental steps or induced by environmental factors, a comparative EST strategy was used [190]. A total of 10,847 ESTs from five libraries, obtained from two different fruit tissues (skin and mesocarp) at four ripening stages (from post-allegation to post-climacteric) and in three different peach genotypes ('OroA', 'Bolero' and 'Suncrest'), were analyzed together with 21,857 peach ESTs available in public databases. The clustering and assembly of both datasets gave 17,858 unigenes and a putative function was assigned to 70.8% of the ESTs. A gene onthology analysis revealed differences between organs and different maturation stages of the same organs in organelles. Moreover, a frequency comparison of the genes putatively involved in the metabolism of some volatiles highlighted the predominant presence of those transcripts in the skin. The metabolic pathways of esters and lactones were selected for further isolation and cloning of candidate key genes.

Chilling injury is a physiological disorder affecting peaches and nectarines stored in cold temperatures to extend their market life, and is a major factor in lower consumption of these fresh fruits compared to others such as apple [191]. A comparative EST transcript profiling approach was performed to study individual and combined biological effects that cold storage and ripening have on both the fruit and fruit quality [192]. A total of 50,625 ESTs from the peach mesocarp of the cv O'Henry, stored at four different postharvest conditions, were sequenced and subsequent to assembling 10,830 unigenes were obtained. The statistical analysis of the EST collection has shown complex expression dynamics, with genes that are differentially expressed during ripening, others in response to cold storage and genes that respond to the combined effects of cold storage and ripening. Pair-wise comparisons revealed at least one significant difference in transcript abundance, between at least two conditions, in 197 contigs, which might be classified into 13 different clusters of gene expression patterns, including groups that increase or decrease transcript abundance during ripening, in response to cold or ripening and cold. These results offer novel candidates to consider in the study of post-harvest diseases.

With the aim of identifying peach candidate genes (CGs) involved in fruit development and quality that can be transferred to other Rosaceae species, a set of EST sequences was developed from two peach cDNA libraries (cv Fantasia) constructed at two stages of development, the first rapid growth stage and the growth plateau [84, 193]. A total of 1,730 peach unigenes was obtained after clustering and 59 CGs involved in sweetness, acidity, and phenolic compound content were selected according to their annotation. Fifty-five primer pairs designed and validated *via* PCR in peach were also tested in strawberry, 36 of which gave amplification products. A high level of marker transportability was observed indicating that designing primers in exons is an efficient strategy, providing a source of CGs that could be of use within the Rosaceae species. Eight CGs were mapped in peach, 14 in strawberry, and four in both species, confirming the pattern of synteny already proposed using comparative mapping approaches [194]. In peach, the CGs are located in three linkage groups (3, 5, 7), and in strawberry they are distributed in all the seven *Fragaria* linkage groups. Interestingly, co-localizations between a few of these CGs and quantitative trait loci for fruit quality traits have also been identified, and deserve further studies to ascertain a possible role of these CGs in the corresponding QTL traits.

To improve knowledge of fruit ripening and also to provide genomic resources for molecular breeding in apricot (*Prunus armeniaca*), 13,006 ESTs were generated from three cDNA libraries of the pericarp at three different critical transition stages during fruit development: green stage (corresponding to the end of stone hardening), half-ripe stage (onset of ethylene production), and ripe stage (maximum in sucrose accumulation) [81]. After clustering, 5,219 (corresponding to 40%) unigenes were obtained. The low interlibrary redundancy indicated that the apricot pericarp EST dataset was far from being saturated. Seventy-six percent of the unigenes displaying homologies with public sequences were clustered into functional categories. The largest proportion of functionally assigned ESTs were related to primary metabolism, stress response, and protein synthesis. Estimates of mRNA abundance from electronic Northern analysis pointed out that stress-related proteins and cell wall modification-related enzymes strongly increased during ripening. Significant homologies in their coding regions (nucleic acid-level isogenes) were found in 186 (42%) out of 448 iso-proteins (amino acid-level isogenes) detected in the unigene set.

In almond (*Prunus dulcis* Mill. L. var. *dulcis*) ESTs were used to obtain a first picture of pistil development. From a cDNA library of almond pistils, over 1,000 clones were selected and sequenced. The assembly process yielded 121 contigs and 595 singletons. Eight of the unique cDNA clones had been previously isolated in *P. dulcis*. A comparison with the NCBI non-redundant protein database showed that ~45.4% of the transcripts had significant similarity to protein coding sequences in the database while 549 of them did not match known protein sequences. Among these latter ones, 56 showed significant similarity to nucleotide sequences in the database. ESTs functional classification allocated the majority of them in the "stress resistance and defense" and "protein synthesis and processing" classes.

Many EST collections have also been used to identify molecular markers, namely Simple Sequence Repeats (SSRs) and Single Nucleotide Polymorphisms (SNPs), in different *Prunus* species, which have been used in genetic mapping studies [50, 56, 61, 195, 196].

The Genomic Database for Rosaceae site (GDR, http://www.rosaceae.org/), is home to a web based repository of all *Prunus* EST sequences [197], which is now in its 5th version. The *Prunus* Unigene v5.0 sequences are of high quality and have been filtered by cross matching public sequences against the NCBI UniVec database and removing species-specific chloroplast, mitochondrial, tRNA, and rRNA sequences using the Basic Local Alignment Sequence Tool (BLAST). Redundancy was also reduced and longer transcripts were generated by the CAP3 program. The final assembly was annotated by the BLAST sequence similarity algorithm searching against the main sequences and proteins databases. To date the repository contains 110,206 EST sequences as shown in Table **4** along with the main features of the *Prunus* unigene v5.0.

Table 4: The GDR *Prunus* EST collection (adapted from http://www.rosaceae.org/)

***Prunus* Unigene v5.0**	
Number of ESTs available	110,806
Number of ESTs available after filtering	106,148
Average Length	589
Number of Contigs (CAP3 Assembly, -p 90)	10,934
Average Length of Contigs	972
Number of Singlets	22,957
Number of Putative Unigenes	33,891

An Italian repository of EST data from *Prunus* species is the ESTree database [198, 199] available at the ESTree Interuniversitary Centre, http://www.itb.cnr.it/. It contains a collection of *Prunus persica* (twelve libraries) and *Prunus amygdalus* (three libraries) EST sequences obtained from different labs all over the world. To date, in its VI version, 75,404 sequences have been deposited. Functional annotation of the sequences based on the Gene Ontology (GO) classification (Table **5**) is also available.

Table 5: ESTree db functional annotation statistics (adapted from http://www.itb.cnr.it/estree)

GO Function	Number and (percentage) over the total of 75,404 sequences
Sequences without a blast annotation	22,762 (30.19%)
GO:0003674 => molecular_function	39,312 (52.14%)
GO:0008150 => biological_process	31,656 (41.98%)
GO:0005575 => cellular_component	21,251(28.18%)

3.1.2. Microarray Analysis

Microarrays are high-throughput platforms used to measure gene expression on a whole genome level and to generate functional data for many genes simultaneously. Microarrays are to be considered a targeted approach for expression profiling since they only measure expression levels of those genes for which a probe, either a clone or a sequence, is available [200]. In *Prunus* species, microarray technology has been utilized to perform expression-profiling studies mainly related to fruit quality traits and, to a lesser extent, flower compatibility, bud dormancy and *in vitro* regeneration [188]. The first microarray developed for a large-scale transcriptome analysis was the μPEACH1.0 [201]. It consisted of 70-mer oligo-probes for a set of 4,806 unigenes obtained after assembling an initial dataset of 11,201 sequences: 2,394 deriving from cDNA libraries of the mesocarp of the peach cultivars 'Fantasia', 'Redhaven', 'OroB' and 'Bolero' at S3 and S4 ripening stages, and 8,807 retrieved from the NCBI GenBank and GDR. The μPEACH1.0 was used to compare S3 and S4 fruits and showed a dramatic change in the mesocarp transcriptome, with hundreds of genes, many belonging to the cell wall and chloroplast compartments, up- and down-regulated [202]. Subsequently, several other studies in peach used the same tool for addressing different aspects related to fruit ripening [203-205]. The same μPEACH1.0 tool was also applied to expression profiling studies in apricot fruit development [206]. Three different developmental stages of 'Goldrich' apricot fruit, corresponding to immature green (6 weeks before fully ripe stage), mature firm ripe (change of peel colour, 1 week before fully ripe stage) and fully ripe were assayed. Apricot target cDNAs showed significant hybridization with an average of 43% of spotted probes, validating the use of μPEACH1.0 to profile the transcriptome of apricot fruit and confirming the high degree of sequence conservation among *Prunus* species [65, 207]. Moreover, results comparison of microarray analysis conducted separately on peach and apricot fruit development, showed that 70% of genes had the same expression pattern in both species. Among those showing a species specific transcript profile, a considerable number appeared to diverge during the last stages of on-tree fruit development suggesting the presence of diverse mechanism regulating ripening in these two phylogenetically closely related species.

The same approach was also used to elucidate mechanisms regulating plum (*Prunus salicina*) fruit ripening in genotypes differing for ethylene production during ripening, namely an ethylene suppressed ('Shiro') and an ethylene producing ('Santa Rosa') cultivar [208]. Both cultivars harvested at commercial maturity stage were allowed to ripen at room temperature for a further 4 days. Surprisingly, no differences were detected between the two cultivars in the transcript levels of genes involved in auxin metabolism, antioxidant system and stress response. The same was observed for transcripts related to ethylene biosynthesis, primarily 1-aminocyclopropane-1-carboxylate synthase (ACS), which appeared to increase, suggesting induction of the ethylene biosynthetic pathway also in plum cultivars in which the burst of ethylene is undetectable.

A specialized EST collection database (ChillPeach) [83], was produced to target genes expressed during chilling injury (CI) and was used to develop a microarray platform. A set of 7,862 high-quality ESTs (comprising 4,468 unigenes) was obtained, 48% of which had never been reported for *Prunus* in the available databases including NCBI, TAIR, PlantTA, ESTree, and GDR. A microarray containing 4,261 probes derived from the ChillPeach unigenes was developed and used in a pilot experiment to identify differentially expressed genes in cold-treated mesocarp tissues compared to control mesocarp tissues, and in vegetative compared to mesocarp tissues. The analysis highlighted 287 up-regulated and 112 down-regulated genes in cold stored fruits. A website containing all the information on the ChillPeach database was also created (http://bioinfo.ibmcp.upv.es/genomics/ChillPeachDB).

A combination of suppression subtractive hybridization (SSH) and microarray was used to study the molecular and physiological mechanisms underlying maintenance and release of seasonal bud dormancy in peach [209]. Four SSH libraries were constructed and enriched in cDNAs highly expressed in dormant buds, in dormancy-released buds and in cultivars with different chilling requirement, sampled after dormancy release. A microarray developed with approximately 2,500 clones picked from the four libraries was used to validate the clones selected with the SSH method at the end of the procedure. Four hundred positive clones were sequenced, and found to correspond to 101 different unigenes with diverse functional annotations. Transcription factors belonging to the MADS-box class, known to be related to growth cessation and terminal bud formation in the ever-growing mutant of peach, were isolated together with others never correlated with bud dormancy before.

The μPEACH2.0 microarray platform has recently been designed to study gene expression differences that might be associated with the aroma profiles of two peach varieties ('Bolero' and 'OroA') differing in their composition of Volatile Organic Compounds (VOCs) [210]. It is based on the 12K Combimatrix platform and contains 4,776 oligonucleotides probes, 1,951 of which are in common with μPEACH1.0. Transcriptomic profiling at three developmental stages has so far allowed the identification of 12 genes putatively involved in VOC biosynthesis.

3.1.3. RNA-seq

Advances made in sequencing platforms with the introduction of next generation technologies, have dramatically reduced the costs of sequencing projects and increased the speed and amount of data that can be obtained. High-throughput sequencing approaches applied to the transcriptome are the latest frontier for gene functional analysis. Deep sequencing of the transcriptome, the so called RNA-seq approach, enables the sampling of almost all RNA species and has some major advantages over other high-throughput approaches, including microarray analysis: unlike hybridization based approaches, no prior knowledge of the genome sequence is required and sequence variations (for example SNPs) can also be detected; this renders RNA-seq particularly useful for non-model genomeless organisms. Moreover, since it is based on the direct sequencing of cDNAs, no background noise is observed; RNA-seq does not have an upper limit for quantification, which correlates with the number of sequences obtained; there are high levels of reproducibility, for both technical and biological replicates; and since there are no cloning steps less RNA sample is required [211].

In *Prunus* species RNA-seq transcriptome profiling approaches are starting to be applied.

In peach, the whole transcriptome sequencing approach was used to investigate host plant defence responses to *Xanthomonas arboricola* pv. *pruni*, a quarantine bacterial pathogen that attacks peach production causing necrotic spots on leaves and fruits, severely reducing yields [212]. Differential gene expression was analysed at 2 and 12 hours post inoculation (hpi), through treated *vs.* control sample comparison. Out of 19,781 known peach genes that were expressed at all time points and in all conditions, 34 were found to be differentially expressed (up and down regulated) at 2 hpi and 263 at 12 hpi. Functional annotation according to the Gene Ontology vocabulary revealed that genes involved in stress response were particularly represented at 2 hpi. Pathogen-associated molecular pattern (PAMP) receptors, disease resistance genes including several RPM1-like and pathogenesis related thaumatin encoding genes were found among differentially expressed genes, together with others involved in photosynthesis, cell wall reorganization, hormone signalling pathways and cytochrome encoding. This work is the starting point to elucidate peach defence mechanisms at the very early stages of infection with a bacterial pathogen in a compatible interaction.

In sweet cherry, *Prunus avium,* RNA-seq has been used to identify candidate genes involved in cuticle formation [213]. The cuticular membrane of cherry fruits is a barrier against pathogens and uncontrolled water loss/uptake and is subjected to micro-cracks impairing its function. Deep sequencing of the transcriptome was carried out in the exocarp tissue and revealed the presence of 18 genes potentially involved in cuticle formation. Expression profiling of a few selected genes was also performed. Results suggested that the cessation of cuticle deposition could be attributable to down regulation of genes involved in this process. A few genes were identified for further analyses.

In Chinese plum (*Prunus mume*) identification through genome-wide analysis of *late embryogenesis abundant* (*LEA*) genes, playing important roles in plant desiccation tolerance, was coupled with expression profile analysis employing an RNA-seq approach on five organs: young leaves, stems, young roots, flowers and immature fruits [214]. In this work 30 *P. mume LEA* genes were found that could be divided into eight groups including the dehydrin and seed maturation protein (SMP) group. Most *Pm*LEA genes were highly expressed in flower and up-regulated by abscissic acid (ABA) treatment. A similar approach was also used to identify in *P. mume* transcription factors of the APETALA 2/ethylene-responsive element [215].

In apricot, an integrated approach using genomic, proteomic and RNA-seq gene expression profiling, has been undertaken to study the molecular basis of resistance to plum pox virus (PPV), one of the major stone fruit diseases [116].

RNA-seq is also being used in *Prunus persica* as a supportive approach to study cytokinin (a phytohormone playing a major role in plant development) response pathway during fruit development [216]. RNA-seq has been performed on peach fruits treated with exogenously applied t-zeatin. Comparison with orthologs of the *Arabidopsis* cytokinin signaling pathway (Arr1 and Arr6) and a cytokinin/auxin crosstalk factor (Shy2) suggests pathway conservation of the cytokinin response between *Arabidopsis* and Peach.

RNA-seq deep sequencing technology is also being applied to *Prunus* species for the isolation of micro RNAs. MicroRNAs (miRNAs) are short non-coding RNA molecules involved in gene expression regulation with major roles in plant development and stress response [217].

An extensive survey of the entire repertoire of miRNAs, including isomiRs (the population of variants of miRNAs deriving from the same precursors) was carried out in three replicates of five biological samples arising from a set of different organs and/or phenological stages of peach [218]. Three hundred and ninety-two isomiRs, corresponding to 26 putative miRNA coding loci were detected, contributing to a deeper understanding of the miRNome complexity.

Another work focused on the characterization of conserved and non-conserved miRNA components from two tissues representing non-dormant peach leaves and dormant peach leaf buds [219]. Among the non-conserved miRNAs identified, 205 appeared to be specific to peach. Within the conserved miRNA families, five (miR1446, miR473, miR479, miR3629, and miR3627) were reported only in tree species (*Populus trichocarpa*, *Citrus trifolia*, and *Prunus persica*). This study also identified several cold-responsive miRNAs and their predicted gene targets in QTLs for chilling requirement and bud dormancy thus connecting miRNA regulatory activities to these two aspects.

Finally, transcriptome deep sequencing of different tissues was used to help functional annotation of the peach genomic sequence [4]. For this purpose RNA was extracted from peach fruit (endocarp, mesocarp and epicarp at S1-S2-S3-S4 developmental stages), roots, fully expanded leaves, embryos and cotyledons, and from two bulks of different seed tissues.

4. GENOMICS: THE PEACH GENOME AS A CRUCIAL TOOL FOR BREEDING

The ability to view the whole DNA sequence (coding and non-coding) of a cell type or an organism is the genomics archetype. Access to the genome sequence, and how genes are encoded and organized within the genome, is a crucial tool in understanding the biological processes that define quality productions. Since 2000, the availability of several plant genome sequences has provided the opportunity to apply new strategies to breeding (genotyping by sequencing, re-sequencing) or speed up consolidated procedures like marker assisted breeding (MAB). At the beginning of 2013 the high quality draft genome of peach (*Prunus persica*) was published [4]. The assembly was obtained from a double haploid genotype of the peach cv Lovell by a Sanger-sequencing shotgun approach with a coverage level of 8.47 fold. The genome was assembled in 391 major scaffolds covering 226.6 Mb. Forty scaffolds were anchored on an updated version of the *Prunus* reference map [51] using 827 molecular markers (SSRs, SNPs), and organized in eight pseudomolecules (peach 2n = 16) covering 96.1% of the Peach v1.0 assembly. On the genome sequence, repeated sequences were annotated and genes were predicted. The percentage of the repeated fraction of the peach genome (29.60%) was lower, as expected, than other plant species with greater genome size, according to the theory that the genome size is directly dependent on the amount of repetitive elements. A total of 27,852 genes were predicted by homology with 88,000 ESTs of peach and 424,000 ESTs of related species. Gene content is similar to those of *A. thaliana* (27,416) and grape (30,434), but lower compared to poplar (45,654) and apple (57,386). This evidence can be explained by the recent whole genome duplication (WGD) that occurred in these two species [4, 220, 221].

A genome can tell different stories depending on what we are looking for; in the case of Peach v1.0 the authors identified signatures of domestication and patterns of genome evolution characteristic of the species. The re-sequencing analysis of 15 accessions belonging to the *Prunus* genus (11 peach accessions and one each for *P. ferganensis*, *P. kansuensis*, *P. davidiana* and *P. mira*) individuated ~ 1,000,000 unique SNPs in *P. persica* accessions and in *P. ferganensis*. These SNPs highlight a low level of nucleotide diversity in the cultivated peach ($\pi = 1.5 \times 10^{-3}$) compared to that of the wild species *Prunus davidiana* ($\pi = 4.8 \times 10^{-3}$).

Using a whole-genome SNP analysis *P. persica* accessions and *P. ferganensis* were grouped according to their geographical origin, with the exception of 'Quetta', a Pakistani cultivar, which likely originated after its dispersion to the west (Fig. **1**) [4]. The logic to include *P. ferganensis* in the peach group is supported by the results of whole-genome sequence comparison. *P. ferganensis* is indistinguishable from *P. persica* accessions. Its geographical origin (Fergana valley, Central Asia), the whole genome analysis and some undomesticated traits allow for *P. ferganensis* to be defined as a wild undomesticated peach or an intermediate genome in peach domestication. The effect of domestication events on the nucleotide diversity (π) is also clear comparing the wild relative *P. davidiana* with the domesticated Asian peach genotype *P. ferganensis*. As reported in literature, the center of origin of cultivated peaches is China, the first domestication bottleneck dated to 2,000-3,000 BC. This event was confirmed by examining the π value of *P. davidiana* (4.8×10^{-3}) that was three times higher than eastern peaches (1.6×10^{-3}). A second bottleneck occurred during the dispersion of peach to the west through Persia and breeding activities after the introduction of peaches in the United States, when a reduced pool of varieties was largely used for modern breeding. This strong founder effect could explain the nucleotide diversity value of western cultivars (1.1×10^{-3}) compared to that of the Asian ones. The slow decay of linkage disequilibrium (LD) r^2 value, compared with other plant species [222, 223], and the large linkage blocks present in peaches provide further molecular evidence of these bottlenecks. High local LD may result from selective sweeps related to domestication and breeding; quantitative trait loci for fruit size, a typical domestication trait, have been mapped in regions showing LD peaks on pseudomolecule 4 (at ~2 Mb, ~8 Mb and ~20 Mb) and pseudomolecule 5 [90, 111, 139].

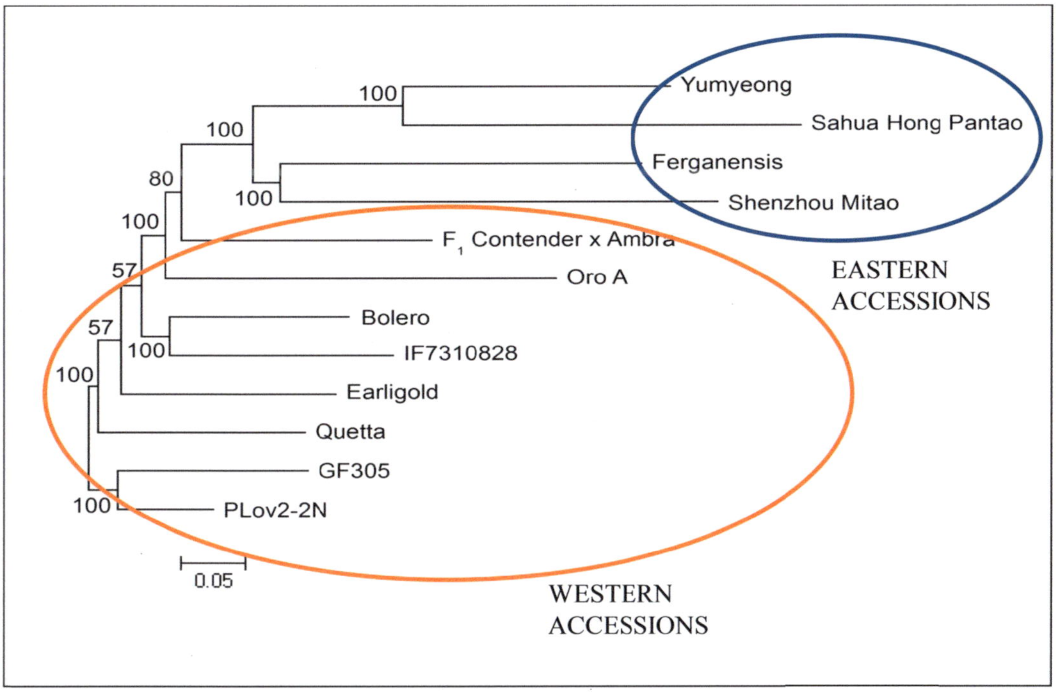

Figure 1: A neighbor joining phylogenetic tree constructed using genome-wide SNP data. From Verde and colleagues [4].

As reported by Dirlewanger and colleagues [65] the genome of the *Prunus* diploid species, peach, almond, apricot, cherry, *P. davidiana*, *P. cerasifera*, and *P. ferganensis*, is essentially syntenic and collinear. This evidence makes the peach genome a crucial tool for stone fruit breeding, giving the opportunity to have a landscape on genus and a favored starting point for all the molecular assisted procedures. Indeed a remarkable number of scientific works report the use of Peach v1.0 as a research starting point.

Using 101-fold next generation sequencing data a first draft of the *Prunus mume* genome was published [224]. The authors assembled a 280Mb genome and anchored 83.9% of scaffolds to eight chromosomes of a genetic map constructed by restriction-site-associated DNA sequencing.

4.1. Gene Candidate Approaches Using Peach v1.0

The candidate gene (CG) approach has been applied in plant genetics over the last two decades for the characterization and cloning of Mendelian and quantitative trait loci (QTLs). The CG analysis is based on the hypothesis that known-function genes (the candidate genes) could correspond to loci controlling traits of interest. Candidate genes either refer to cloned genes presumed to affect a given trait ('functional CGs') or to genes suggested due to their close proximity on linkage maps to loci controlling the trait ('positional CGs') [225]. When a candidate is finally individuated, the finishing validation will be provided through physiological analyses, genetic transformation and/or sexual complementation. As already discussed above in this chapter many linkage maps, with a number of traits localized, are available in stone fruits, and this knowledge coupled with the peach genome sequence has allowed the CG approach to be successfully applied in *Prunus* spp.

This approach has been extensively used in stone fruits since the long inter fertile period and orchard dimensions require the use of marker assisted selection (MAS) and marker assisted breeding (MAB). In sweet cherry (*Prunus avium*), a rosaceous fruit crop cultivated in temperate regions of the world for its highly valued fruits, large fruits are highly desirable by the consumers and the development of new cherry varieties with this characteristic is a major breeding goal. A targeted approach was applied to saturate two fruit-size QTLs located on linkage groups 2 and 6 (G2 and G6) which had been previously identified [169]. A total of 14 simple sequence repeats (SSRs) were developed from the syntenic peach genomic regions and used to further narrow the sweet cherry QTL regions [226]. Comparison of these syntenic regions with the Peach v1.0 identified cell number regulator genes (CNR), as candidate genes for both the G2 and G6 fruit size QTLs [227]. The CNR gene family, first described in maize based upon homology to the tomato fruit weight gene *fw2.2*, is hypothesized to contain genes involved in cell number regulation that ultimately affect plant growth and organ size [228]. Sequencing of the CNR genes from the parents of the segregant populations and a panel of diverse sweet cherry selections identified 3 and 2 allelic variants, for G2 and G6 respectively [226]. Comparisons of sweet cherry fruit weights based on their CNR genotypes further supported the hypothesis that the phenotypic differences in fruit size associated with the G2 and G6 QTLs were caused by allelic variants of the CNR genes [226].

The peach genome release v1.0 was used in a candidate gene approach for the isolation of the root-knot nematode (RKN) resistance genes in apricot and peach trees grown in France. These two cultivations are mostly grafted on the peach seedling rootstock 'Montclar' which is sensitive to damage by *Meloidogyne* spp. The genetic resistance to RKN was conferred by the *RMia* gene from 'Nemared' (homozygous resistant), that had previously been localized on the linkage group 2 (G2) of the *Prunus* reference map. SSR markers, in the gene region, were selected from the peach genome, and two of these, polymorphic between 'Nemared' and other peach accessions, were shown to flank the *RMia* gene in an interval of ~30 kb in the Peach v1.0. From the 949 total progenies used, four recombinant individuals were still available for a further reduction of this interval. For all the 945 non-recombinant hybrids the predicted phenotype was confirmed on young seedlings by inoculation tests with the RKN *M. incognita*. As a result, both flanking SSR markers were available for MAB of the *RMia* gene. This strategy will be applied in order to obtain a seedling rootstock homozygous for *RMia* [229].

Sharka disease, caused by the plum pox virus (PPV), is one of the major limiting factors for stone fruit production all over the world. Introgression of PPV resistance for crop improvement is therefore one of the most important goals in *Prunus* breeding programs. The major bottleneck in the breeding pipeline is the sharka phenotyping procedure which is time- and labor-consuming. In this context, screening of seedlings at early stages of development and marker-assisted selection (MAS) would provide the best solution for enhancing breeding efficiency [230]. Forty-two simple sequence repeat (SSR) markers were generated from the peach genome assembly v1.0 and from an apricot bacterial artificial chromosome (BAC) clone identified in the physical map of the PPV resistance locus [231]. SSR markers tightly linked to PPV resistance trait were localized in a bi-parental population. Three SSR markers, PGS1.21 PGS1.23 and PGS1.24, showed allelic variants associated with PPV resistance with no recombinants in the crosses analyzed. These markers unambiguously discriminated resistant from susceptible accessions in different genetic backgrounds. These results are the first successful application of MAS for breeding resistance in *Prunus* species [230].

A comprehensive study was performed to elucidate the genetic determinism of flowering and maturity dates in three *Prunus* species (peach, apricot and sweet cherry). Several QTLs for these traits were highly stable and this suggested that they are probably not affected by climatic variations. For flowering date, major QTLs were detected on linkage group 4 (G4) for apricot and sweet cherry and on G6 for peach. QTLs were identified on G2, G3, G4 and G7 for the three species. For maturity date, a major QTL was detected on G4 in the three species [232]. Using the peach genome sequence data, candidate genes underlying the major QTLs on G4 and G6 were investigated and key genes were identified. Millions of bases, forming the QTLs, were scanned looking for candidate genes (CGs) with functions compatible with the studied traits. On G4, the *Arabidopsis* ethylene responsive transcription factor 4 (ERF4) was proposed as a good candidate for controlling the ripening date in peach and apricot; it could be a major gene in agreement with the inheritance of this trait. For flowering date, 16 CGs belonging to the auxin efflux carrier family protein, cell division control protein 2 homolog C, protein Scarecrow, glycosyl transferase Quasimodo1, transcription factor MYB86, anaphase-promoting complex [232], were identified in the two regions. Flowering date is a much more complex trait than maturity date and in fact, most of CGs cross interact. These results can be the starting point for future research with the aim of understanding plant plasticity and adaptation to climate exchange.

Recently, Dardick and colleagues [154], using a next generation sequence-based mapping approach, identified a candidate gene (*PpeTAC1*) for the *br* mutation in peach associated with vertically oriented growth of branches referred to as “pillar” or “broomy”.

All these outcomes showed how the peach genomic sequence is a powerful tool for breeding and genetic studies in all stone fruits.

4.2. Whole Genome Approaches (Re-Sequencing Strategies, SNP Discovery, SNP-Chip use and Applications, Genotyping-by-Sequencing)

Whole genome strategies can be defined as a synergic network required for molecular plant breeding at the whole genome level. The final goal of these strategies is to help the best combinations of genotypes/genes, haplotypes, linkage disequilibrium (LD) blocks emerge into breeding products with desirable phenotypes [233]. Breeding for complex traits needs to understand the genetic, physiological and molecular bases of the characters including the interactions between the environment and the traits. Nowadays, innovation of the sequence-by-synthesis technology (or next generation sequencing, NGS) that dramatically reduces the cost and the time to generate sequences can sequence a genome in very little time (not including the huge bioinformatics efforts) and at a reduced cost [234]. The availability of the peach genome sequence coupled with NGS technologies has rapidly changed the plant genome analysis paradigm. Alignment of the short NGS reads obtained from diverse and unrelated varieties to a reference genome allows the identification of large numbers of Single Nucleotide Polymorphisms (SNPs) and small Deletion/Insertion Polymorphisms (DIPs) [25]. In this scenario, an international consortium (The International Peach SNP Consortium; IPSC) has pursued a coordinated effort to perform genome-scale SNP discovery in peach using next generation sequencing platforms to develop and characterize a high-throughput Illumina Infinium II SNP genotyping array platform. Whole genome re-sequencing of 56 peach breeding accessions was performed. Polymorphism detection algorithms identified a total of 1,022,354 SNPs. Conservative filtering was applied to arrive at a set of 8,144 SNPs that were included on the IPSC peach SNP array v1, distributed over all eight peach chromosomes with an average spacing of 26.7 kb between SNPs. The use of this platform to screen a total of 709 accessions of peach in two separate evaluation panels identified a total of 6,869 (84.3%) polymorphic SNPs [25]. The IPSC peach SNP array v1 was used worldwide for genetic studies in peach and related stone fruit and nut species. Some recent results are given below. Using the peach genome to anchor NGS reads, a 6K SNP array (Illumina Infinium® II system) was developed for diploid sweet cherry (*Prunus avium*) and allotetraploid sour cherry (*P. cerasus*) [96].

Volatile organic compounds (VOCs) in plants are involved in aroma and pest resistance. The knowledge of molecular, genetic, and physiological mechanisms underlying aroma formation is limited despite their importance in determining peach fruit quality. A recent research was conducted in order to identify peach QTLs for fruit VOCs and to understand their genetic basis using an F_1 population of 126 seedlings deriving from the cross between 'Bolero' and 'OroA', two peach cultivars differing in their aroma profile [130]. Dense single nucleotide polymorphism (SNP) and SSR maps covering the eight linkage groups of the peach genome were constructed by genotyping with the International Peach SNP Consortium peach SNP array v1, and data for 23 VOCs with high or unknown "odor activity value" were obtained by gas chromatography-mass spectrometry analysis of fruits. A total of 72 QTLs were identified for the 23 VOCs studied. Co-locations between candidate genes and major QTLs were identified taking advantage of the peach genome sequence: genes encoding two putative terpene synthases and one lipoxygenase (Lox) might be involved in the biosynthesis of linalool and p-menth-1-en-9-al, and nonanal, respectively [130]. These data could open opportunities for aroma MAS in peach.

Bacterial spot, caused by *Xanthomonas arboricola* pv. *pruni* (Xap), is a severe disease that causes defoliation and blemishing of fruits, seriously affecting the production. In peach, it is necessary to understand the molecular basis of its tolerance and susceptibility. To elucidate the genetics response to Xap, an F_2 segregating population derived from the cross of two peach cultivars, 'Clayton' (Xap resistant) and 'O'Henry' (highly susceptible to Xap) was obtained. After the population phenotyping, 63 individuals exhibiting high tolerance/resistance to Xap were genotyped with the IPSC 9K peach SNP array v1 and 1,341 SNPs were used to construct a genetic linkage map [129]. This map covers a genetic distance of 421.4 cM and was used for mapping QTLs responsible for Xap resistance in peach. Fourteen QTLs involved in Xap resistance were identified. One major QTL, on G4, was associated with Xap resistance in leaf, and two major QTLs, on G1 and G6, were associated with Xap resistance in fruit. In addition, one major QTL, on G5, was associated with Xap resistance on both leaf and fruit [129]. These results represent a potential for marker-assisted selection for Xap resistance in peach.

Molecular markers could improve the efficiency of breeding. Nowadays many Mendelian traits are mapped in peach and SSR markers have been found to be linked to some of the key major genes. However, there is insufficient linkage between the markers and the genes, and this still limits the use of molecular markers in tree breeding programs. To address this limitation, about 1,300 peach cultivars were genotyped with the 9K peach SNP chip in the frame of the

FruitBreedomics project [26]. This germplasm was chosen as representative of the genetic diversity present in five different collections (from Europe and China). Approximately 4,300 SNPs were positively genotyped and used for further analysis. The average number of heterozygous loci in the genotyped accessions was 1,186 (from 13 to 2,775). Preliminary results of the population structure revealed three main subpopulations and the presence of a high number of admixed individuals. LD seems to decay at a distance longer than ca. 1 Mb. These results will be instrumental for implementing LD-based mapping of QTLs and genes in peach [26].

A combined genetic and genomic approach was performed to elucidate molecular mechanism controlling bud dormancy release and flowering time in peach. Exploiting the high quality genome assembly and the robust QTL mapping results [117], 2 high- *vs.* 2 low-chill individuals were resequenced using the Illumina platform [235]. Reads were aligned against the reference genome represented by a high-chill doubled haploid Lovell genotype. The haplotype configuration of individuals allowed reconstruction of the two homozygous alleles at the three strongest QTLs (G1, G4 and G7) that collectively explain up to 70% of phenotypic variation for chilling requirement (CR) and bloom date. Using bioinformatic pipelines SNPs, DIPs and structural variants in the QTL intervals in high- *vs.* low-chill haplotypes were scanned to create a prioritized list of low-chill specific polymorphisms at the three most significant quantitative trait loci. The SNPs and DIPs were verified by wet methodologies (Sanger sequencing, and electrophoresis) [235]. These results are in agreement with the hypothesis of the potential network of candidate genes for control of the CR and bloom date in peach.

The two fundamental steps for genotyping germplasm are: i) finding DNA sequence polymorphisms for marker discovery and ii) assay design and genotyping for testing markers across the populations. An important strength of sequence-based genotyping (GBS) approaches is that marker discovery and genotyping are completed at the same time [236]. This facilitates exploration of new germplasm sets or even new species without the upfront effort of discovering and characterizing polymorphisms. In the past, the first application using NGS technologies was the discovery of single nucleotide polymorphisms (SNPs) and presence-absence variations (PAVs) in different populations with and without reference genomes [237-241]. The key objective of the GBS approach, therefore, is not merely to discover polymorphisms and then transfer these to a fixed assay, but to simultaneously discover polymorphisms and obtain genotypic information across the whole population of interest [236]. As GBS can be readily used for *de-novo* discovery and application of new molecular polymorphisms, it is particularly powerful for new germplasm sets and uncharacterized species. For breeding applications, informative polymorphisms can be discovered as novel germplasm is introduced into the breeding pool [236].

Currently, some research in stone fruits regard the use of GBS for different purposes such as for the saturation of the 'Texas' x 'Earligold' (TxE) *Prunus* reference map. For this aim, Randomly Amplified DNA Fingerprinting (RAF) fragment libraries were sequenced by a Roche platform and a software was developed (Bin Mapping Program) in order to identify SNPs and DIPs, selecting the most informative ones and genotyping and bin mapping the segregating RAF fragments. In total, 5,712 DIPs and 6,678 SNPs were detected, revealing a frequency of 1 SNP/170 bp. A total of 1,127 RAF contigs were bin mapped on the *Prunus* reference map. The combination of NGS, bioinformatics and bin mapping has proven to be an efficient strategy for SNPs and DIPs discovery and linkage map saturation in peach in order to develop tools for marker assisted selection [242].

As already mentioned, blooming date (BD) and chilling requirement (CR) are economically important traits for breeding fruit and nut tree species adapted to local climate conditions. An F_2 mapping population obtained by the self-crossing of an F_1 ('Hakuho' x 'UFGold') was analyzed [243]. Initially, 37 polymorphic microsatellite (SSR) markers were localized across all linkage groups. In order to saturate the linkage map a genotyping-by-sequencing (GBS) approach was performed to discover the single nucleotide polymorphisms (SNPs) present in the population. Genomic DNA was ApeKI restricted, ligated to barcoded adaptors, and pooled and amplified for multiplex sequencing on the Illumina platform. Approximately 160 million sequence reads were processed using TASSEL 3.0. SNPs were filtered to retain not only loci homozygous within each parent, but also polymorphic between parents. The linkage map consisting of 37 SSRs and 380

SNPs in eight linkage groups was constructed using JoinMap v4.1. Several quantitative trait loci (QTLs) were detected for BD and CR on G4 and G7 [243].

The GBS approach was also applied to study Sharka disease in apricot. Fine mapping of resistance has been performed [230]; F_1, F_2 and BC_1 populations originating from different donors have been tested in different environments (Czech republic, Italy and France) to check for stability and durability of the resistance. This research has also allowed refining the genetic location of the host actor(s) underlying the resistance trait [244]. These data are being integrated into a whole genome association genetics study to explore the broad diverse apricot germplasm as a substrate for candidate gene(s) identification for Sharka resistance as well as for other important agronomical traits in *Prunus* [244]. The GBS approach seems to be perfect for plant breeding, and preliminary studies on genomic selection show a great acceleration rate for developing new improved varieties.

Finally, the availability of a genome sequence is a powerful tool that makes performing whole-genome genotyping possible using different approaches: re-sequencing, GBS, or high-density chip. The availability of high-density markers, due to the decreasing sequencing cost, will allow marker assisted breeding for many traits that were previously inaccessible and more efficiently than ever. Moreover, the very high quality of the peach genome sequence make it likely to become the reference not only for the *Prunus* genus, but also for all the tree species as demonstrated by Kubisiak and colleagues [245]. They used Peach v.1.0 to construct a transcriptome-based genetic map of Chinese chestnut identifying segmental regions homologous to peach.

5. FUTURE PERSPECTIVES

As widely discussed in this chapter, the availability of the peach genome sequence [4] coupled with the release of important genomic tools in this species [25, 96] is rapidly providing the opportunity to fill the gap between genomics and breeding, not only in peach but in all *Prunus* spp. These tools, together with recent studies estimating the extent of LD decay and population stratification in *Prunus*, will foster Genome Wide Association Studies (GWAS) [4, 53, 246, 247]. Several initiatives have recently been set up worldwide. Notably, the Fruitbreedomics project funded by the EU within the Seventh Framework Programme (FPPVII) is using the whole genome SNP array [25] on germplasm collections of four peach countries (China, Italy, Spain and France) in order to set up a set of national core collections and perform GWAS for several important agronomic traits [26]. Whole genome association genetic study is underway to explore the broad diverse apricot germplasm as a substrate for candidate gene(s) identification for Sharka resistance, as well as for other important agronomic traits in *Prunus* [244], using a whole genome re-sequencing approach (genotyping-by-sequence, GBS) and the peach genome sequence as reference. Many other researches are being conducted in the *Prunus* spp, taking advantage of the latest technical advances to address major issues of stone fruits cultivation, which will make genetic and genomic tools for the breeding of new, eco-sustainable varieties available in the near future.

ACKNOWLEDGEMENTS

Declared none.

CONFLICT OF INTEREST

The authors confirm that this chapter contents have no conflict of interest.

REFERENCES

[1] FAO (1997), State of the world's plant genetic resources for food and agriculture. Food and Agriculture Organization of the United Nations (FAO), Rome Italy.

[2] FAO (2010), Second report on the state of the world's plant genetic resources for food and agriculture. Food and Agriculture Organization of the United Nations (FAO), Rome, Italy.

[3] Faust, M, Timon, B (1995), Origin and dissemination of peach. *Hortic Rev (Am Soc Hortic Sci)* 17, 331-379.

[4] Verde, I, Abbott, AG, Scalabrin, S, Jung, S, *et al.* (2013), The high-quality draft genome of peach (*Prunus persica*) identifies unique patterns of genetic diversity, domestication and genome evolution. *Nat Genet* 45, 487-94. doi: 10.1038/ng.2586.

[5] Faust, M, Surányi, D, Nyujto, F (1998), Origin and dissemination of apricot. *Hortic Rev Then NewYork* 22, 225-260.

[6] Potter, D (2012), Basic Information on the Stone Fruit Crops. In: Kole C, Abbott AG (eds) *Genetics, Genomics and Breeding of Stone fruits*. CRC Press Taylor & Francis Group, pp 1-21.

[7] Faust, M, Surányi, D (1999), Origin and dissemination of plums. *Hort Rev* 23, 179-231.
[8] Faust, M, Surányi, D (1997), Origin and dissemination of cherry. *Hortic Rev (Am Soc Hortic Sci)* 19, 263-317.
[9] Li, YX, Li, TH, Zhang, HL, Qi, YW (2007), Sampling strategy for a primary core collection of peach (*Prunus persica* (L.) Batsch.) germplasm. *Eur J Hortic Sci* 72, 268.
[10] Wang, Y, Zhang, J, Sun, H, Ning, N, *et al.* (2011), Construction and evaluation of a primary core collection of apricot germplasm in China. *Sci Hortic (Amsterdam)* 128, 311-319.
[11] Engelmann, F (2010), Use of biotechnologies for the conservation of plant biodiversity. *Vitr Cell Dev Biol - Plant* 47, 5-16. doi: 10.1007/s11627-010-9327-2.
[12] Frankel, OH, Brown, AHD (1984), Plant genetic resources today: a critical appraisal. In: Holden JHW, Williams J (eds) *Crop genetic resources: conservation and evaluation.* pp 249-257.
[13] Frankel, OH, Brown, AHD (1984), Current plant genetic resources - a critical appraisal. *Proceedings XV International Congress of Genetics.* Oxford and IBH Publishing Co, New Delhi, pp 1-13.
[14] FAO (1996), Global plan of action for the conservation and sustainable utilization of plant genetic resources for food and agriculture. Food and Agriculture Organization of the United Nations (FAO), Rome Italy.
[15] Van Hintum, TJL, Brown, AHD, Spillane, C, Hodgkin, T (2000), Core collections of plant genetic resources. IPGRI Tech. Bullettin.
[16] Brown, AHD (1989), *The case for core collections. use plant Genet Resour Cambridge Univ Press Cambridge* 136-156.
[17] Upadhyaya, HD, Ortiz, R (2001), A mini core subset for capturing diversity and promoting utilization of chickpea genetic resources in crop improvement. *Theor Appl Genet* 102, 1292-1298.
[18] McKhann, HI, Camilleri, C, Bérard, A, Bataillon, T, *et al.* (2004), Nested core collections maximizing genetic diversity in *Arabidopsis thaliana. Plant J* 38, 193-202.
[19] Escribano, P, Viruel, MA, Hormaza, JI (2008), Comparison of different methods to construct a core germplasm collection in woody perennial species with simple sequence repeat markers. A case study in cherimoya (*Annona cherimola*, Annonaceae), an underutilised subtropical fruit tree species. *Ann Appl Biol* 153, 25-32.
[20] Bataillon, TM, David, JL, Schoen, DJ (1996), Neutral genetic markers and conservation genetics: simulated germplasm collections. *Genetics* 144, 409-417.
[21] Kim, KW, Chung, HK, Cho, GT, Ma, KII, *et al.* (2007), PowerCore: a program applying the advanced M strategy with a heuristic search for establishing core sets. *Bioinformatics* 23, 2155-2162.
[22] Le Cunff, L, Fournier-Level, A, Laucou, V, Vezzulli, S, *et al.* (2008), Construction of nested genetic core collections to optimize the exploitation of natural diversity in *Vitis vinifera* L. subsp. sativa. *BMC Plant Biol* 8, 31.
[23] Barnaud, A, Lacombe, T, Doligez, A (2006), Linkage disequilibrium in cultivated grapevine, *Vitis vinifera* L. *Theor Appl Genet* 112, 708-16. doi: 10.1007/s00122-005-0174-1.
[24] Li, TH, Li, YX, Li, ZC, Zhang, HL, *et al.* (2008), Simple sequence repeat analysis of genetic diversity in primary core collection of peach (*Prunus persica*). *J Integr Plant Biol* 50, 102-110.

[25] Verde, I, Bassil, N, Scalabrin, S, Gilmore, B (2012), Development and evaluation of a 9K SNP array for peach by internationally coordinated SNP detection and validation in breeding germplasm. *PLoS One* 7, e35668.

[26] Micheletti, D, Aramini, V, Arús, P, Banchi, E, *et al.* (2012), Genetic variability description in a wide germplasm of *domestica*ted peach through high throughput genotyping. *6th Rosaceous Genomics Conference (RGC6).* Mezzocorona (Trento), Italy, p 46.

[27] Krichen, L, Audergon, JM, Trifi-Farah, N (2012), Relative efficiency of morphological characters and molecular markers in the establishment of an apricot core collection. *Hereditas* 149, 163-172.

[28] Botstein, D, White, R (1980), Construction of a genetic linkage map in man using restriction fragment length polymorphisms. *Am J Hum Genet* 32, 314-331.

[29] Eldredge, L, Ballard, R, Baird, W (1992), Application of RFLP analysis to genetic linkage mapping in peaches. *Hortic Sci* 27, 160-163.

[30] Arús, P, Messeguer, R, Viruel, M, Tobutt, K, *et al.* (1994), The European *Prunus* mapping project. *Euphytica* 77, 97-100.

[31] Quarta, R, Dettori, MT, Verde, I, Laino, P, *et al.* (1996), Restriction fragment length polymorphism analysis in peach. *Acta Hortic* 374, 61-70.

[32] Dirlewanger, E, Pronier, V, Parvery, C, Rothan, C, *et al.* (1998), Genetic linkage map of peach [*Prunus persica* (L.) Batsch] using morphological and molecular markers. *Theor Appl Genet* 97, 888-895. doi: 10.1007/s001220050969.

[33] Viruel, M, Messeguer, R (1995), A linkage map with RFLP and isozyme markers for almond. *Theor Appl Genet* 91, 964-971.

[34] Foolad, M, Arulsekar, S, Becerra, V, Bliss, F (1995), A genetic map of *Prunus* based on an interspecific cross between peach and almond. *Theor Appl Genet* 91, 964-971.

[35] Joobeur, T, Viruel, M (1998), Construction of a saturated linkage map for *Prunus* using an almond× peach F2 progeny. *Theor Appl Genet* 97, 1034-1041.

[36] Rajapakse, S, Belthoff, L, He, G (1995), Genetic linkage mapping in peach using morphological, RFLP and RAPD markers. *Theor Appl Genet* 90, 503-510.

[37] Quarta, R, Dettori, MT, Verde, I, Marchesi, U, *et al.* (2001), Characterization and evaluation of genetic diversity in peach germplasm using RAPD and RFLP markers. *Acta Hortic* 546, 489-496.

[38] Quarta, R, Dettori, MT, Verde, I, Gentile, A, *et al.* (1998), Genetic analysis of agronomic traits and genetic linkage mapping in a BC_1 peach polulation using RFLPs and RAPDs. *Acta Hort* 465, 51-60.

[39] Dettori, M, Quarta, R, Verde, I (2001), A peach linkage map integrating RFLPs, SSRs, RAPDs, and morphological markers. *Genome* 44, 783-790.

[40] Morgante, M, Salamini, F (2003), From plant genomics to breeding practice. *Curr Opin Biotechnol* 14, 214-219. doi: 10.1016/S0958-1669(03)00028-4.

[41] Cipriani, G, Lot, G, Huang, WG, Marrazzo, MT, *et al.* (1999), AC/GT and AG/CT microsatellite repeats in peach [*Prunus persica* (L) Batsch]: isolation, characterisation and cross-species amplification in *Prunus*. *Theor Appl Genet* 99, 65-72. doi: 10.1007/s001220051209.

[42] Testolin, R, Marrazzo, T, Cipriani, G, Quarta, R (2000), Microsatellite DNA in peach (*Prunus persica* L. Batsch) and its use in fingerprinting and testing the genetic origin of cultivars. *Genome* 43, 512-520.

[43] Sosinski, B, Gannavarapu, M, Hager, LD, Beck, LE, *et al.* (2000), Characterization of microsatellite markers in peach [*Prunus persica* (L.) Batsch]. *Theor Appl Genet* 101, 421-428. doi: 10.1007/s001220051499.

[44] Dirlewanger, E, Cosson, P, Tavaud, M, Aranzana, J, *et al.* (2002), Development of microsatellite markers in peach [*Prunus persica* (L.) Batsch] and their use in genetic diversity analysis in peach and sweet cherry (*Prunus avium* L.). *Theor Appl Genet* 105, 127-138. doi: 10.1007/s00122-002-0867-7.

[45] Aranzana, MJ, Garcia-Mas, J, Carbo, J, Arus, P, *et al.* (2002), Development and variability analysis of microsatellite markers in peach. *Plant Breed* 121, 87-92. doi: 10.1046/j.1439-0523.2002.00656.x.

[46] Yamamoto, T, Kimura, T, Shoda, M, Imai, T, *et al.* (2002), Genetic linkage maps constructed by using an interspecific cross between Japanese and European pears. *Theor Appl Genet* 106, 9-18. doi: 10.1007/s00122-002-0966-5.

[47] Georgi, L, Wang, Y, Yvergniaux, D (2002), Construction of a BAC library and its application to the identification of simple sequence repeats in peach. *Theor Appl Genet* 105, 1151-1158.

[48] Wang, Y, Georgi, LL, Zhebentyayeva, TN, Reighard, GL, *et al.* (2002), High-throughput targeted SSR marker development in peach (*Prunus persica*). *Genome* 45, 319-328. doi: 10.1139/G01-153.

[49] Verde, I, Vendramin, E, Dettori, MT, Quarta, R, *et al.* (2004), Target SSR development in peach and SSR mapping in a peach BC_1 progeny. *Acta Hort* 63-68.

[50] Vendramin, E, Dettori, MT, Giovinazzi, J, Micali, S, *et al.* (2007), A set of EST-SSRs isolated from peach fruit transcriptome and their transportability across *Prunus* species. *Mol Ecol Notes* 7, 307-310. doi: 10.1111/j.1471-8286.2006.01590.x.

[51] Howad, W, Yamamoto, T, Dirlewanger, E, Testolin, R, *et al.* (2005), Mapping with a few plants: using selective mapping for microsatellite saturation of the *Prunus* reference map. *Genetics* 171, 1305-9. doi: 10.1534/genetics.105.043661.

[52] Aranzana, MJ, Carbó, J, Arús, P (2003), Microsatellite variability in peach [*Prunus persica* (L.) Batsch]: cultivar identification, marker mutation, pedigree inferences and population structure. *Theor Appl Genet* 106, 1341-52. doi: 10.1007/s00122-002-1128-5.

[53] Aranzana, MJ, Abbassi, E-K, Howad, W, Arús, P (2010), Genetic variation, population structure and linkage disequilibrium in peach commercial varieties. *BMC Genet* 11, 69. doi: 10.1186/1471-2156-11-69.

[54] Mnejja, M, Garcia-Mas, J, Howad, W, Badenes, ML, *et al.* (2004), Simple-sequence repeat (SSR) markers of Japanese plum (*Prunus salicina* Lindl.) are highly polymorphic and transferable to peach and almond. *Mol Ecol Notes* 4, 163-166. doi: 10.1111/j.1471-8286.2004.00603.x.

[55] Mnejja, M, Garcia-Mas, J, Howad, W, Arús, P (2005), Development and transportability across *Prunus* species of 42 polymorphic almond microsatellites. *Mol Ecol Notes* 5, 531-535. doi: 10.1111/j.1471-8286.2005.00977.x.

[56] Decroocq, V, Favé, MG, Hagen, L, Bordenave, L, *et al.* (2003), Development and transferability of apricot and grape EST microsatellite markers across taxa. *Theor Appl Genet* 106, 912-22. doi: 10.1007/s00122-002-1158-z.

[57] Hagen, LS, Chaib, J, Fady, B, Decroocq, V, *et al.* (2004), Genomic and cDNA microsatellites from apricot (*Prunus armeniaca* L.). *Mol Ecol Notes* 4, 742-745. doi: 10.1111/j.1471-8286.2004.00802.x.

[58] Lopes, MS, Sefc, KM, Laimer, M, Da Camara Machado, A, *et al.* (2002), Identification of microsatellite loci in apricot. *Mol Ecol Notes* 2, 24-26. doi: 10.1046/j.1471-8286.2002.00132.x.

[59] Messina, R, Lain, O, Marrazzo, MT, Cipriani, G, *et al.* (2004), New set of microsatellite loci isolated in apricot. *Mol Ecol Notes* 4, 432-434. doi: 10.1111/j.1471-8286.2004.00674.x.

[60] Vilanova, S, Soriano, JM, Lalli, DA, Romero, C, *et al.* (2006), Development of SSR markers located in the G1 linkage group of apricot (*Prunus armeniaca* L.) using a bacterial artificial chromosome library. *Mol Ecol Notes* 6, 789-791. doi: 10.1111/j.1471-8286.2006.01346.x.

[61] Li, X, Shangguan, L, Song, C, Wang, C, *et al.* (2010), Analysis of expressed sequence tags from *Prunus mume* flower and fruit and development of simple sequence repeat markers. *BMC Genet* 11, 66. doi: 10.1186/1471-2156-11-66.

[62] Maghuly, F, Fernandez, EB, Ruthner, S, Pedryc, A, *et al.* (2005), Microsatellite variability in apricots (*Prunus armeniaca* L.) reflects their geographic origin and breeding history. *Tree Genet Genomes* 1, 151-165. doi: 10.1007/s11295-005-0018-9.

[63] Dondini, L, Lain, O, Geuna, F, Banfi, R, *et al.* (2007), Development of a new SSR-based linkage map in apricot and analysis of synteny with existing *Prunus* maps. *Tree Genet Genomes* 3, 239-249. doi: 10.1007/s11295-006-0059-8.

[64] Ohta, S, Nishitani, C, Yamamoto, T (2005), Chloroplast microsatellites in *Prunus*, Rosaceae. *Mol Ecol Notes* 5, 837-840. doi: 10.1111/j.1471-8286.2005.01080.x.

[65] Dirlewanger, E, Graziano, E, Joobeur, T, Garriga-Calderé, F, *et al.* (2004), Comparative mapping and marker-assisted selection in Rosaceae fruit crops. *Proc Natl Acad Sci U S A* 101, 9891-6. doi: 10.1073/pnas.0307937101.

[66] Cantini, C, Iezzoni, AF, Lamboy, WF, Boritzki, M, *et al.* (2001), DNA fingerprinting of tetraploid cherry germplasm using simple sequence repeats. *J Amer Soc Hort Sci* 126, 205-209.

[67] Downey, SL, Iezzoni, AF (2000), Polymorphic DNA markers in black cherry (*Prunus serotina*) are identified using sequences from sweet cherry, peach, and sour cherry. *J Amer Soc Hort Sci* 125, 76-80.

[68] Struss, D, Ahmad, R, Southwick, SM, Boritzki, M (2003), Analysis of sweet cherry (*Prunus avium* L.) cultivars using SSR and AFLP markers. *J Amer Soc Hort Sci* 128, 904-909.

[69] Clarke, JB, Tobutt, KR (2003), Development and characterization of polymorphic microsatellites from *Prunus avium*'Napoleon'. *Mol Ecol Notes* 3, 578-580. doi: 10.1046/j.1471-8286.2003.00517.x.

[70] Vaughan, SP, Russell, K (2004), Characterization of novel microsatellites and development of multiplex PCR for large-scale population studies in wild cherry, *Prunus avium*. *Mol Ecol Notes* 4, 429-431. doi: 10.1111/j.1471-8286.2004.00673.x.

[71] Dirlewanger, E, Cosson, P, Boudehri, K, Renaud, C, *et al.* (2006), Development of a second-generation genetic linkage map for peach [*Prunus persica* (L.) Batsch] and characterization of morphological traits affecting flower and fruit. *Tree Genet Genomes* 3, 1-13. doi: 10.1007/s11295-006-0053-1.

[72] Yamamoto, T, Yamaguchi, M, Hayashi, T (2005), An integrated genetic linkage map of peach by SSR, STS, AFLP and RAPD. *J Jpn Soc Hort Sci* 74, 204-213.

[73] Sawa, A, Kaneko, S, Isagi, Y, Mariko, S, *et al.* (2010), Development and characterization of microsatellite markers for *Prunus verecunda* and *Prunus grayana* (Rosaceae). *Conserv Genet* 11, 1167-1169. doi: 10.1007/s10592-009-9909-x.

[74] Ai, C, Yu, X, Dahlenburg, A, Zhao, HJ, *et al.* (2008), Development and characterization of SSR markers in Chinese cherry (*Prunus pseudocerasus* Lindl.). *Eur J Hortic Sci* 73, 104.

[75] Testolin, R, Messina, R, Lain, O, Marrazzo, MT, *et al.* (2004), Microsatellites isolated in almond from an AC-repeat enriched library. *Mol Ecol Notes* 4, 459-461. doi: 10.1111/j.1471-8286.2004.00700.x.

[76] Gunderson, KL, Steemers, FJ, Lee, G, Mendoza, LG, *et al.* (2005), A genome-wide scalable SNP genotyping assay using microarray technology. *Nat Genet* 37, 549-54. doi: 10.1038/ng1547.

[77] Lijavetzky, D, Cabezas, JA, Ibáñez, A, Rodríguez, V, *et al.* (2007), High throughput SNP discovery and genotyping in grapevine (*Vitis vinifera* L.) by combining a re-sequencing approach and SNPlex technology. *BMC Genomics* 8, 424. doi: 10.1186/1471-2164-8-424.

[78] Pindo, M, Vezzulli, S, Coppola, G, Cartwright, DA, *et al.* (2008), SNP high-throughput screening in grapevine using the SNPlex genotyping system. *BMC Plant Biol* 8, 12. doi: 10.1186/1471-2229-8-12.

[79] Gunderson, KL (2009), Whole-genome genotyping on bead arrays. *Methods Mol Biol* 529, 197-213. doi: 10.1007/978-1-59745-538-1_13.

[80] Rafalski, A (2002), Applications of single nucleotide polymorphisms in crop genetics. *Curr Opin Plant Biol* 5, 94-100.

[81] Grimplet, J, Romieu, C, Audergon, J-M, Marty, I, *et al.* (2005), Transcriptomic study of apricot fruit (*Prunus armeniaca*) ripening among 13 006 expressed sequence tags. *Physiol Plant* 125, 281-292. doi: 10.1111/j.1399-3054.2005.00563.x.

[82] Horn, R, Lecouls, AC, Callahan, A, Dandekar, A, *et al.* (2005), Candidate gene database and transcript map for peach, a model species for fruit trees. *Theor Appl Genet* 110, 1419-28. doi: 10.1007/s00122-005-1968-x.

[83] Ogundiwin, E, Martí, C, Forment, J, Pons, C, *et al.* (2008), Development of ChillPeach genomic tools and identification of cold-responsive genes in peach fruit. *Plant Mol Biol* 68, 379-97. doi: 10.1007/s11103-008-9378-5.

[84] Le Dantec, L, Cardinet, G, Bonet, J, Fouché, M, *et al.* (2010), Development and mapping of peach candidate genes involved in fruit quality and their transferability and potential use in other Rosaceae species. *Tree Genet Genomes* 6, 995-1012. doi: 10.1007/s11295-010-0308-8.

[85] Silva, C, Garcia-Mas, J, Sánchez, a M, Arús, P, *et al.* (2005), Looking into flowering time in almond (*Prunus dulcis* (Mill) D. A. Webb): the candidate gene approach. *Theor Appl Genet* 110, 959-68. doi: 10.1007/s00122-004-1918-z.

[86] Ogundiwin, E, Peace, CP, Gradziel, TM, Parfitt, DE, *et al.* (2009), A fruit quality gene map of *Prunus*. *BMC Genomics* 10, 587. doi: 10.1186/1471-2164-10-587.

[87] Testone, G, Condello, E, Verde, I (2009), The peach (*Prunus persica* Batsch) homeobox gene *KNOPE3*, which encodes a class 2 knotted-like transcription factor, is regulated during leaf development and triggered by sugars. *Mol Genet Genomics* 282, 47-64.

[88] Testone, G, Condello, E, Verde, I, Nicolodi, C, *et al.* (2012), The peach (*Prunus persica* L. Batsch) genome harbours 10 *KNOX* genes, which are differentially expressed in stem development, and the class 1 *KNOPE1* regulates elongation and lignification during primary growth. *J Exp Bot* 63, 5417-35. doi: 10.1093/jxb/ers194.

[89] Chen, L, Zhang, S, Illa, E, Song, L, *et al.* (2008), Genomic characterization of putative allergen genes in peach/almond and their synteny with apple. *BMC Genomics* 9, 543. doi: 10.1186/1471-2164-9-543.

[90] Illa, E, Eduardo, I, Audergon, JM, Barale, F, *et al.* (2011), Saturating the *Prunus* (stone fruits) genome with candidate genes for fruit quality. *Mol Breed* 28, 667-682. doi: 10.1007/s11032-010-9518-x.

[91] Tavassolian, I, Rabiei, G, Gregory, D, Mnejja, M, *et al.* (2010), Construction of an almond linkage map in an Australian population Nonpareil x Lauranne. *BMC Genomics* 11, 551. doi: 10.1186/1471-2164-11-551.

[92] Cabrera, A, Kozik, A, Howad, W, Arus, P (2009), Development and bin mapping of a Rosaceae Conserved Ortholog Set (COS) of markers. *BMC Genomics* 10, 562.

[93] Sargent, DJ, Fernandéz-Fernandéz, F, Ruiz-Roja, JJ, Sutherland, BG, *et al.* (2009), A genetic linkage map of the cultivated strawberry (*Fragaria* × *ananassa*) and its comparison to the diploid *Fragaria* reference map. *Mol Breed* 24, 293-303. doi: 10.1007/s11032-009-9292-9.

[94] Illa, E, Sargent, DJ, Girona, EL, Bushakra, J, *et al.* (2011), Comparative analysis of rosaceous genomes and the reconstruction of a putative ancestral genome for the family. *BMC Evol Biol* 11, 9.

[95] Ahmad, R, Parfitt, DE, Fass, J, Ogundiwin, E, *et al.* (2011), Whole genome sequencing of peach (*Prunus persica* L.) for SNP identification and selection. *BMC Genomics* 12, 569. doi: 10.1186/1471-2164-12-569.

[96] Peace, C, Bassil, N, Main, D, Ficklin, S, *et al.* (2012), Development and evaluation of a genome-wide 6K SNP array for diploid sweet cherry and tetraploid sour cherry. *PLoS One* 7, e48305. doi: 10.1371/journal.pone.0048305.

[97] Hesse, C (1975), Peaches. In: Janick J, Moore JN (eds) *Advances in fruit breeding Temperate fruits*. Purdue University Press, W. Lafayette, Indiana, pp 285-335.

[98] Monet, R, Guye, A, Roy, M, Dachary, N (1996), Peach mendelian genetics: a short review and new results. *Agronomie* 16, 321-329. doi: 10.1051/agro:19960505.

[99] Chaparro, JX, Werner, DJ, O'Malley, D, Sederoff, RR (1994), Targeted mapping and linkage analysis of morphological isozyme, and RAPD markers in peach. Theor Appl Genet. doi: 10.1007/BF00221132.

[100] Byrne, DH (1990), Isozyme Variability in Four Diploid Stone Fruits Compared with Other Woody Perennial Plants. *J Hered* 81, 68-71.

[101] Mnejja, M, Garcia-Mas, J, Audergon, J-M, Arús, P (2010), *Prunus* microsatellite marker transferability across rosaceous crops. *Tree Genet Genomes* 6, 689-700. doi: 10.1007/s11295-010-0284-z.

[102] Scorza, R, Mehlenbacher, S, Lightner, G (1985), Inbreeding and coancestry of freestone peach cultivars of the eastern United States and implications for peach germplasm improvement. *J Am Soc Hortic Sci* 110, 547-552.

[103] Bliss, FA, Arulsekar, S, Foolad, MR, Becerra, V, *et al.* (2002), An expanded genetic linkage map of *Prunus* based on an interspecific cross between almond and peach. *Genome* 45, 520-529. doi: 10.1139/G02-011.

[104] Jáuregui, B, de Vicente, MC, Messeguer, R, Felipe, A, *et al.* (2001), A reciprocal translocation between 'Garfi' almond and 'Nemared' peach. *Theor Appl Genet* 102, 1169-1176. doi: 10.1007/s001220000511.

[105] Arús, P, Verde, I, Sosinski, B, Zhebentyayeva, T, *et al.* (2012), The peach genome. *Tree Genet Genomes* 8, 531-547. doi: 10.1007/s11295-012-0493-8.

[106] Aranzana, MJ, Pineda, A, Cosson, P, Dirlewanger, E, *et al.* (2003), A set of simple-sequence repeat (SSR) markers covering the *Prunus* genome. *Theor Appl Genet* 106, 819-25. doi: 10.1007/s00122-002-1094-y.

[107] Arús, P, Yamamoto, T, Dirlewanger, E, Abbott, A (2005), Synteny in the Rosaceae. In: Janick J (ed) *Plant Breeding Reviews*. John Wiley, London, UK, pp 175-211.

[108] Pozzi, C, Vecchietti, A (2009), Peach structural genomics. In: Folta KM, Gardiner SE (eds) *Genetics and genomics of Rosaceae*. Springer, New York, pp 235-257.

[109] Dirlewanger, E, Illa, E, Howad, W (2012), Molecular Linkage Maps: Strategies, Resources and Achievements. In: Kole C, Abbott AG (eds) *Genetics, Genomics and Breeding of Stone Fruits*. CRC Press Taylor & Francis Group, pp 76-104.

[110] Foulongne, M, Pascal, T, Arús, P, Kervella, J (2003), The potential of *Prunus davidiana* for introgression into peach [*Prunus persica* (L.) Batsch] assessed by comparative mapping. *Theor Appl Genet* 107, 227-38. doi: 10.1007/s00122-003-1238-8.

[111] Quilot, B, Wu, BH, Kervella, J, Génard, M, *et al.* (2004), QTL analysis of quality traits in an advanced backcross between *Prunus persica* cultivars and the wild relative species P. *davidiana. Theor Appl Genet* 109, 884-97. doi: 10.1007/s00122-004-1703-z.

[112] Decroocq, V, Foulongne, M, Lambert, P, Gall, O Le, *et al.* (2005), Analogues of virus resistance genes map to QTLs for resistance to sharka disease in *Prunus davidiana. Mol Genet genomics* 272, 680-9. doi: 10.1007/s00438-004-1099-0.

[113] Verde, I, Lauria, M, Dettori, MT, Vendramin, E, *et al.* (2005), Microsatellite and AFLP markers in the *Prunus persica* [L. (Batsch)]xP. ferganensis BC(1)linkage map: saturation and coverage improvement. *Theor Appl Genet* 111, 1013-21. doi: 10.1007/s00122-005-0006-3.

[114] Blenda, A, Verde, I, Georgi, LL, Reighard, GL, *et al.* (2007), Construction of a genetic linkage map and identification of molecular markers in peach rootstocks for response to peach tree short life syndrome. *Tree Genet Genomes* 3, 341-350. doi: 10.1007/s11295-006-0074-9.

[115] Marandel, G, Pascal, T, Candresse, T, Decroocq, V (2009), Quantitative resistance to Plum pox virus in *Prunus davidiana* P1908 linked to components of the eukaryotic translation initiation complex. *Plant Pathol* 58, 425-435. doi: 10.1111/j.1365-3059.2008.02012.x.

[116] Martínez-Gómez, P, García-Ibarra, A, Hernández, JA, Rubio, M (2011), An integrated genomic, proteomic and transcriptomic approach to plum pox virus resistance in *Prunus*. *Acta Hort* 899, 81-86.

[117] Fan, S, Bielenberg, DG, Zhebentyayeva, TN, Reighard, GL, *et al.* (2010), Mapping quantitative trait loci associated with chilling requirement, heat requirement and bloom date in peach (*Prunus persica*). *New Phytol* 185, 917-30. doi: 10.1111/j.1469-8137.2009.03119.x.

[118] Joobeur, T, Periam, N, de Vicente, MC, King, GJ, *et al.* (2000), Development of a second generation linkage map for almond using RAPD and SSR markers. *Genome* 43, 649-55.

[119] Sánchez-Pérez, R, Howad, W, Garcia-Mas, J, Arús, P, *et al.* (2010), Molecular markers for kernel bitterness in almond. *Tree Genet Genomes* 6, 237-245. doi: 10.1007/s11295-009-0244-7.

[120] Hurtado, A, Romero, C, Vilanova, S, Abbott, G, *et al.* (2002), Genetic linkage maps of two apricot cultivars (*Prunus armeniaca* L.), and mapping of PPV (sharka) resistance. *Theor Appl Genet* 105, 182-191. doi: 10.1007/s00122-002-0936-y.

[121] Vilanova, S, Romero, C, Abbott, a G, Llácer, G, *et al.* (2003), An apricot (*Prunus armeniaca* L.) F2 progeny linkage map based on SSR and AFLP markers, mapping plum

pox virus resistance and self-incompatibility traits. *Theor Appl Genet* 107, 239-47. doi: 10.1007/s00122-003-1243-y.

[122] Lambert, P, Hagen, LS, Arus, P, Audergon, JM (2004), Genetic linkage maps of two apricot cultivars (*Prunus armeniaca* L.) compared with the almond Texas x peach Earlygold reference map for *Prunus*. *Theor Appl Genet* 108, 1120-30. doi: 10.1007/s00122-003-1526-3.

[123] Olukolu, BA, Trainin, T, Fan, S, Kole, C, *et al.* (2009), Genetic linkage mapping for molecular dissection of chilling requirement and budbreak in apricot (*Prunus armeniaca* L.). *Genome* 52, 819-28. doi: 10.1139/g09-050.

[124] Wang, D, Karle, R, Brettin, TS, Iezzoni, AF (1998), Genetic linkage map in sour cherry using RFLP markers. *Theor Appl Genet* 97, 1217-1224.

[125] Olmstead, JW, Sebolt, AM, Cabrera, A, Sooriyapathirana, SS, *et al.* (2008), Construction of an intra-specific sweet cherry (*Prunus avium* L.) genetic linkage map and synteny analysis with the *Prunus* reference map. *Tree Genet Genomes* 4, 897-910. doi: 10.1007/s11295-008-0161-1.

[126] Clarke, JB, Sargent, DJ, Bošković, RI, Belaj, A, *et al.* (2009), A cherry map from the inter-specific cross *Prunus avium* 'Napoleon' × P. nipponica based on microsatellite, gene-specific and isoenzyme markers. *Tree Genet Genomes* 5, 41-51. doi: 10.1007/s11295-008-0166-9.

[127] Dirlewanger, E, Cosson, P, Howad, W, Capdeville, G, *et al.* (2004), Microsatellite genetic linkage maps of myrobalan plum and an almond peach hybrid location of root-knot nematode resistance genes. *Theor Appl Genet* 109, 827-38. doi: 10.1007/s00122-004-1694-9.

[128] Martínez-García, PJ, Parfitt, DE, Ogundiwin, EA, Fass, J, *et al.* (2013), High density SNP mapping and QTL analysis for fruit quality characteristics in peach (*Prunus persica* L.). *Tree Genet Genomes* 9, 19-36. doi: 10.1007/s11295-012-0522-7.

[129] Yang, N, Reighard, G, Ritchie, D, Okie, W, *et al.* (2013), Mapping quantitative trait loci associated with resistance to bacterial spot (*Xanthomonas arboricola* pv. *pruni*) in peach. *Tree Genet Genomes* 9, 573-586. doi: 10.1007/s11295-012-0580-x.

[130] Eduardo, I, Chietera, G, Pirona, R, Pacheco, I, *et al.* (2013), Genetic dissection of aroma volatile compounds from the essential oil of peach fruit: QTL analysis and identification of candidate genes using dense SNP maps. *Tree Genet Genomes* 9, 189-204. doi: 10.1007/s11295-012-0546-z.

[131] Klagges, C, Campoy, JA, Quero-García, J, Guzmán, A, *et al.* (2013), Construction and comparative analyses of highly dense linkage maps of two sweet cherry intra-specific progenies of commercial cultivars. *PLoS One* 8, e54743. doi: 10.1371/journal.pone.0054743.

[132] Dirlewanger, E, Arus, P (2007), *Molecular Marker Systems in Plant Breeding and Crop Improvement (Google eBook)*. 2007, 279-302.

[133] Abbott, A, Arús, P, Scorza, R (2007), Peach. In: Kole C (ed) *Genome mapping and molecular breeding in plants*. Springer, Berlin, pp 137-156.

[134] Pascal, T, Pfeiffer, F, Kervella, J (2010), Powdery Mildew Resistance in the Peach Cultivar Pamirskij 5 Is Genetically Linked with the Gr Gene for Leaf Color. *HortScience* 45, 150-152.

[135] Bailey, J, French, A (1933), The inheritance of certain characteristics in the peach. *Proc Am Soc Hortic Sci* 29, 127-130.

[136] Bailey, J, French, A (1949), The inheritance of certain fruit and foliage characteristics in the peach. *Mass Agr Exp St* 452.

[137] Monet R (1989), Peach genetics: past, present and future. *Acta Hortic* 254, 49-53.

[138] Peace, CP, Ahmad, R, Gradziel, TM, Dandekar, AM, *et al.* (2005), The use of molecular genetics to improve peach and nectarine post- storage quality. *Acta Hort* 682, 403-410.

[139] Cantín, CM, Crisosto, CH, Ogundiwin, EA, Gradziel, T, *et al.* (2010), Chilling injury susceptibility in an intra-specific peach [*Prunus persica* (L.) Batsch] progeny. *Postharvest Biol Technol* 58, 79-87.

[140] Peace, CP, Callahan, AM, Ogundiwin, EA, Potter, D, *et al.* (2007), Endopolygalacturonase Genotypic Variation in *Prunus*. *Acta Hort* 738, 639-645.

[141] Peace, C, Norelli, JL (2009), *Genetics and Genomics of Rosaceae*. 19-53. doi: 10.1007/978-0-387-77491-6.

[142] Boudehri, K, Bendahmane, A, Cardinet, G, Troadec, C, *et al.* (2009), Phenotypic and fine genetic characterization of the *D* locus controlling fruit acidity in peach. *BMC Plant Biol* 9, 59. doi: 10.1186/1471-2229-9-59.

[143] Arus, P, Ballester, J, Jauregui, B, Joobeur, T, *et al.* (1998), The European *Prunus* mapping project: update on marker development in almond. *Acta Hort* 484, 331-338.

[144] Ballester, J, Socias I Company, R, Arus, P, De Vicente, MC (2001), Genetic mapping of a major gene delaying blooming time in almond. *Plant Breed* 120, 268-270. doi: 10.1046/j.1439-0523.2001.00604.x.

[145] Badenes, ML, Hurtado, MA, Sanz, F, Archelos, DM, *et al.* (2000), Searching for molecular markers linked to male sterility and self-compatibility in apricot. *Plant Breed* 119, 157-160. doi: 10.1046/j.1439-0523.2000.00463.x.

[146] Salava, J, Wang, Y, Krska, B, Polak, J, *et al.* (2002), Identification of molecular markers linked to resistance of apricot (*Prunus armeniaca* L.) to plum pox virus. *J Plant Dis Protec* 9, 64-67.

[147] Lalli, DA, Abbott, AG, Zhebentyayeva, TN, Badenes, ML, *et al.* (2008), A genetic linkage map for an apricot (*Prunus armeniaca* L.) BC_1 population mapping plum pox virus resistance. *Tree Genet Genomes* 4, 481-493. doi: 10.1007/s11295-007-0125-x.

[148] Soriano, JM, Vera-Ruiz, EM, Vilanova, S, Martínez-Calvo, J, *et al.* (2008), Identification and mapping of a locus conferring plum pox virus resistance in two apricot-improved linkage maps. *Tree Genet Genomes* 4, 391-402. doi: 10.1007/s11295-007-0118-9.

[149] Soriano, J, Domingo, M, Zuriaga, E, Romero, C, *et al.* (2011), Genetic tools for selecting resistance to sharka disease in apricot. *Acta Hort* 966, 255-258.

[150] Cachi, AM, Wünsch, A (2011), Characterization and mapping of non-S gametophytic self-compatibility in sweet cherry (*Prunus avium* L.). *J Exp Bot* 62, 1847-56. doi: 10.1093/jxb/erq374.

[151] Falchi, R, Vendramin, E, Zanon, L, Scalabrin, S, *et al.* (2013), Three distinct mutational mechanisms acting on a single gene underpin the origin of yellow flesh in peach. *Plant J* 76, 175-87. doi: 10.1111/tpj.12283.

[152] Adami, M, Franceschi, P, Brandi, F, Liverani, A, *et al.* (2013), Identifying a Carotenoid Cleavage Dioxygenase (ccd4) Gene Controlling Yellow/White Fruit Flesh Color of Peach. *Plant Mol Biol Report* 31, 1166-1175. doi: 10.1007/s11105-013-0628-6.

[153] Vendramin, E, Pea, G, Dondini, L, Pacheco, I, *et al.* (2014), A unique mutation in a MYB gene cosegregates with the nectarine phenotype in peach. *PLoS One* 9, e90574. doi: 10.1371/journal.pone.0090574.

[154] Dardick, C, Callahan, A, Horn, R, Ruiz, KB, *et al.* (2013), *PpeTAC1* promotes the horizontal growth of branches in peach trees and is a member of a functionally conserved gene family found in diverse plants species. *Plant J* 75, 527-709. doi: 10.1111/tpj.12234.

[155] Tanksley, SD (1993), Mapping polygenes. *Annu Rev Genet* 27, 205-33. doi: 10.1146/annurev.ge.27.120193.001225.

[156] Arús, P, Howad, W, Mnejja, M (2005), Marker development and marker assisted selection in temperate fruit trees. In: Tuberosa R, Phillips R, Gale M (eds) *Proceedings of the international congress: "In the wake of the double helix: from the green revolution to the Gene Revolution" 27-31 May 2003, Bologna, Italy*. Avenue media, Bologna, Italy, pp 309-325.

[157] Abbott, A, Rajapakse, S, Sosinski, B, Lu, Z, *et al.* (1998), Construction of saturated linkage maps of peach crosses segregating for characters controlling fruit quality, tree architecture and pest resistance. *Acta Hort* 465, 41-49.

[158] Dirlewanger, E, Moing, A, Rothan, C, Svanella, L, *et al.* (1999), Mapping QTLs controlling fruit quality in peach (*Prunus persica* (L.) Batsch). *Theor Appl Genet* 98, 18-31. doi: 10.1007/s001220051035.

[159] Etienne, C, Rothan, C, Moing, A, Plomion, C, *et al.* (2002), Candidate genes and QTLs for sugar and organic acid content in peach [*Prunus persica* (L.) Batsch]. *Theor Appl Genet* 105, 145-159. doi: 10.1007/s00122-001-0841-9.

[160] Verde, I, Quarta, R, Cedrola, C, Dettori, M (2002), QTL analysis of agronomic traits in a BC_1 peach population. *Acta Hort* 592, 291-297.

[161] Foulongne, M, Pascal, T, Pfeiffer, F, Kervella, J (2003), QTLs for powdery mildew resistance in peach × *Prunus davidiana* crosses : consistency across generations and environments. *Mol Breed* 12, 33-50.

[162] Bielenberg, DG, Wang, Y (Eileen), Li, Z, Zhebentyayeva, T, *et al.* (2008), Sequencing and annotation of the evergrowing locus in peach [*Prunus persica* (L.) Batsch] reveals a cluster of six MADS-box transcription factors as candidate genes for regulation of terminal bud formation. *Tree Genet Genomes* 4, 495-507. doi: 10.1007/s11295-007-0126-9.

[163] Peace, C, Crisosto, C, Garner, D, Dandekar, A, *et al.* (2006), Genetic control of internal breakdown in peach. *Acta Hortic* 713, 489-496.

[164] Ogundiwin, EA, Peace, CP, Gradziel, TM, Dandekar, AM, *et al.* (2007), Molecular genetic dissection of chilling injury in peach fruit. *Acta Hortic* 738, 633-638.

[165] Frett, TJ, Reighard, GL, Okie, WR, Gasic, K (2014), Mapping quantitative trait loci associated with blush in peach [*Prunus persica* (L.) Batsch]. Tree Genet Genomes. doi: 10.1007/s11295-013-0692-y.

[166] Pirona, R, Eduardo, I, Pacheco, I, Da Silva Linge, C, *et al.* (2013), Fine mapping and identification of a candidate gene for a major locus controlling maturity date in peach. *BMC Plant Biol* 13, 166. doi: 10.1186/1471-2229-13-166.

[167] Sooriyapathirana, SS, Khan, A, Sebolt, AM, Wang, D, *et al.* (2010), QTL analysis and candidate gene mapping for skin and flesh color in sweet cherry fruit (*Prunus avium* L.). *Tree Genet Genomes* 6, 821-832. doi: 10.1007/s11295-010-0294-x.

[168] Wang, D, Karle, R, Iezzoni, AF (2000), QTL analysis of flower and fruit traits in sour cherry. *Theor Appl Genet* 100, 535-544. doi: 10.1007/s001220050070.

[169] Zhang, G, Sebolt, AM, Sooriyapathirana, SS, Wang, D, *et al.* (2010), Fruit size QTL analysis of an F_1 population derived from a cross between a *domestica*ted sweet cherry

cultivar and a wild forest sweet cherry. *Tree Genet Genomes* 6, 25-36. doi: 10.1007/s11295-009-0225-x.

[170] Cervellati, C, Paetz, C, Dondini, L, Tartarini, S, *et al.* (2012), A qNMR approach for bitterness phenotyping and QTL identification in an F_1 apricot progeny. *J Biotechnol* 159, 312-9. doi: 10.1016/j.jbiotec.2011.09.004.

[171] Lambert, P, Dicenta, F, Rubio, M, Audergon, JM (2007), QTL analysis of resistance to sharka disease in the apricot (*Prunus armeniaca* L.) 'Polonais' × 'Stark Early Orange' F_1 progeny. *Tree Genet Genomes* 3, 299-309. doi: 10.1007/s11295-006-0069-6.

[172] Sánchez-Pérez, R, Howad, W, Dicenta, F, Arús, P, *et al.* (2007), Mapping major genes and quantitative trait loci controlling agronomic traits in almond. *Plant Breed* 126, 310-318. doi: 10.1111/j.1439-0523.2007.01329.x.

[173] Wang, Y, Georgi, LL, Reighard, GL, Scorza, R, *et al.* (2002), Genetic mapping of the evergrowing gene in peach [*Prunus persica* (L.) Batsch]. *J Hered* 93, 352-8.

[174] Jáuregui, B (1998), Identification of molecular markers linked to agronomic characters in an interspecific almond x peach progeny. University of Barcelona, Spain.

[175] Scorza, R, Melnicenco, L, Dang, P, Abbott, AG (2002), Testing a microsatellite marker for selection of columnar growth habit in peach [*Prunus persica* (L.) Batsch]. *Acta Hort* 592, 285-289.

[176] Joobeur, T (1998), Construction of a molecular marker map and genetic analysis of agronomic characters in *Prunus*. University of Lleida, Spain.

[177] Yamamoto, T, Shimada, T, Imai, T, Yaegaki, H, *et al.* (2001), Characterization of morphological traits based on a genetic linkage map in peach. *Breed Sci* 51, 271-278.

[178] Quarta, R, Dettori, MT, Sartori, A, Verde, I (2000), Genetic linkage map and QTL analysis in peach. *Acta Hort* 521, 233-242.

[179] Warburton, ML, Becerra-Velásquez, VL, Goffreda, JC, Bliss, FA (1996), Utility of RAPD markers in identifying genetic linkages to genes of economic interest in peach. *Theor Appl Genet* 93, 920-925.

[180] Rubio, M, Pascal, T, Bachellez, A, Lambert, P (2010), Quantitative trait loci analysis of Plum pox virus resistance in *Prunus davidiana* P1908: new insights on the organization of genomic resistance regions. *Tree Genet Genomes* 6, 291-304. doi: 10.1007/s11295-009-0249-2.

[181] Claverie, M, Bosselut, N, Lecouls, AC, Voisin, R, *et al.* (2004), Location of independent root-knot nematode resistance genes in plum and peach. *Theor Appl Genet* 108, 765-73. doi: 10.1007/s00122-003-1463-1.

[182] Lu, ZX, Reighard, GL, Nyczepir, AP, Beckman, TG, *et al.* (1998), Inheritance of resistance to root-knot nematodes in peach rootstocks. *Acta Hort* 465, 111-116.

[183] Viruel, MAM, Madur, D, Dirlewanger, E, Pascal, T, *et al.* (1998), Mapping quantitative trait loci controlling peach leaf curl resistance. *Acta Hort* 465, 79-88.

[184] Houle, D, Govindaraju, DR, Omholt, S (2010), Phenomics: the next challenge. *Nat Rev Genet* 11, 855-66. doi: 10.1038/nrg2897.

[185] Arumuganathan, K, Earle, E (1991), Nuclear DNA content of some important plant species.*Plant Mol Biol Report* 9, 208-218.

[186] Trainotti, L, Cagnin, S, Forcato, C, Bonghi, C, *et al.* (2012), Functional genomics: transcriptomics. In: Kole C, Abbott AG (eds) *Genetics, Genomics and Breeding of Stone Fruits*. CRC Press Taylor & Francis Group, pp 292-322.

[187] Nagaraj, SH, Gasser, RB, Ranganathan, S (2007), A hitchhiker's guide to expressed sequence tag (EST) analysis. *Brief Bioinform* 8, 6-21. doi: 10.1093/bib/bbl015.

[188] Martínez-Gómez, P, Crisosto, CCH, Bonghi, C, Rubio, M (2011), New approaches to *Prunus* transcriptome analysis. *Genetica* 139, 755-69. doi: 10.1007/s10709-011-9580-2.

[189] Trainotti, L, Zanin, D, Casadoro, G (2003), A cell wall-oriented genomic approach reveals a new and unexpected complexity of the softening in peaches. *J Exp Bot* 54, 1821-32. doi: 10.1093/jxb/erg198.

[190] Vecchietti, A, Lazzari, B, Ortugno, C (2009), Comparative analysis of expressed sequence tags from tissues in ripening stages of peach (*Prunus persica* L. Batsch). *Tree Genet genomes* 5, 377-391.

[191] Crisosto, CH, Crisosto, G, Neri, F (2006), Understanding Tree Fruit Quality Based on Consumer Acceptance. *SHS Acta Hortic* 712, 183-190.

[192] Vizoso, P, Meisel, L a, Tittarelli, A, Latorre, M, *et al.* (2009), Comparative EST transcript profiling of peach fruits under different post-harvest conditions reveals candidate genes associated with peach fruit quality. *BMC Genomics* 10, 423. doi: 10.1186/1471-2164-10-423.

[193] Etienne, C, Moing, A, Dirlewanger, E, Raymond, P, *et al.* (2002), Isolation and characterization of six peach cDNAs encoding key proteins in organic acid metabolism and solute accumulation: involvement in regulating peach fruit acidity. *Physiol Plant* 114, 259-270. doi: 10.1034/j.1399-3054.2002.1140212.x.

[194] Vilanova, S, Sargent, DJ, Arús, P, Monfort, A (2008), Synteny conservation between two distantly-related Rosaceae genomes: *Prunus* (the stone fruits) and *Fragaria* (the strawberry). *BMC Plant Biol* 8, 67. doi: 10.1186/1471-2229-8-67.

[195] Yamamoto, T, Mochida, K, Imai, T, Shi, YZ, *et al.* (2002), Microsatellite markers in peach [*Prunus persica* (L.) Batsch] derived from an enriched genomic and cDNA libraries. *Mol Ecol Notes* 2, 298-301. doi: 10.1046/j.1471-8286.2002.00242.x.

[196] Jung, S, Abbott, A, Jesudurai, C, Tomkins, J, *et al.* (2005), Frequency, type, distribution and annotation of simple sequence repeats in Rosaceae ESTs. *Funct Integr Genomics* 5, 136-43. doi: 10.1007/s10142-005-0139-0.

[197] Jung, S, Jesudurai, C, Staton, M, Du, Z, *et al.* (2004), GDR (Genome Database for Rosaceae): integrated web resources for Rosaceae genomics and genetics research. *BMC Bioinformatics* 5, 130. doi: 10.1186/1471-2105-5-130.

[198] Lazzari, B, Caprera, A, Vecchietti, A, Stella, A, *et al.* (2005), ESTree db: a tool for peach functional genomics. *BMC Bioinformatics* 6 Suppl 4, S16. doi: 10.1186/1471-2105-6-S4-S16.

[199] Lazzari, B, Caprera, A, Vecchietti, A, Merelli, I, *et al.* (2008), Version VI of the ESTree db: an improved tool for peach transcriptome analysis. *BMC Bioinformatics* 9 Suppl 2, S9. doi: 10.1186/1471-2105-9-S2-S9.

[200] Rensink, WA, Buell, CR (2005), Microarray expression profiling resources for plant genomics. *Trends Plant Sci* 10, 603-9. doi: 10.1016/j.tplants.2005.10.003.

[201] EST Consortium (2005), Development of an oligo-based microarray(μPEACH 1.0) for genomics studies in peach fruit. *Acta Hort* 682, 263-268.

[202] Trainotti, L, Bonghi, C, Ziliotto, F, Zanin, D (2006), The use of microarray μPEACH1.0 to investigate transcriptome changes during transition from pre-climacteric to climacteric phase in peach fruit. *Plant Sci* 170, 606-613.

[203] Trainotti, L, Tadiello, A, Casadoro, G (2007), The involvement of auxin in the ripening of climacteric fruits comes of age: the hormone plays a role of its own and has an intense interplay with ethylene in ripening peaches. *J Exp Bot* 58 (12):, 3299-3308.

[204] Ziosi, V, Bonghi, C, Bregoli, AM, Trainotti, L, *et al.* (2008), Jasmonate-induced transcriptional changes suggest a negative interference with the ripening syndrome in peach fruit. *J Exp Bot* 59, 563-73. doi: 10.1093/jxb/erm331.

[205] Ziliotto, F, Begheldo, M, Rasori, A, Bonghi, C, *et al.* (2008), Transcriptome profiling of ripening nectarine (*Prunus persica* L. Batsch) fruit treated with 1-MCP. *J Exp Bot* 59, 2781-91. doi: 10.1093/jxb/ern136.

[206] Manganaris, GA, Rasori, A, Bassi, D, Geuna, F, *et al.* (2011), Comparative transcript profiling of apricot (*Prunus armeniaca* L.) fruit development and on-tree ripening. *Tree Genet Genomes* 7, 609-616. doi: 10.1007/s11295-010-0360-4.

[207] Jung, S, Main, D, Staton, M, Cho, I, *et al.* (2006), Synteny conservation between the *Prunus* genome and both the present and ancestral *Arabidopsis* genomes. *BMC Genomics* 7, 81. doi: 10.1186/1471-2164-7-81.

[208] Manganaris, GA, Bonghi, C, Ramina, A, Ziosi, V, *et al.* (2010), A Preliminary Transcriptomic Approach to Elucidate Post Harvest Ripening of Plum Fruit. *Acta Hort* 874, 99-106.

[209] Leida, C, Terol, J, Martí, G, Agustí, M, *et al.* (2010), Identification of genes associated with bud dormancy release in *Prunus persica* by suppression subtractive hybridization. *Tree Physiol* 30, 655-66. doi: 10.1093/treephys/tpq008.

[210] Pirona, R, Vecchietti, A, Lazzari, B, Caprera, A, *et al.* (2013), Expression profiling of genes involved in the formation of aroma in two peach genotypes. *Plant Biol* 15, 443-51. doi: 10.1111/j.1438-8677.2012.00666.x.

[211] Wang, Z, Gerstein, M, Snyder, M (2009), RNA-seq: a revolutionary tool for transcriptomics. *Nat Rev Genet* 10, 57-63. doi: 10.1038/nrg2484.

[212] Socquet-Juglard, D, Kamber, T, Pothier, JFJ, Christen, D, *et al.* (2013), Comparative RNA-seq analysis of early-infected peach leaves by the invasive phytopathogen *Xanthomonas arboricola* pv. *pruni*. *PLoS One* 8, e54196. doi: 10.1371/journal.pone.0054196.

[213] Alkio, M, Jonas, U, Sprink, T, van Nocker, S, *et al.* (2012), Identification of putative candidate genes involved in cuticle formation in *Prunus avium* (sweet cherry) fruit. *Ann Bot* 110, 101-112.

[214] Du, D, Zhang, Q, Cheng, T, Pan, H, *et al.* (2013), Genome-wide identification and analysis of late embryogenesis abundant (LEA) genes in *Prunus mume*. *Mol Biol Rep* 40, 1937-46. doi: 10.1007/s11033-012-2250-3.

[215] Du, D, Hao, R, Cheng, T, Pan, H, *et al.* (2012), Genome-Wide Analysis of the *AP2/ERF* Gene Family in *Prunus mume*. *Plant Mol Biol Report* 31, 741-750. doi: 10.1007/s11105-012-0531-6.

[216] Avendano, C, Rodriguez, F, Barindell, E, Mujica, K, *et al.* (2013), The Cytokinin Response Pathway During *Prunus persica* Fruit Development. Plant Anim. Genome XXI Conf.

[217] Jones-Rhoades, MW (2012), Conservation and divergence in plant microRNAs. *Plant Mol Biol* 80, 3-16. doi: 10.1007/s11103-011-9829-2.

[218] Colaiacovo, M, Bernardo, L, Centomani, I, Crosatti, C, *et al.* (2012), A Survey of MicroRNA Length Variants Contributing to miRNome Complexity in Peach (*Prunus persica* L.). *Front Plant Sci* 3, 165. doi: 10.3389/fpls.2012.00165.

[219] Barakat, A, Sriram, A, Park, J, Zhebentyayeva, T, *et al.* (2012), Genome wide identification of chilling responsive microRNAs in *Prunus persica*. *BMC Genomics* 13, 481. doi: 10.1186/1471-2164-13-481.

[220] Tuskan, GA, DiFazio, S, Jansson, S, Bohlmann, J, *et al.* (2006), The genome of black cottonwood, *Populus trichocarpa* (Torr. & Gray). *Science (80-)* 313, 1596-1604.

[221] Velasco, R, Zharkikh, A, Affourtit, J, Dhingra, A, *et al.* (2010), The genome of the *domestica*ted apple (*Malus* × *domestica* Borkh.). *Nat Genet* 42, 833-9. doi: 10.1038/ng.654.

[222] Branca, A, Paape, TD, Zhou, P, Briskine, R, *et al.* (2011), Whole-genome nucleotide diversity, recombination, and linkage disequilibrium in the model legume *Medicago truncatula*. *Proc Natl Acad Sci U S A* 108, E864-70. doi: 10.1073/pnas.1104032108.

[223] Cao, J, Schneeberger, K, Ossowski, S, Günther, T, *et al.* (2011), Whole-genome sequencing of multiple *Arabidopsis thaliana* populations. *Nat Genet* 43, 956-63. doi: 10.1038/ng.911.

[224] Zhang, Q, Chen, W, Sun, L, Zhao, F, *et al.* (2012), The genome of *Prunus mume*. *Nat Commun* 3, 1318. doi: 10.1038/ncomms2290.

[225] Pflieger, S, Lefebvre, V, Causse, M (2001), The candidate gene approach in plant genetics: a review. *Mol Breed* 7, 275-291. doi: 10.1023/A:1011605013259.

[226] Iezzoni, A, Cabrera, A, De Franceschi, P, Van der Knaap, E, *et al.* (2012), The genetic control of fruit size in cherry (*Prunus*): from phenotype to candidate genes. *6th Rosaceous Genomics Conference,*. Mezzocorona (Trento), Italy, p 32.

[227] Guo, M, Rupe, MA, Dieter, JA, Zou, J, *et al.* (2010), Cell Number Regulator1 affects plant and organ size in maize: implications for crop yield enhancement and heterosis. *Plant Cell* 22, 1057-73. doi: 10.1105/tpc.109.073676.

[228] Frary, A, Nesbitt, TC, Grandillo, S, Knaap, E, *et al.* (2000), fw2.2: a quantitative trait locus key to the evolution of tomato fruit size. *Science* 289, 85-8.

[229] Duval, H, Hoerter, M, Masse, M, Moretti, A, *et al.* (2012), Marker assisted breeding for peach seedling rootstocks resistant to rootknotnematodes. *6th Rosaceous Genomics Conference*. Mezzocorona (Trento), Italy, p 61.

[230] Soriano, JM, Domingo, ML, Zuriaga, E, Romero, C, *et al.* (2012), Identification of simple sequence repeat markers tightly linked to plum pox virus resistance in apricot. *Mol Breed* 30, 1017-1026. doi: 10.1007/s11032-011-9685-4.

[231] Zuriaga, E, Soriano, JM, Zhebentyayeva, T, Romero, C, *et al.* (2013), Genomic analysis reveals MATH gene(s) as candidate(s) for Plum pox virus (PPV) resistance in apricot (*Prunus armeniaca* L.). *Mol Plant Pathol* 14, 663-677. doi: 10.1111/mpp.12037.

[232] Dirlewanger, E, Quero-García, J, Le Dantec, L, Lambert, P, *et al.* (2012), Comparison of the genetic determinism of two key phenological traits, flowering and maturity dates, in three *Prunus* species: peach, apricot and sweet cherry. *Heredity (Edinb)* 109, 280-92. doi: 10.1038/hdy.2012.38.

[233] Xu, P, Wu, X, Wang, B, Luo, J, *et al.* (2012), Genome wide linkage disequilibrium in Chinese asparagus bean (Vigna. unguiculata ssp. *sesquipedialis*) germplasm: implications for *domestica*tion history and genome wide association studies. *Heredity (Edinb)* 109, 34-40. doi: 10.1038/hdy.2012.8.

[234] Metzker, ML (2010), Sequencing technologies - the next generation. *Nat Rev Genet* 11, 31-46.

[235] Zhebentyayeva, T, Fan, S, Olukolu, B, Hughes-Murphree, S, Chandra, A, *et al.* (2012), Deep resequencing the QTL intervals for chilling requirement and bloom date in peach, a model genome system for woody tree species genetics. *6th Rosaceous Genomics Conference*. Mezzocorona (Trento), Italy, p 134.

[236] Poland, JA, Rife, TW (2012), Genotyping-by-Sequencing for Plant Breeding and Genetics. *Plant Genome J* 5, 92. doi: 10.3835/plantgenome2012.05.0005.

[237] Gore, MA, Chia, J-M, Elshire, RJ, Sun, Q, *et al.* (2009), A first-generation haplotype map of maize. *Science* 326, 1115-7. doi: 10.1126/science.1177837.

[238] Deschamps, S, la Rota, M, Ratashak, JP, Biddle, P, *et al.* (2010), Rapid Genome-wide Single Nucleotide Polymorphism Discovery in Soybean and Rice *via* Deep Resequencing of Reduced Representation Libraries with the Illumina Genome Analyzer. *Plant Genome J* 3, 53. doi: 10.3835/plantgenome2009.09.0026.

[239] Hyten, DL, Choi, I-Y, Song, Q, Specht, JE, *et al.* (2010), A High Density Integrated Genetic Linkage Map of Soybean and the Development of a 1536 Universal Soy Linkage Panel for Quantitative Trait Locus Mapping. *Crop Sci* 50, 960. doi: 10.2135/cropsci2009.06.0360.

[240] Nelson, JC, Wang, S, Wu, Y, Li, X, *et al.* (2011), Single-nucleotide polymorphism discovery by high-throughput sequencing in sorghum. *BMC Genomics* 12, 352. doi: 10.1186/1471-2164-12-352.

[241] Byers, RL, Harker, DB, Yourstone, SM, Maughan, PJ, *et al.* (2012), Development and mapping of SNP assays in allotetraploid cotton. *Theor Appl Genet* 124, 1201-14. doi: 10.1007/s00122-011-1780-8.

[242] Meneses, C, Vera, G, Tonda, R, Arús, P, *et al.* (2012), Saturation of the 'Texas' x 'Earligold' *Prunus* reference map using genotyping by sequencing. *6th Rosaceous Genomics Conference*. Mezzocorona (Trento), Italy, p 110.

[243] Rauh, B, Fan, S, Abbott, A, Gasic, K, *et al.* (2012), Use of genotypingbysequencing for quantitative trait loci mapping of chilling requirement and bloom date in peach',. *6th Rosaceous Genomics Conference*. Mezzocorona (Trento), Italy, p 79.

[244] Decroocq, V, Zhebentyayeva, T, Lambert, P, Geuna, F, Salava, J, *et al.* (2012), Integrated approaches in *Prunus armeniaca* for resistance to Sharka disease. *6th Rosaceous Genomics Conference*. Mezzocorona (Trento), Italy, p 62.

[245] Kubisiak, TL, Nelson, CD, Staton, ME, Zhebentyayeva, T, *et al.* (2012), A transcriptome-based genetic map of Chinese chestnut (*Castanea mollissima*) and identification of regions of segmental homology with peach (*Prunus persica*). *Tree Genet Genomes* 9, 557-571. doi: 10.1007/s11295-012-0579-3.

[246] Mariette, S, Tavaud, M, Arunyawat, U, Capdeville, G, *et al.* (2010), Population structure and genetic bottleneck in sweet cherry estimated with SSRs and the gametophytic self-incompatibility locus. *BMC Genet* 11, 77. doi: 10.1186/1471-2156-11-77.

[247] Arunyawat, U, Capdeville, G, Decroocq, V, Mariette, S (2012), Linkage disequilibrium in French wild cherry germplasm and worldwide sweet cherry germplasm. *Tree Genet Genomes* 8, 737-755. doi: 10.1007/s11295-011-0460-9.

Chapter 6: Stone Fruits Production, Postharvest Storage, Processing and Nutrition

CHAPTER 6

Stone Fruits Production, Postharvest Storage, Processing and Nutrition

Muhammad Siddiq*

Department of Food Science and Human Nutrition, Michigan State University, East Lansing, MI 48824, USA (Currently: Food Science Consultant, Windsor, Canada)

Abstract: Apricots, cherries, peaches, nectarines, and plums (all belonging to genus *Prunus*), are commonly referred to as "stone fruits" due to the fact that their seed is enclosed in a hard (stone-like) endocarp. Stone fruits are produced in almost all agro-ecological regions of the world, with commercial production reported in over 80 countries. All stone fruits are healthy and nutritious, being low in calories and rich in many nutrients and bioactive compounds (antioxidants). Stone fruits, except for sour cherries, are enjoyed fresh due to their rich flavors and aromas. Stone fruits are processed into a variety of products, *e.g.*, canned, frozen, dried, puree, juice, concentrate, jam, and jelly. This chapter provides an overview of stone fruits production, postharvest physiology and storage, processing/products, nutritional profile, and health benefits.

Keywords: *Prunus*, postharvest physiology, storage, processing, antioxidants, canned peaches, dried apricots, prune juice, plum paste, cherry concentrate, apricot jam, Maraschino cherries.

INTRODUCTION

Stone fruits, so called since their seed is enclosed in a hard (stone-like) endocarp, include apricots (*Prunus armeniaca*), cherries (sweet: *Prunus avium* L.; sour/tart: *Prunus cerasus* L.), peaches and nectarines (*Prunus persica*), and plums (*Prunus domestica*). Almonds are another stone fruit but they are categorized under nuts and therefore not covered in this chapter. Stone fruits are produced in almost all agro-ecological regions of the world, although production varies greatly in different countries. Commercial production of most stone fruits, except sour

***Corresponding author Muhammad Siddiq:** Department of Food Scienceand Human Nutration, Michigan State University, East Lansing, MI 48824, USA; Tel: +15173558474; E-mail: siddiq@msu.edu

Innocenzo Muzzalupo & Sabrina Micali (Eds.)

cherries, is reported in over 80 countries, whereas sour cherries are grown in about 30 countries [1].

The total world production of all stone fruits was 38.21 million metric tons in 2010 (Fig. **1**). Peaches and nectarines represent the biggest share (53.73%) of total production with 20.38 million metric tons, while cherries (sweet and sour) account for only 8.45% or 3.22 million metric tons. Production of peaches and nectarines has seen a tremendous growth since 1990, with 2010 production almost 120% higher than that of 1990. In the same time period, plums (79.80%), apricots (58.13%), sweet cherries (47.74%) and sour cherries (34.06%) have also experienced significant increases.

All stone fruits are healthy and nutritious, being low in calories and rich in many nutrients and bioactive compounds (antioxidants). Stone fruits, except for sour cherries, are enjoyed fresh due to their rich flavors and aromas. Stone fruits can also be processed into a variety of packaged products, *e.g.* canned, frozen, dried, puree, juice, concentrate, jam, and jelly. This chapter provides a review of production, postharvest physiology and storage, processing/products, nutritional profile, and health benefits of stone fruits.

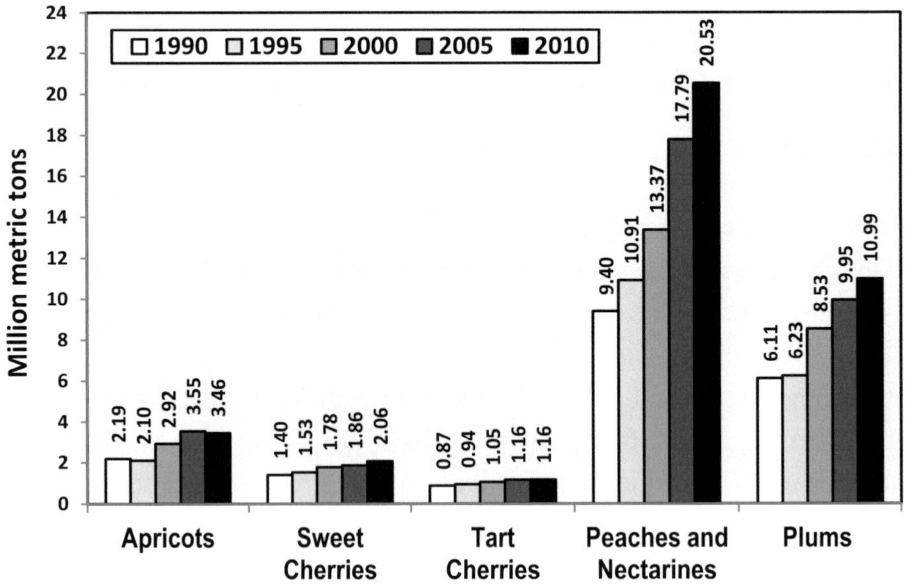

Figure 1: World production of different stone fruits for selected years from 1990 to 2010. Adapted from FAOSTAT [1].

APRICOTS

Apricot (*Prunus armeniaca)* belongs to the genus *Prunus* in the Rosaceae family [2]. This fruit is native to temperate Asia, first discovered growing wild on the mountain slopes of China, and later cultivated in Armenia. Apricot, like most stone fruits, thrives in a Mediterranean climate of long, hot summers and cool, wet winters; the fruits mature primarily in early summer making them one of the earliest available summer fruits [3]. The apricot tree is of medium size, usually held under 18 feet by pruning. The fruit is generally globose to slightly oblong, 1.25 to 2.5 inches in diameter; the fruit flesh is yellow and the skin is yellow or blushed red. The early apricot bloom is highly susceptible to injury by frost. In addition, fruit is subjected to cracking in humid climates, thus limiting commercial production in the U.S. to the states west of the Rocky Mountains [2, 4].

Table 1: Leading apricot producing countries, by quantity—million metric tons (1990–2010).

Country/Year	1990	1995	2000	2005	2010
Turkey	300,000	281,000	579,000	894,000	476,132
Iran	85,474	192,922	262,432	275,578	400,000
Uzbekistan	--[1]	55,000	68,000	170,000	325,000
Italy	184,710	104,685	201,372	232,882	252,892
Algeria	34,979	41,233	56,354	145,097	239,700
Pakistan	81,000	190,634	125,889	197,239	220,000
France	110,432	100,639	138,944	176,950	139,569
Morocco	73,700	78,000	119,600	103,600	132,398
China	--	48,022	88,317	90,937	94,995
Syria	72,700	30,392	78,873	65,513	93,700
Egypt	38,000	53,948	62,613	73,000	92,704
Japan	--	--	121,200	123,000	92,400

[1]not reported
From FAOSTAT [1]

Major apricot producing countries are listed in Table **1**. Turkey was the leading apricot producing country in 2010, followed by Iran and Uzbekistan; these three countries accounted for 35% share of the total world production of 3.46 million metric tons. Apricot production in Turkey has declined in recent years. Major apricot exporting countries (by quantity) were France, Spain, Turkey, Italy,

Uzbekistan, whereas, the Russian Federation, Germany, Italy, Austria, and Netherlands were the major importing countries in 2010 [1]. In the U.S., the first major production of apricots was recorded in California in 1792. California continues to be the leading apricot producing state in the United States. Over 90% of apricots in the U.S. are grown in California, with much smaller production operations in Washington and Utah [5].

Production and Postharvest Physiology

Production, Maturity and Fruit Quality

Mediterranean-type mild climates, where danger of spring frost is minimal, are most-suited for apricot production. Apricot trees exhibit moderate tolerance to high pH soils and salinity but are intolerant to water-logging. Deep, fertile, well-drained soils are best suited to produce tall and healthy trees. Apricots are pruned fairly heavily as they bear too many fruits. Generally, most new growth and interfering wood is removed each year, exposing the spurs to maximal sunlight [6]. Perez-Gonzales [7] reported that the most important morphological traits of the apricot trees that correlated with fruit weight were tree growth habit, apical and basal diameter of fruiting spurs, and bud and leaf size. Timing and type of fertilizer application are known to improve external and internal fruit quality and storage ability, reduce production costs, maintain soil fertility, avoid nutrient deficiency or excess, and control tree vigor [8].

Fruit maturity is determined mainly based on changes in the skin ground color (green to yellow); the exact yellowish-green color is cultivar-dependent. Since apricots are susceptible to high bruising when fully ripe and soft, picking is recommended when the fruit is still firm. Important quality criteria include fruit size, shape, freedom from defects (including gel breakdown and pit burn) and decay. Soluble solids content of over 10% and a moderate titratable acidity of 0.7-1.0% are critical for high consumer acceptance. Fruit with flesh firmness of 2-3 lb force (8.9-13.3 N) is ready for fresh consumption [9]. Femenia and colleagues [10] studied the ripening-related changes in the cell wall of apricot fruit and concluded that the decrease in pectic galactans and inhibition of cross-linking within the pectic backbone were related to the softening process that occurs during apricot fruit ripening.

Harvest and Storage

Manual picking of apricots for fresh consumption and processing is most common and trees are usually picked over 2-3 times each, depending on the optimal fruit firmness. Apricots are generally handled in half bins, tray-packed in single/double layers, and shipped in shallow containers to prevent crushing and bruising. Apricots should be uniform in size, and by count, no more than 5% apricots in each container may vary more than 6 mm when measured at the widest part of the cross section. Gomez and Ledbetter [11] reported that full flavor never develops when apricots ripen off the tree.

Though not stored in large quantities like apples, apricots generally keep well for 1-2 weeks at –0.5 to 0 °C with RH of 90-95%. Susceptibility to freezing injury depends on the soluble solids content, which varies from 10 to 14%. Cultivars that are sensitive to chilling, develop and express chilling injury symptoms (gel breakdown, flesh browning), and loss of flavor, more rapidly at 5 °C than at 0 °C. Controlled atmosphere (CA) storage conditions of 2-3% O_2 + 2 to 3% CO_2 are suggested to retain fruit firmness and ground color. The development of off-flavors may hasten at O_2 exposure of < 1% and CO_2 exposure in excess of 5% for over 2 weeks can cause flesh browning and loss of flavor. Pre-storage treatment with 20% CO_2 for 2 days may reduce incidence of decay during subsequent transport and/or storage in CA or air. The respiration rate at 10 °C is more than double compared to 0 °C. The ethylene production rate increases significantly with temperature increase; from < 0.1 $\mu L\ kg^{-1}h^{-1}$ at 0 °C to 4-6 $\mu L\ kg^{-1}h^{-1}$ at 20 °C for firm-ripe apricots and even higher for soft-ripe apricots. The greatest hazard in handling and shipping apricots is decay, mainly brown rot and rhizopus rot, and accelerated ethylene production can hasten the development of such decay. In order to delay ripening, softening and decay, quick cooling to 4 °C or less and storage at temperatures as close to 0 °C as possible is recommended [9].

A variety of methods and treatments have been investigated to extend shelf-life and preserve flavor, firmness, and other quality attributes of apricots. In recent years, the use of ethylene inhibitor 1-methylcyclopropene (1-MCP) to delay ripening and prolong storage life of apricots has shown good results for ethylene inhibition, decline in pectin methyl esterase activity, and preserving fruit firmness and flavor [12, 13]. Zhang and colleagues [14] studied the effects of 1-MCP treatment (1 mg/kg for 24 hours) on the postharvest quality deterioration of fresh apricots stored at 22 ºC and 4 ºC. Evaluation of respiration rate, ethylene release, and weight-loss showed that 1-MCP was effective in delaying ripening and maintaining quality.

Martinez-Romero and colleagues [15] studied the effects of post-harvest putrescine treatment on the physicochemical properties and physiological response to mechanical damage of apricots (cv. Maurizio) during storage. Apricots harvested at the commercial ripening stage were treated with 1 mM putrescine by pressure infiltration, then mechanically damaged with a 25 N force and stored at 10°C for 6 days. Putrescine treatment increased fruit firmness and reduced bruising due to mechanical damage. In putrescine-treated fruit, color changes, weight loss, ethylene emission and respiration rates were significantly reduced.

Physiological Disorders

Chilling injury, also known as gel breakdown, and pit burn are the two main physiological disorders in apricots. Chilling injury, which is characterized by the formation of water-soaked pockets that subsequently turn brown, usually develops at 2.2-7.6 ºC during extended cold storage [9]. Tissue breakdown is sometimes accompanied by sponginess and gel formation. In addition to a short market life, fruit stored at elevated temperatures also loses flavor. "Pit burn" occurs as a result of softening of the flesh that turns brown around the stone when apricots are exposed to temperatures above 38 ºC before harvest. Tagliavini and Marangoni [8] recommend early diagnosis of bitter pit for guiding applications of calcium sprays. Bussi *et al.* [16] evaluated the effects of N and K fertilization on the incidence of pit burn defect (PD), yields and fruit quality of apricot (cv. Bergeron). They reported that fruit N increased with N fertilization level; high fruit N was shown to be a predisposing factor for PD, whereas higher Ca contents tended to reduce PD during storage.

Processed Products

About 20% of apricots produced are consumed fresh and the rest are processed into dried, canned, frozen, jam, juice, and pureed products [5]. Haciseferogullari and colleagues [17] indicate that physical properties, such as fruit length and diameter, mass, fruit volume, geometric mean diameter, sphericity, bulk density, fruit density, porosity, projected area, static and dynamic coefficient of friction are important in designing processing equipment. The processing information in this section mainly relates to apricots produced and processed in the U.S., unless noted otherwise.

Dried Apricots

Turkey produces almost half of the world's total dried apricots. Dried apricots are processed from the pitted fruit, with a final moisture content of about 25%. Dried apricots may be treated with SO_2 to retain a characteristic color prior to packing.

The fruit is harvested at fully ripe stage and treated with SO_2 to preserve the color of the finished product. Drying is either natural, in the sun, or in large dehydrators commonly used for other fruits, such as plums [6]. Apricots are dried to a final moisture content of around 25%. California apricots are sun-dried in halves while Turkish (Mediterranean) dried apricots are whole with the pit squeezed out [5].

In Turkey, apricots are traditionally sun-dried after pretreatment with SO_2 (obtained by burning sulfur in specially constructed rooms) to moisture content of 23-28%. Sun-drying (Fig. **2**) produces a product with a rich orange color, a translucent appearance and a desirable texture.

Figure 2: Sun drying of apricots in open fields in Turkey. From http://en.wikipedia.org/wiki/Apricot.

However, the long drying time, depending on weather, and manual labor requirements are some of the disadvantages of sun-drying. In the U.S., almost all dried apricots are produced in California. Cultivars that retain their color and flavor during processing and storage are best suited for drying.

To preserve the color quality during storage, sulfites are the most widely used chemicals in the production of dried apricots. Traditionally, sulfited apricots can have SO_2 contents ranging from 1,000-6,000 ppm. However, the maximum legal limit allowed in most countries is 2,000 ppm. Various methods of desulfiting apricots, in order to meet legal limits for SO_2, have been reported by Ozkan and Cemeroglu [18]. Piga and colleagues [20] reported that sulfiting pretreatment can reduce ascorbic acid contents in the dried apricot. Nisar-Alizai and Ahmad [21] reported that apricot products treated with polyphenol oxidase (PPO) inhibitors prior to dehydration were of better color and textural quality than those prepared with sulfite treatment alone. However, shelf-life after drying was restricted, suggesting that PPO inhibitors usually do not provide the same extent of protection as sulfiting does.

Canned Apricots

Though processed on a much smaller scale than the dried form, canned apricots are convenient and easy to use in recipes such as baked goods, salads and sauces. For quick ways to add flavor and nutrients, canned apricots can be added to plain oatmeal, cottage cheese, ice cream or yogurt.

Fruit for canning is selected at the peak of ripeness, which helps to retain the highest quality and nutrient levels. The optimum harvesting period is very short; if the fruit is green, the flavor is astringent and if it is over ripe, it is too soft to handle during processing. After delivery to the processing plant, apricots must be canned without any delay so that the processed fruit maintains better nutritional and flavor quality [6-22]. Typical processing steps for apricot canning [23] are given below:

- ***Grading and Washing:*** Since the fruit received at the processing plant varies in size, the first operation is grading for size by running them over screens of 40/32, 48/32, 56/32, 64/32 and 68/32 inches. After grading, apricots are washed under spray water and any bad smutty spots are trimmed.

- ***Pitting and Cutting:*** Apricots are then split into halves (or other styles) and pitted using mechanical pitters and cutters; splitting is done on the natural dividing line in order to make symmetrical pieces and to have the pit come out easily. The pitters make a preliminary sorting by placing the prime fruit in one receptacle, the hard and irregular in another, and the soft in a third one.

- ***Filling:*** The filling of cut apricots is done by hand or by hand pack fillers. Standard of Identity requires a minimum cut-out Brix (detail given below). Therefore, in order to estimate the strength of the syrup to use, it is important to know soluble solids of fruit, and also how much fruit goes into each can. The syrup is added hot by means of rotary and straight-line syrupers.

- ***Exhausting:*** Apricots contain considerable trapped air, which presents a danger of can pin holing if not driven out. Cans must be either thoroughly exhausted prior to closing, or closed in a steam-flow machine. Depending on the fruit texture, an exhaust procedure up to 10 minutes at 180-190 ºF is used. If necessary, additional syrup is added before the cans are closed.

- ***Processing and Cooling**:* To attain a commercially sterile product, canned apricots should be processed sufficiently long enough for the center temperature of the product to reach a minimum of 190 ºF or 195 ºF for air and water-cooling, respectively. Most canned apricots are processed in continuous retorts or reel cookers at 212 ºF for 17 to 30 minutes, depending on the size of the cans and the texture of the fruit. The cans are generally partially cooled in water and then air-cooled on trays.

The most important quality attribute of canned apricots is texture or firmness; if the texture is too soft, product acceptability is very low even if color and flavor are of an acceptable quality. The textural quality of canned apricots has been researched extensively [24-26]. Chitarra and colleagues [27] studied fruit softening and cell wall pectin solubilization in canned 'Patterson' apricots and reported that canning of slightly immature fruit led to immediate softening of the halves, to the extent that many were of an unacceptable texture. The results showed that such softening occurred in conjunction with a substantial solubilization of (presumably pectic) polysaccharides, rich in uronic acid, arabinose and galactose, in the canning liquid.

Generally, canning apricots are not peeled; however, the lye peeling method, the same as that used for peaches, is employed for those apricots that are peeled before canning. Toker and Bayindirli [28] investigated enzymatic peeling of apricots at 20 ºC, 35 ºC, and 50 ºC. They concluded that enzymatic peeling could be an alternative to chemical or mechanical peeling of stone fruits due to (1) better quality, as fruit retained their structural integrity and fresh fruit properties, (2) reduced heat treatment, and (3) less industrial waste.

Frozen Apricots

The majority of frozen apricots are bulk-packed for use in the food applications listed above. Retail marketing of frozen apricots has not gained popularity on a scale similar to some other frozen fruits. Apricots to be processed as frozen should have even ripening, good color and flavor, low browning tendency, a tender and smooth skin, and firm texture. In California, the principal variety is ‘Patterson’, followed by ‘Blenheim’, ‘Tilton’, and ‘Modesto’, which are processed on a smaller scale. Apricot halves are preferred to sliced fruit and can be processed with or without peeling. Halves and slices are sold to bakers, ice cream makers, and frozen dessert makers, whereas multiple-scored apricots (called machine-pitted) are primarily used for jam, jelly, and preserve making [29, 30].

An apricot freezing method adapted from Boyle and colleagues [29] and Scorza and Hui [30] is described here briefly. Upon arrival at the processing plant, apricots are graded and inspected on a conveyor belt and then passed through a halving and pitting machine. Apricots to be processed as “machine-pitted” (for jam, jelly, and preserves) are pitted using Elliott Pitters. After separation of the pits, halves, slices, or dices are inspected and washed to remove small fruit pieces or skin. They are then treated to prevent browning before they are syruped, packed, and frozen. One of the following two methods are used for browning control: (1) the fruit may be blanched in hot water or steam, or (2) treated with an anti-browning agent, like ascorbic acid. Blanching is done in a single layer on a mesh belt for 3 to 4 minutes in steam to give satisfactory results for firm fruit. In the case of a softer fruit batch, it is best to treat it with ascorbic acid or other anti-browning agents at a level of 0.05% or more in the syrup used for packing. The packing medium can vary from 15 °Brix syrup to dry sugar (fruit to sugar ratio is usually 3:1), which may be sprinkled on the fruit as it is filled into the containers or lightly mixed for even distribution.

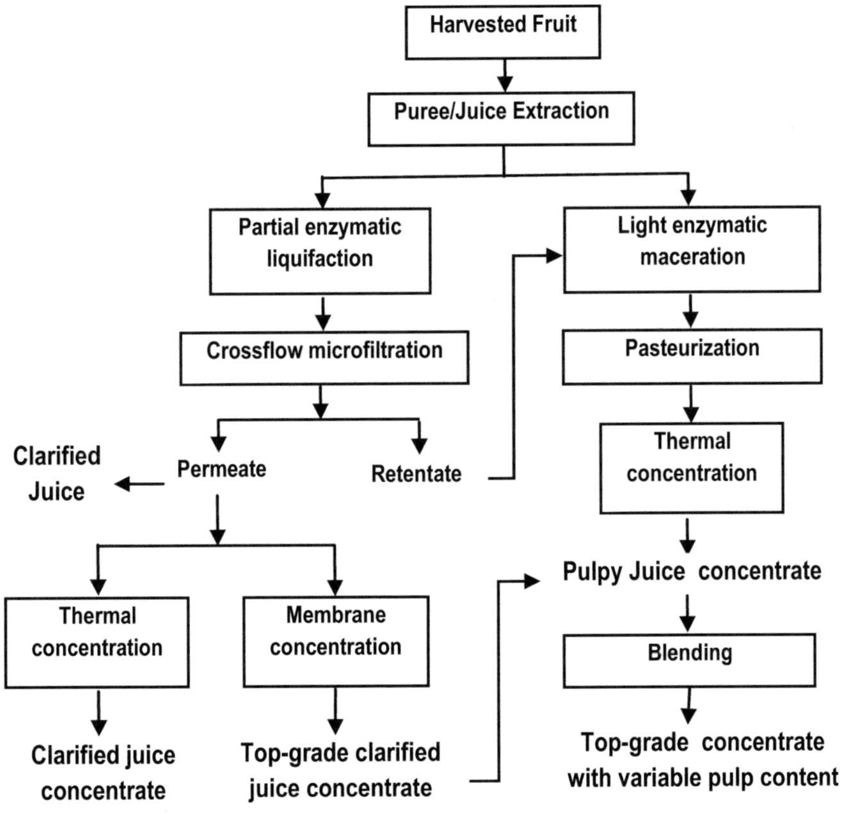

Figure 3: Proposed flowchart for producing clarified juice and juice concentrate from pulpy juices on an industrial scale. From Vaillant and colleagues [34].

Another process for frozen apricots, "osmodehydrofreezing" (osmotic dehydration followed by air drying and freezing) has been proposed for the production of intermediate moisture apricot ingredients without SO_2 application, with a natural and agreeable color [31]. Ten-mm apricot cubes were osmodeyhydrated in sucrose, maltose, or sorbitol syrups at 56% and 3:1 fruit/syrup ratio. Ascorbic acid (1%) and sodium chloride (0.1%) were added as antioxidants. Osmodehydration was carried out at 25 °C for 15 min under vacuum (700mm Hg), and for 45 and 120 minutes at atmospheric pressure. Drying was conducted in an upward air-circulated drier, at a dry-bulb temperature of 65 °C and an air speed of 1.5 m/s, to achieve a soluble solids content corresponding to a_w of 0.86. Osmo-air-dehydrated apricot cubes were then frozen in an air-blast tunnel at -40 °C and an air speed of 4 m/s. Results showed that incorporation of different sugars into the apricot cubes affected their low-temperature phase transitions and the percentage of sugars distribution. Ascorbic acid retention during air-drying was affected by the composition of the syrup. Maltose exhibited the greatest protective effect on color stability during the subsequent 8-months of frozen storage.

Apricot Juice and Concentrate

The apricot juice production method is the same as described later for peaches (see "Peaches and Nectarines" section). The use of pectinolytic enzymes (pectinesterase and polygalacturonase) aids in the liquefaction of apricot pulp and juice extraction [32]. Juices thus obtained have higher total soluble solids, total sugars and acidity, but lower crude fiber and vitamin C. Apricot juice can be processed into concentrate for subsequent various food applications. The use of apricot juice in juice blends/cocktails is also gaining popularity. Zhang and colleagues [33] reported on the development of a juice blend using juices from Japanese apricots, grapes, Russian olives, and dates. Juice blends were evaluated for optimum formula, stability, and flavor.

The use of appropriate enzymes (pectinase and cellulase) during juice extraction is common for maximizing yields. The enzyme treatment of puree is a critical step in apricot juice production. In addition to the combination and rate of enzymes used, treatment time and temperature combination play a significant role in the final consistency of the juice. Vaillant and colleagues [34] reported that the enzymes applied at 0.14 g/L at 27-30 °C for 30 min produced the highest juice yield. These researchers also proposed a microfiltration process for producing clarified juices and concentrates from pulp-rich juices (such as apricot); this process outline is shown in Fig. **3**. These researchers suggested that the proper enzyme treatment of diluted fruit purees/pulps can increase the efficiency of the membrane filtration process.

Apricot Puree

Apricot puree use is gaining popularity as a new substitute for oil or water in many high-calorie and high-fat recipes. Unlike prunes, which can darken the color of some baked goods, or applesauce, which may cause recipes to be diluted, apricot puree reduces the fat content and adds a touch of flavor without any negative effects [5]. McHugh and colleagues [35] investigated the use of apricot puree as an edible film and concluded that fruit puree edible barriers may be used in food systems, not only for favorable sensory characteristics, but also to control mass transfer, improve product quality and extend shelf-life. Fruit barrier films, which are good oxygen barriers, can be used for low to intermediate moisture food systems such as nuts, confections and baked goods.

Apricot Jam/Preserve

A typical formulation for making apricot jam or preserve, as described by Lopez [36], is shown in Table **2**. Benamara and colleagues [37] prepared a 'light' apricot jam and studied the effects of partial substitution of aspartame for sucrose on the quality of jam. Substitution of aspartame for 90 or 100% of the sugar did not give satisfactory results; 80% level was the maximum substitution possible. Changes in the color of apricot jam during storage increased with the increasing concentration of added aspartame; this was probably due to gradual breakdown of the aspartame to release free aspartic acid and phenylalanine, which underwent a browning reaction with sugars.

Apricot by-Products

A number of by-products from apricot processing have been reported in the literature. It is to be noted that the commercial processing of these by-products vary from country to country. Seeds or kernels of the apricot grown in central Asia and around the Mediterranean are so sweet that they are sometimes substituted for almonds, especially in sweets or desserts [38]. The Italian liqueur *amaretto* and *amaretti biscotti* are flavored with the extract of apricot kernels rather than almonds; furthermore, oil pressed from these cultivars has been used as cooking oil [38].

Table 2: Formulation for apricot jam and preserve

Ingredient	**Unit**	**Higher Quality (50/50)**	**Standard Quality (45/55)**
Water	Pounds	20	20
Fruit	Pounds	100	82
Pectin, Rapid set	Ounces	5.50-6.75	5.50-6.75
Sugar	Pounds	100	100
Acid Solution[1]	Fl. Ounces	Approx. 16	Approx. 14
Cooking Temperature to finish[2]	°F	221	221
Soluble Solids	%	65	65
Yield	Pounds	160	157
pH		3.3	3.3

[1] The acid solution is prepared by mixing 1 pound of citric acid with 1 pint of hot water. A sufficient quantity is added to adjust to the indicated pH.
[2] The finishing temperature applies to cooking at or near sea level. At a higher elevation, the temperatures are corrected for the difference between 212 °F and the boiling point of water.
From Lopez [36]

The kernel is an important source of dietary protein as well as oil and fiber [39]. The use of apricot pits/kernels has been explored to extract oil, benzaldehyde, and as an alternative source of proteins [40, 41]. The oil from the apricot kernel is very rich in essential fatty acids, especially linoleic acid. Conditions for extraction of oil from high quality apricot kernels were optimized by Fathollahzadeh and colleagues [42], whose study focused on the effects of moisture content and compression axis on the amount of force needed to crack apricot pits and kernels. It was observed that, with an increasing moisture content of the apricot pits and kernels, the energy for cracking decreased. Their study also revealed that cracking an apricot pit requires higher rupture force and energy when compressed along its length and that cracking an apricot kernel requires a higher rupture force when compressed along its thickness.

The use of apricot kernel flour as a fat replacement (up to 40% level) in cookies was studied by Seker and colleagues [39]. The addition of apricot kernel flour

(AKF) decreased the spread ratio and increased the hardness of the cookies; however, a sensory evaluation revealed that cookies were acceptable to the panelists at all concentrations of AKF from 10% to 40%, though 10-20% fat replacement with AKF was found best for physical quality attributes.

Innovative Processing Technologies

Patrignani and colleagues [43] studied the effect of high-pressure homogenization (HPH) on *Saccharomyces cerevisiae* inactivation and physical-chemical features in apricot juice; their results demonstrated that repeated HPH passes at 100 MPa allowed a significant inactivation of the spoilage caused by yeast inoculated in juices. However, the inactivation of the considered strain was greatly affected by the food matrix. Furthermore, refrigeration of the treated samples prevented cell recovery and, in some cases, induced a further decrease in cell viability in the inoculated sample, thereby allowing a further increase in the juice shelf-life.

The pulsed electric field (PEF) treatment for preservation of different apricot products has been reported recently. Evrendilek and colleagues [44] reported the use of a bench-scale PEF system – OSU-4 (Fig. **4**) for the inactivation of *Penicillum expansum* in apricot nectar.

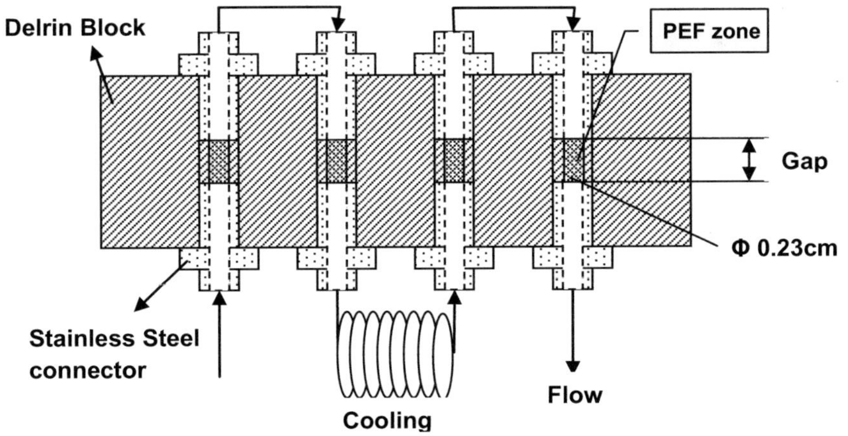

Figure 4: Diagram of co-field flow treatment chambers for OSU-4 PEF systems. From Evrendilek and colleagues [44], provided by the authors.

After inoculation of apricot nectar with *P. expansum* spores at the level of 105-106cfu/mL, the samples were processed by the OSU-4 PEF system as a function of differing electric field strengths (0, 13, 17, 20, 23, 27, 30 and 34kV/cm). The results showed that with an increase in processing time and strength of electric field, germination tube elongation (GTE) and spore germination were inhibited completely. Furthermore, light and scanning electron microscopy (SEM) observations revealed considerable morphological alterations in fungal conidia such as cytoplasmic coagulation, vacuolations, shrinkage and protoplast leakage. This was one of the earliest studies reporting on the inactivation of *P. expansum* using PEF technology.

Evrendilek and colleagues [45] investigated the effectiveness of PEF treatment for inactivation of *Botrytis cinerea* that was inoculated into apricot nectar. Based on the rate of spore germination, measurement of GTE, and light and SEM microscopic observations, PEF processing proved to be effective in inactivating *B. cinerea*; they concluded that PEF treatment can prevent product loss in fruit juice/nectars due to *B. cinerea* contamination.

Nutritional Profile and Dietary Benefits

Apricots are a good source of dietary fiber, which is important to a healthy diet and can both help control weight and lower cholesterol levels. Dried apricots are a concentrated source of fiber and one of the most highly nutrient-dense dried fruits [6]. Apricots and their processed products are low in fat, especially saturated ones, and are a rich source of some important nutrients. The chemical and nutritional composition of fresh apricots and their processed products is shown in Table **3** [46]. Apricots are an excellent source of potassium, iron and magnesium. It should be noted that varietal differences could also contribute to variations in the composition of raw and finished products.

In comparison with other major fruits, apricots contain significantly higher amounts of flavonoids, especially catechin and epicatechin up to 4.40 and 20.20 mg/100 g fruit, respectively. A variety of dietary flavonoids have been found to inhibit tumor development [47]. Jimenez and colleagues [48] studied the effects of canning and freezing on the antioxidant properties compared to raw apricots. The results showed that raw apricots exhibited the highest inhibition of oxidation (evaluated by the lipid peroxidation assay); the freezing process had a slight loss of antioxidant activity, whereas canned apricots completely lost their antioxidant capacity. Furthermore, the capacity of raw apricot to scavenge superoxide radicals was higher than that of butylated hydroxyanisole (BHA) and butylated hydroxytoluene (BHT). Canned apricots showed higher 2,2'-azino-bis(3-ethylbenzothiazoline-6-sulphonic acid) or ABTS radical scavenging capacity than the raw fruit [48].

Apricot seeds were used to treat tumors as early as 502 A.D., and apricot oil was used against tumors and ulcers in England in the 1600s [6]. The American Heart Association continues to recommend an overall healthy dietary pattern that is rich in fruits, vegetables, whole grains, low-fat dairy products, lean meats, poultry and fish. Diets rich in fruits, vegetables, whole grains, and fish have been associated, in many studies, with a lower risk of cardiovascular disease and strokes [49]. Apricots are rich in beta-carotenes. Since beta-carotene turns to vitamin A in the body, it is often referred to as vitamin A on food labels. Beta-carotenes play a critical role in fighting disease and infections by maintaining strong immunity, protecting the eyes, helping to keep skin, hair, gums, and various glands healthy, and helping to build bones and teeth.

Table 3: Composition of raw and processed apricots (per 100 g edible portion)

	Unit	Raw	Canned[1]	Frozen[2]	Dried[3]	Nectar[4]
Proximate:						
Water	g	86.35	82.56	73.3	30.89	84.87
Energy	kcal	48	63	98	241	56
Protein	g	1.4	0.53	0.7	3.39	0.37
Total lipid (fat)	g	0.39	0.05	0.1	0.51	0.09
Ash	g	0.75	0.37	0.8	2.57	0.29
Carbohydrate, by difference	g	11.12	16.49	25.1	62.64	14.39
Fiber, total dietary	g	2	1.6	2.2	7.3	0.6
Sugars, total	g	9.24	14.89	--[5]	53.44	13.79
Minerals:						
Calcium	mg	13	11	10	55	7
Iron	mg	0.39	0.39	0.9	2.66	0.38
Magnesium	mg	10	8	9	32	5
Phosphorus	mg	23	13	19	71	9
Potassium	mg	259	138	229	1162	114
Sodium	mg	1	4	4	10	3

Table 3: contd…

Zinc	mg	0.2	0.11	0.1	0.39	0.09
Copper	mg	0.078	0.079	0.064	0.343	0.073
Manganese	mg	0.077	0.052	0.05	0.235	0.032
Selenium	µg	0.1	0.1	0.4	2.2	11.9
Vitamins:						
Vitamin A	IU	1926	1322	1680	3604	1316
Vitamin C, total ascorbic acid	mg	10	2.7	9	1	0.6
Thiamin	mg	0.03	0.016	0.02	0.015	0.009
Riboflavin	mg	0.04	0.02	0.04	0.074	0.014
Niacin	mg	0.6	0.304	0.8	2.589	0.26
Pantothenic acid	mg	0.24	0.092	0.2	0.516	0.096
Vitamin B-6	mg	0.054	0.054	0.06	0.143	0.022
Folate, total	µg	9	2	2	10	1
Carotene, beta	µg	1094	789	--	2163	786
Vitamin E (α-tocopherol)	mg	0.89	0.6	--	4.33	0.31
Lutein + zeaxanthin	µg	89	26	--	0	14

[1]in light syrup pack, with skin, solids and liquids; [2]frozen, sweetened; [3]sulfured, uncooked; [4]canned
[5]not reported
From USDA [46]

CHERRIES (SWEET AND SOUR/TART)

Cherries belong to the *Rosaceae* family, subfamily *Prunoideae*. There are two different types of cherries: "Sweet" cherry (*Prunus avium* L.) and "Sour" or "Tart" cherry (*Prunus cerasus* L.) [50]. The cherry fruits in commerce are usually obtained from a limited number of species, including specialty cultivars of the wild cherry, *Prunus avium*; the name 'cherry' also refers to the cherry tree [51]. Sweet cherries originated in the area between the Black and Caspian seas. Sweet cherries came to the U.S.A. with English Colonists in 1629, and were later introduced to California by Spanish Missionaries. Washington, Oregon, and California are the main sweet cherry producing states in the U.S.A. The sour cherry most likely developed from a natural cross between *P. avium* and *P. fruticosa*; it originated in Turkmenistan and later spread to Europe. English settlers introduced sour cherry to the U.S.A. [50]. In the U.S, major sour cherry producing states are Michigan, Colorado, New York, Oregon, Pennsylvania, Utah, Washington, and Wisconsin [52]; Michigan alone contributes about 75% of the total U.S. production.

Major cherry cultivars, as described by Rieger [50], are: *Sweet Cherry Cultivars* – There are less than 100 sweet cherry cultivars grown around the world. 'Bing,' 'Napoleon' ('Royal Ann'), 'Ranier,' and 'Lambert' are the most important cultivars in North America (pollinizers for 'Bing' cherries are often 'Early Burlat,' 'Black Tartarian,' and 'Van'). *Sour Cherry Cultivars* – 'Montmorency' is the main sour cherry in the U.S.A. and Canada, accounting for 99% of all production; whereas in Europe, 'Schattenmorelle' is the leading cultivar. Sour or tart cherries differ from sweet cherries, in that they are sourer and also have a different nutrient profile. The flavor profile, and therefore usage of the two cherries also varies. Sweet cherries have a more candy-like flavor, and are therefore canned with added syrup, or eaten raw. Whereas, most tart cherries are used exclusively for processing [52].

Cherries comprised about 3% of the total stone fruit production of 38.21 million metric tons in 2010 [1]. The 2010 world production was 2.06 and 1.16 million metric tons for sweet and sour cherries, respectively. As shown in Table **4**, Turkey, U.S.A., Iran, Italy, and Spain were the top five producers of sweet cherries; Turkey alone accounted for 20% of the total world production. Turkey also led in sour cherry production, followed by the Russian Federation, Ukraine, Poland, and Iran (Table **4**). The top five sweet cherry exporters in 2010 were Turkey, U.S.A., Chile, Spain, and Austria while the Russian Federation, Canada, Germany, Austria, and China led the list of importers. Hungary, Poland, the Czech Republic, Spain, and France were the major exporters of sour cherries and Germany, the Russian Federation, Hungary, Austria, and Belgium were the major importers [1].

Cherries, being a rich source of antioxidants, offer many benefits to human health [53, 54]. With an increased consumer awareness of the many health benefits of consuming cherries or processed cherry products, the production and marketing of cherry juice concentrate has grown consistently. The use of cherry juice concentrate has also been gaining support in functional foods as its nutraceutical potential is just beginning to be understood.

Table 4: Leading sweet and sour cherry producing countries, by quantity—million metric tons (1990–2010)

Country/Year	1990	1995	2000	2005	2010
Sweet Cherries:					
Turkey	143,000	186,000	230,000	280,000	417,905
U.S.A.	142,180	150,140	185,070	227,522	284,130
Iran	85,411	156,755	216,313	224,892	255,500
Italy	100,470	120,167	145,672	101,295	115,476
Spain	54,900	55,500	112,900	95,726	80,300
Uzbekistan	--[1]	18,000	19,800	22,000	75,000
Ukraine	--	46,600	76,200	100,200	73,000
Romania	67,700	60,500	73,700	117,859	70,290
Russian Federation	--	62,000	85,000	93,000	66,700
Chile	13,700	20,000	31,050	32,000	59,000
Syria	19,400	40,800	56,285	53,441	58,100
France	82,422	62,994	66,494	69,024	45,905
Sour (Tart) Cherries:					
Turkey	90,000	92,000	106,000	140,000	194,989
Russian Federation	--	109,000	200,000	230,000	165,000
Ukraine	--	154,300	155,300	181,800	154,500
Poland	77,464	144,382	139,595	139,851	147,238
Iran	19,260	32,000	48,859	48,670	103,232
U.S.A.	94,710	179,440	127,640	122,651	86,364
Serbia	--	--	--	--	66,224
Hungary	61,175	47,600	48,894	48,082	51,870
Belarus	--	11,200	16,000	27,616	51,089
Uzbekistan	--	3,500	4,000	16,500	30,000
Germany	118,380	33,313	35,613	24,571	18,265
Azerbaijan	--	--	13,000	21,316	18,049

From [1] FAOSTAT.

Production and Harvest

The sweet cherry is grown almost everywhere in temperate zones; this genus needs warm and light conditions, but can also tolerate cold winters and requires 550–600 mm (22-24 inch) precipitation [55]. Cherries grow on trees, which need to be well maintained, in order to provide a well ripened, undamaged fruit during harvesting. Trees strive well in deep, well-drained, gravelly to sandy loam soils. Both sweet and sour cherry trees are typically maintained at about 12-15 ft in height. While sour cherries are self-fertile, pollination is necessary for the production of sweet cherries. Pruning and training of trees is common to maximize vigor and fruit yield. Both types of cherries require only about 2-3 months for fruit development [50].

Sweet cherries are harvested manually whereas sour cherries are mechanically-harvested. Typically, color change and soluble solids content are used as maturity indices. However, fruit removal force, which is a more reliable indicator, has been used increasingly [50]. The harvesting and processing schedule of tart cherries is very strenuous, because fruits can only be held in a cold water bath for up to eight hours before their quality begins to deteriorate. The harvested cherries fall onto a landing area lined with large polyethylene sheets. After falling from the tree, cherries are then conveyed or carried to a cold water bath. It is important to leave unripe cherries on the tree to allow them to ripen and supply the farmer with a greater harvest [50, 55]. Preharvest treatment of gibberelic acid is common for sweet cherries. Typically, these sprays are done 3 weeks prior to harvesting and it results in firmer and larger sized fruits, which are also less prone to pitting after the treatment.

Table 5: Fruit weight, fruit firmness, total soluble solids, and titratable acidity of Michigan tart cherry selections

Tart Cherry Selection	**Fruit Weight (g)**	**Fruit Firmness (Force-N)**	**Total Soluble Solids (°Brix)**	**Titratable Acidity (% malic acid)**
Montmorency	4.82	0.52	13.74	1.13
Balaton	8.17	0.77	17.99	1.36
Erdi Jubileum	7.00	0.69	20.20	1.41
Tamaris	5.59	0.63	17.47	1.21
25-14(20)	3.95	0.54	17.97	1.21
27-10(50)	5.07	0.59	15.81	1.28

Adapted from Siddiq and colleagues [56]

Both the sweet and sour cherry fruit is a "drupe" with a size ranging from ½ inch (13 mm) to 1¼ inch (31 mm). Sour cherries have lower sugars and higher organic acid contents than sweet cherries; they are generally bright red in color, and exhibit less color variation than sweet cherries. Typically, the skin color of sweet cherries skin is deep-red or purple, yellow (has a red check or shade), or rarely white, whereas flesh color varies from white to dark-red [57]. Varietal/cultivar differences can have significant effect on various physical-chemical attributes of the fruit [56], as shown in Table **5** for some Michigan grown sour (tart) cherries.

Postharvest Physiology and Storage

Cherries should be pre-cooled immediately after harvesting for optimum shelf-life and to minimize postharvest losses. Pre-cooling is also necessary for cherries intended for processing, especially if there are any delays between harvesting and processing. Sweet cherries are not sensitive to chilling temperatures and should be stored as cold as possible without freezing. Sweet cherries produce very low amounts of ethylene but will respond to exogenous or wound-induced ethylene with increased respiration and quality loss. Cherries are a low respiring fruit; the respiration rate (mg CO_2/kg/hr) at different temperatures is reported to be 6-10 (0 °C), 16-28 (5 °C), 20-36 (10 °C), 28-64 (15 °C), and 40-90 (20 °C). Sweet cherries maintain good quality for 2-4 weeks under recommended storage conditions of –1 to 0 °C with > 95% RH.

Controlled atmosphere (CA) and modified atmosphere packaging (MAP) are commonly used methods to improve the shelf-life quality of cherries. Wei and colleagues [59] investigated the effect of cold storage and 1-methylcyclopropene (1-MCP) treatment combined with 0 °C storage on the quality, cell wall composition and degrading enzymes activity of 'Summit' sweet cherries. In comparison with room temperature storage, both cold storage and 1-MCP/0 °C treatments significantly maintained better fruit quality and firmness, reduced respiration rate, prevented degradation of the total soluble solids and soluble proteins and extended shelf life of sweet cherries.

The effect of edible coating (sodium alginate: 1%, 3% or 5% w/v) in sweet cherry fruits harvested at commercial maturity stage was studied by Diaz-Mula and colleagues [60]. Edible coatings were effective in delaying the evolution of the parameters related to postharvest ripening, such as color, softening and loss of acidity, and in reducing the respiration rate. Moreover, the coatings showed a positive effect on maintaining higher levels of total phenolics and antioxidant activity.

Cherry pitting and bruising are common quality problems caused by harvest injury as well as rough postharvest handling. Fruit pitting is due to subsurface damage that develops into sunken areas close to the fruit surface, while bruising can occur from excess compression, drops or large impacts during harvest, transport or packing. Additionally, sweet cherries are also prone to shriveling and water loss due to the lack of a well-developed cuticle. Linke and colleagues [61] reported that changes in color and structure (shrinkage, bending) of fruit peduncles (initially green, turgid and metabolically fully active at harvest) are good indicators of freshness loss in sweet cherries.

Cherry Processing and Processed Products

Major processed products are frozen, canned, and dried cherries and cherry juice/concentrate. Almost 99% of tart cherries produced in the U.S.A. are processed; 71% as frozen cherries, 22% as canned cherries, and 7% as juice, wine, brined, and dried products. Michigan produces 75% of tart cherry production in the United States; 95% of this yield is the Montmorency cultivar [62]. A general outline of the various steps involved in the processing of different cherry products is shown in Fig. **5**.

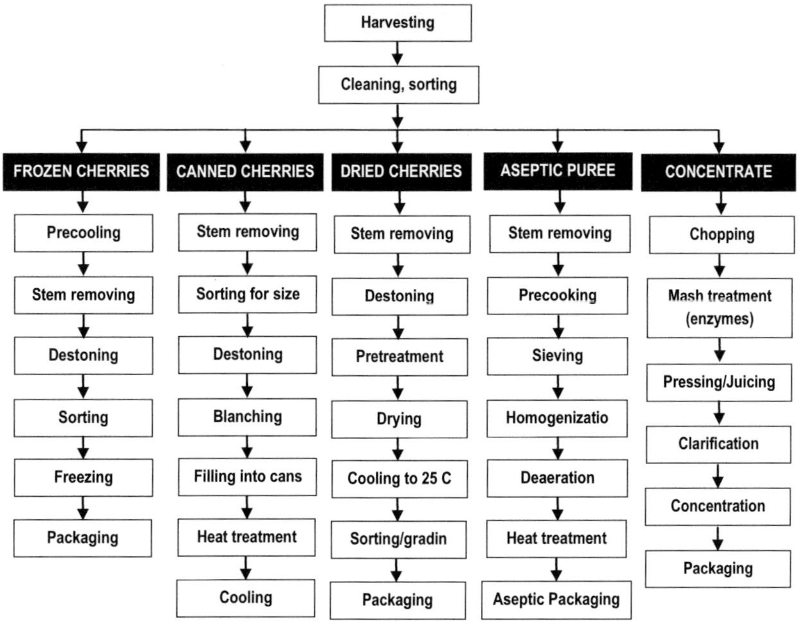

Figure 5: Processing steps for different cherry products. Adapted from Steger-Mate [54].

Pre-Processing Preparation Steps

As described above, the first step after harvest is pre-cooling of cherries to remove field heat. This step also serves as one of the first methods of cleaning the fruit. At the processing plant, cherries are washed and cleaned in water; cherries will settle in the water bath, where sticks, twigs, steams, undesirable cherries, and other debris will float on top of the bath, thus allowing it to be skimmed off. Subsequently, the cherries are sorted, washed, pitted, and processed further depending on their final product form (frozen, canned, dried, *etc.*). Following washing, sorting is performed; sorting techniques vary depending upon the end-product. Sorting options include the use of an eliminator, which is a machine that removes small or crushed cherries based upon preset parameters; other size grading methods may be used, such as a belt where fruits less than 5/8 inch diameter fall through the belt [63].

A popular and effective method of cleaning berries, including tart cherries is the McLauchlan air-and-water cleaner [63]. This cleaner includes an air cushion that suspends the product, while removing outer debris, and then the product is further processed using a vacuum chamber where air currents remove contamination, finally the fruit moves into a water bath, which is used to remove dirt [63]. Many of the quality losses in the processing of tart cherries are due to factors such as temperature, force, acidity, and mineral content of water in the washing step.

Pitting is performed using a mechanical pitter. The pitting process of tart cherry processing is crucial to the product yield. The FDA requirements indicate that no more than one pit can be present in a 20-oz can of cherries [52]. Therefore, if the pitting equipment is not maintained properly it is possible that the processor could have adulterated products. According to McLellan and Padilla-Zakour [52] processors continue to research and attempt to develop efficient pitting equipment. Presently, the loss of product during the pitting operation is 15%. Different methods for pitting have also been explored, and some have found that pitting a practically frozen product yields the greatest benefits [52]. Once the cherries are pitted they are then further processed according to their end use.

Frozen Cherries

Among different cherry products, frozen cherries offer better color quality, improved flavor, and firmer texture when compared to canned cherries. The processing of tart cherries into a frozen product uses the same initial steps (sorting, washing, pitting) after which cherries are blanched at 100 °C for 40-60 seconds [52]. Blanching inactivates enzymes (such as polyphenol oxidase, peroxidase and pectinase) that can deteriorate frozen cherry quality (mushy texture, faded color or browning). Without blanching the enzyme activity in the cherries would be slowed but not stopped completely. After blanching, the product is prepared to be frozen; however, some processors find it desirable to dip the fruit into a calcium chloride solution to retain firmness [52]. Calcium replaces the hydrogen in the cross linkages of pectin, and therefore increases the strength of the linkage and fruit firmness.

Following the pre-freezing processes, the product is then frozen using individually quick frozen (IQF) freezing technique. This process utilizes high velocity air at – 30 °F or below. Following IQF freezing the cherries are packaged into various packages, depending on the retail or bulk market. The final frozen product is then stored in a freezer at 0 °C, until shipping, also done at freezing temperatures. Frozen cherries can also be processed with the addition of chilled sugar syrup (50-60 °Brix); this product is primarily used for bakery applications [52]. The purpose of adding the sugar syrup is to have a protective layer over the fruit.

Frozen products undergo sorting and grading to improve the quality. Frozen cherries are packaged in polyethylene bags or multilayer boxes and the packaged product also passes through a metal detector. Smaller package size products are sold in retail outlets for households, while large packages are used by other food industries (*e.g.*, baking, confectionery, dairy, canning, *etc.*) or food service applications [55].

Canned Cherries

Canned cherries provide a year-round availability of this fruit for a variety of culinary uses. Cleaned and pitted cherries are blanched before being filled into cans. Delays between blanching and filling should be avoided as it can let the product cool down thereby requiring more heat/energy during thermal processing. Depending on the packaging style, there are different categories of filling media used, water or syrup: water (sugar-free), slightly sweet (14-17 °Brix), sweet (17-20 °Brix), strongly sweet (>20 °Brix) sugar syrup [55]. Filling media terminology used in the U.S. is "light," "heavy," or "extra-heavy" syrup. Additives can be used if desired, *e.g.*, citric acid (to adjust acidity), calcium chloride (to firm texture), coloring, and flavoring. Carle and colleagues [64] reported that a calcium brining markedly improved firmness of canned, processed, fresh or frozen cherries. Increased firmness was due partly to the absorption of calcium into fruit tissue, but characterization of enzyme activities suggested the involvement of pectin methylesterase and peroxidase, which were active even at 1-4 °C and pH 4. Added syrup is typically at 60-65 °C, after which the cans are immediately sealed using an automatic sealer. Filled/sealed cans are heat processed at 92-96 °C; the duration of heat treatment is determined by the pH of the fruit, the size of fruit pieces, and the size of the can [55]. Following thermal processing, cans are cooled to <40 °C, dried, labeled, and cased for warehousing and shipment to markets.

Canned cherries must meet a minimum filled weight requirement. For quality evaluation, a scoring system is used for color (20 pts), freedom from pits (20 pts), freedom from defects (30 pts), and product character (30 pts), as per USDA standards for canned cherries [65].

Dried Cherries

Dried fruit has a long shelf life and therefore can provide a good alternative to fresh fruit. Drying is the oldest method of food preservation and is widely used for both fruit and vegetables [66]. Dried sweet or sour cherries can be prepared either from raw or frozen cherries. Like most other fruits, sweet and sour cherries are typically dried in tunnel or tray-based dryers. Even distribution of the fruit on the perforated trays is important for uniform drying and overall product quality. A drying temperature of 68-70 °C is commonly used and cherries are dried to 18–20% final moisture content. The duration of drying is 8–10 hours, depending on the raw material and equipment [55]. Vacuum drying can be used to minimize quality changes; however, the process is more costly than basic hot-air drying [66]. Characteristics of the vacuum-drying technique, such as a high drying rate, a low drying temperature and an oxygen-deficient drying environment, can contribute to maintaining better quality such as color, shape, aroma and flavor and nutritive values of the dried product [67].

The effects of alkali emulsion of ethyl oleate and air temperature (60, 70 and 75 °C) on the drying characteristics of sweet cherries were studied by Doymaz and Ismail [68] using a hot air dryer at a constant air velocity of 2.0 m/sec. It was observed that both the alkali emulsion of ethyl oleate and air temperature affected the drying time. The drying times of pre-treated samples were 19.5-22.6% shorter than those of control samples. The rehydration ratio was significantly affected by pre-treatment and air temperature, moreover, it was found to increase proportionally with the increase in the air drying temperature.

Osmotic dehydration is another method of preservation that is used for drying cherries. This method involves dipping cherries in concentrated sugar syrup. Sucrose is the main sugar used in osmo-dehydration but other materials can also be used, such as fructo-oligosaccharide, concentrated fruit juice, sorbitol, galactosylosorbitol, oligofructose, and inverted sugar [69]. Partially dehydrated cherries prepared in this way can be added to foods such as desserts, yogurt, ice-cream, confectionery, and bakery products. After additional drying, they can also be used as components of cereals or snacks for direct consumption [70]. Although dehydration leads to a loss of nutrients, Torreggiani and Bertolo [70] reported that the dehydrated material becomes impregnated with the osmotic agent that enhances not only its sensory properties, but also has an impact on its dietary value. This method of drying cherries also facilitates the addition of health-beneficial compounds (*e.g.*, probiotics) to the fruit.

Dried cherries are packaged in suitable films with appropriate air/moisture barrier properties. Steger-Mate [55] reported that added value can be created before packaging by procedures such as chopping, size selection to defined fractions, sieving, grinding, and mixture production.

Juice, Juice Concentrate

Sour cherries are the type used for juice extraction and production of concentrate. Cherries should be free of green stalk and any mold growth as these can severely impact the quality of extracted juice. After washing, the cherries are heated (~70 °C) in large vats with mechanical agitation, which initiates maceration of fruits. Adding pectinolytic and cellulytic juice extraction enzymes is common to maximize yield. According to Chaovanalikit and Wrolstad [71], the concentration of bioactive compounds is significantly higher in the fruit skin than the flesh. For maximization of anthocyanins and phenolic compounds during juice extraction, use of cellulolytic and pectinase enzymes is recommended. Juice is filtered and clarified using filtering aids, *e.g.*, China clay and gelatin [55].

Since cherry juice is not typically used for direct consumption, it is further processed into juice concentrate between 55 and 65 °Brix. Vacuum evaporation is the preferred method of concentration for better quality maintenance. Reverse osmosis is another juice concentration method, though it is not commonly used [55]. Cherry juice concentrate is packaged in small containers for the retail market or in bulk containers (up to 55 gallons) for bulk shipments.

Pulsed electric fields (PEF) processing of sour cherry juice was studied by Altuntas and colleagues [72]; these researchers suggested that PEF could be a viable option to process sour cherry juice with a significant amount of microbial inactivation and without adversely affecting important physical and quality parameters.

Cherry concentrate is reconstituted to produce cherry juice that is used in fruit juice blends or cocktails and other applications. Extended storage or fluctuation in temperature during storage can negatively affect the quality of cherry concentrate. Usually, color degradation is the most obvious quality defect during storage manifested by the loss of the characteristic red cherry color and the onset of a darker and dull appearance. Place and colleagues [73] studied the quality of cherry juice concentrate during 24 weeks storage at three temperatures (4, 21, 38 °C). Fig. **6** shows the results of the Hunter color "a" values for both unpasteurized and pasteurized cherry juice concentrate. For all samples, it was found that there was a significant decrease in the Hunter color "a" values. All initial values were positive, representing a red color of the product. During storage, these values dramatically decreased at 21 °C and 38 °C, showing that the longer cherry concentrate is stored at higher temperatures the more it loses its red pigment.

Refrigerated storage of juice concentrate is preferred for minimizing quality changes.

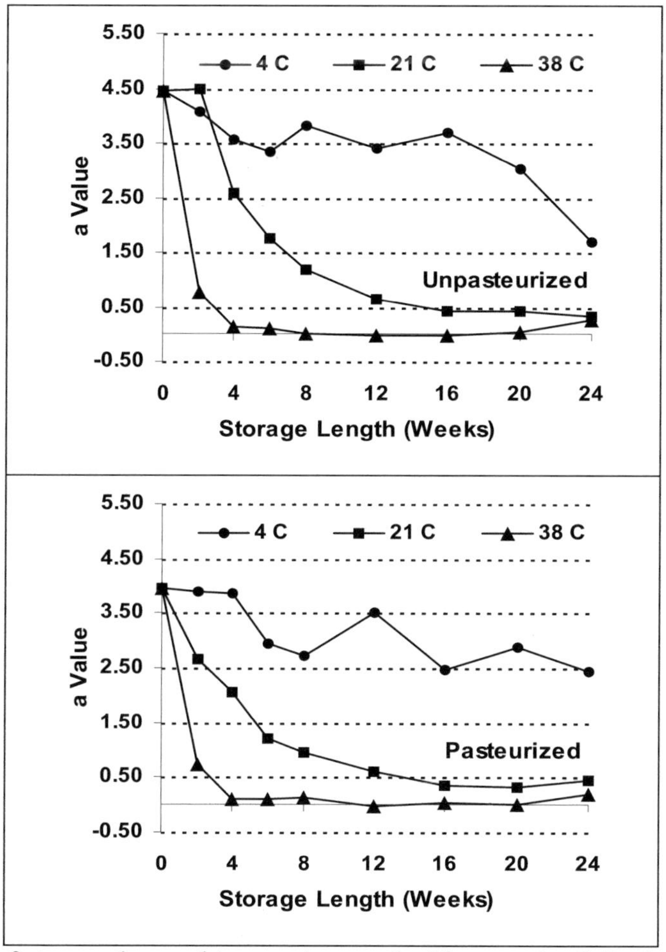

Figure 6: Effects of storage time and temperature on the Hunter color "a" value of unpasteurized and pasteurized cherry juice concentrate. From Place and colleagues [70].

Aseptic Puree

Though not as common as for other fruits (peaches, mango), cherries are processed into puree. Steger-Mate [55] reported that puree can be made from both sweet and sour cherries; it is a natural, semi-finished product that contains pressed fruit flesh without seeds, skin and additives. Puree is used as an ingredient in many fruit-based products, such as baby foods, jams, juices, and fruit sauces. Cleaned and washed cherries are heated and precooked to soften the tissue.

Heating, besides facilitating maceration, also inactivates oxidative and pectinolytic enzymes that can deteriorate puree quality. Pre-cooked cherry macerate is passed through the desired mesh-size sieves to remove skin, pits, *etc.* [55]. The finished puree is bulk packaged in polyethylene bags and sealed, or in polyethylene-lined bulk cartons and drums.

Maraschino Cherries

According to the FDA [74] Maraschino cherries are defined as "cherries which have been dyed red, impregnated with sugar and packed in a sugar syrup flavored with oil of bitter almonds or a similar flavor."

Maraschino cherries are preserved and sweetened cherries, which are typically made from light-colored sweet cherry varieties, such as the 'Royal Ann,' 'Rainier,' or 'Gold.' These days, cherries are first preserved in a brine solution containing SO_2 (to bleach the fruit) and $CaCl_2$ (firming agent), followed by soaking in a suspension of food coloring and sugar syrup. Maraschino cherries are used as an ingredient in many cocktails; as a garnish, they often are used to decorate frozen yogurts, cakes, pastry, parfaits, milkshakes, ice cream sundaes, and ice cream sodas. Canned diced-fruit cocktail typically contains Maraschino cherries [75].

Other Products

Jam is another product made from cherries; however, its market is rather very limited. The jam making process is similar to the one described in the "Apricot" section. In addition to the fruit (cherries *vs.* apricot), the formulation can vary somewhat with respect to added pectin, sugar, or acid. Home jam making is more common than commercial jam production.

Siddiq and colleagues [76] developed a gluten-free extruded snack product using a dry bean flour base and dried fruit (apples, blueberries, cherries, cranberries, and pears). Bean flour and dried fruit bits were extruded at 25% moisture content and 90 °C. Extrudates were dried at 60 °C overnight and physico-chemical and sensory properties evaluated. The cherry-based snack had acceptable quality attributes and was better or comparable to other fruit-based snacks.

Cherry pomace, a by-product of the juice industry, has also been developed into a value-added product in the dried form or powder. Dried cherry pomace is a natural, fiber- and antioxidant-rich product that can be used for added health benefits in various food products. Adil and colleagues [77] used high pressure liquid extraction (HPE) and subcritical fluid extraction (SCE) for the extraction of total phenolic compounds (TPC) from sour cherry pomace. Antiradical efficiency (AE) of the extracts was also determined. TPC and AE at the optimum conditions (54.8-59 MPa, 50.6-54.4 °C, 40 min) for SCE were determined for sour cherry pomace as 0.60 mg GAE/g sample and 2.30 mg DPPH/g sample, respectively.

Popa and colleagues [78] reported that cherry kernels, a substantive byproduct of sour cherry processing, could be a potential source of an edible oil and possibly supplementary protein (analysis of Montmorency cherry pits has shown that the dried pits contained 22.7% core which produced 31.8% oil). The fatty acids profile showed a high content of unsaturated fatty acids: oleic acid (42.9%) and linoleic acid (38.2%), while the dominant saturated acids were palmitic (11 %) and stearic (6.4%).

Nutritional Profile, Bioactive Compounds and Health Benefits

The nutritional profile of sour and sweet raw cherries and selected processed products is given in Table **6**. Cherries and their processed products are a healthy and nutritious food, with total energy values ranging from 46 kcal to 89 kcal/100 g. Cherries are low in sodium but have a higher potassium content, thus making them a better food with respect to the Na-K ratio. Sour and sweet cherries contain all vitamins in low concentrations, except vitamin B12 and vitamin D. The data shown here refer to U.S. grown and processed cherries, some varietal and regional differences are expected for fruit from other parts of the world.

Table 6: Nutritional profile of raw and processed sour and sweet cherries (per 100g)

	Units	Raw		Canned		Frozen	
		Sour	Sweet	Sour	Sweet	Sour[2]	Sweet[3]
Proximate:							
Water	g	86.13	82.25	79.62	81.56	87.2	75.53
Energy	kcal	50	63	75	67	46	89
Protein	g	1	1.06	0.74	0.61	0.92	1.15
Total lipid (fat)	g	0.3	0.2	0.1	0.15	0.44	0.13

Table 6: contd...

Carbohydrate, by diff.	g	12.18	16.01	19.3	17.29	11.02	22.36
Fiber, total dietary	g	1.6	2.1	0.8	1.5	1.6	2.1
Sugars, total	g	8.49	12.82	--[4]	15.79	9.02	20.26
Minerals:							
Calcium	mg	16	13	10	9	13	12
Iron	mg	0.32	0.36	1.32	0.36	0.53	0.35
Magnesium	mg	9	11	6	9	9	10
Phosphorus	mg	15	21	10	18	16	16
Potassium	mg	173	222	95	148	124	199
Sodium	mg	3	0	7	3	1	1
Zinc	mg	0.1	0.07	0.07	0.1	0.1	0.04
Vitamins:							
Vitamin C, total ascorbic acid	Mg	10	7	2	3.7	1.7	1
Thiamin	Mg	0.03	0.027	0.016	0.021	0.044	0.027
Riboflavin	Mg	0.04	0.033	0.039	0.041	0.034	0.047
Niacin	Mg	0.4	0.154	0.17	0.403	0.137	0.177
Vitamin B-6	Mg	0.044	0.049	0.044	0.03	0.067	0.036
Folate, total	µg	8	4	8	4	5	4
Vitamin A	IU	1283	64	726	157	870	189
Vitamin E (α-tocopherol)	Mg	0.07	0.07	--	0.23	0.05	0.07
Vitamin K (phylloquinone)	µg	2.1	2.1	--	1.4	1.5	2.1

[1]in light syrup, solids and liquid; [2] unsweetened; [3] sweetened; [4] not reported
From USDA [46]

The health benefits of cherries have been well documented; cherries are reported to be rich in many non-nutritive components such as flavonoids and phenolic compounds, which provide protection against chronic diseases through multiple effects, including the antioxidant effect [53, 79]. In addition, it has been shown that antioxidants can significantly improve certain immune responses [80]. These beneficial effects may be partially associated with the abundance of anthocyanins, such as the glycosides of cyanidin. When compared to other berries, tart cherries have been shown to have the highest amount of anthocyanins, which are a major contributor to the cherries' antioxidant capacity, commonly referred to as ORAC – oxygen radical absorbance capacity [81]. 'Montmorency' tart cherries have been reported to contain high levels of phytochemicals including melatonin, a molecule critical in regulating the sleep-wake cycle in humans. Howatson and colleagues [82] reported that consumption of tart cherry juice concentrate provides an increase in exogenous melatonin that is beneficial in improving sleep duration and quality in healthy men and women and might be of benefit in managing disturbed sleep.

Processing affects the phytochemical content of cherry products and may consequently affect their related health benefits. Ou and colleagues [83] evaluated processed cherry products (juice concentrate, individually quick-frozen, canned, and dried cherries) for anthocyanins, ORAC values, and total phenolics (Table **7**, Fig. **7**). On a per serving basis, total anthocyanins were highest in frozen cherries and total proanthocyanidins were highest in juice concentrate. Total phenolics were highest in juice concentrate. On a per serving basis, juice concentrate was superior to other tart cherry products. Juice concentrate had the highest oxygen radical absorbance capacity (ORAC).

Table 7: Anthocyanin content and ORAC values of tart cherry products

Cherry Product	**Cyanidin-3-glucosylrutinoside**[1]	**Cyanidin-3-rutinoside**[1]	**Peonidin-3-rutinoside**[1]	**Total Anthocyanins**[1]	**ORAC**[2]
Juice concentrate	0.05 ± 0.001	0.02 ± 0.001	0.002 ± 0.0001		128 ± 8
-Per serving[3]	1.44 ± 0.1	0.47 ± 0.1	0.07 ± 0.1	2.0 ± 0.03	3622 ± 226
Frozen cherries	0.047 ± 0.002	0.02 ± 0.003	0.003 ± 0.001		20.3 ± 0.44
-Per serving	3.14 ± 0.04	1.32 ± 0.05	0.23 ± 0.01	4.7 ± 0.09	1362.3 ± 29.8
Canned cherries	0.008 ± 0.001	0.002 ± 0.0	0.0003 ± 0.001		17 ± 0.7
-Per serving	0.91 ± 0.03	0.26 ± 0.01	0.04 ± 0.001	1.2 ± 0.05	2057 ± 80.7
Dried cherries	0.02 ± 0.002	0.006 ±.001	0.001 ± 0.0		68 ± 2.7
-Per serving	1.02 ± 0.01	0.27 ± 0.01	0.05 ± 0.001	1.4 ± 0.06	3060 ± 120

[1]mg/g, fresh weight; [2] Trolox equiv., fresh-weight;
[3] One serving of cherry concentrate = 1 ounce or 28.3 g, frozen cherries = ½ cup or 67 g, canned cherries = ½ cup or 121 g, dried cherries = ¼ cup or 45 g
Adapted from Ou and colleagues [83].

Research has shown that anthocyanins, acting primarily as antioxidants, have anti-inflammatory, anti-carcinogenic, and anti-aging properties [84]. A broad spectrum of *in vitro*, *in vivo*, and human studies have demonstrated that anthocyanins promote antioxidant status, healthy vision, urinary tract health, and dermal health as well as exhibiting potential health benefits including cardiovascular, neuroprotective, anticarcinogenic potential, and anti-diabetic properties [85]. Kirakosyan and colleagues [87], in their study of the chemical profile and antioxidant capacities of tart cherry products, showed a similar relationship between the ORAC of products made from 'Montmorency' and 'Balaton'. Furthermore, their study on antioxidant activities of crude extracts of ten tart cherry products showed that these products preserve their antioxidant capacities after processing and storage. Besides anthocyanin-rich pigment, tart cherries are known to have other kinds of flavonoids that are common in many edible berries. The main colorless polyphenols of tart cherries are neochlorogenic acid, 3-coumaroylquinic acid, and chlorogenic acid [88], though these compounds are relatively less bioactive compared to anthocyanins.

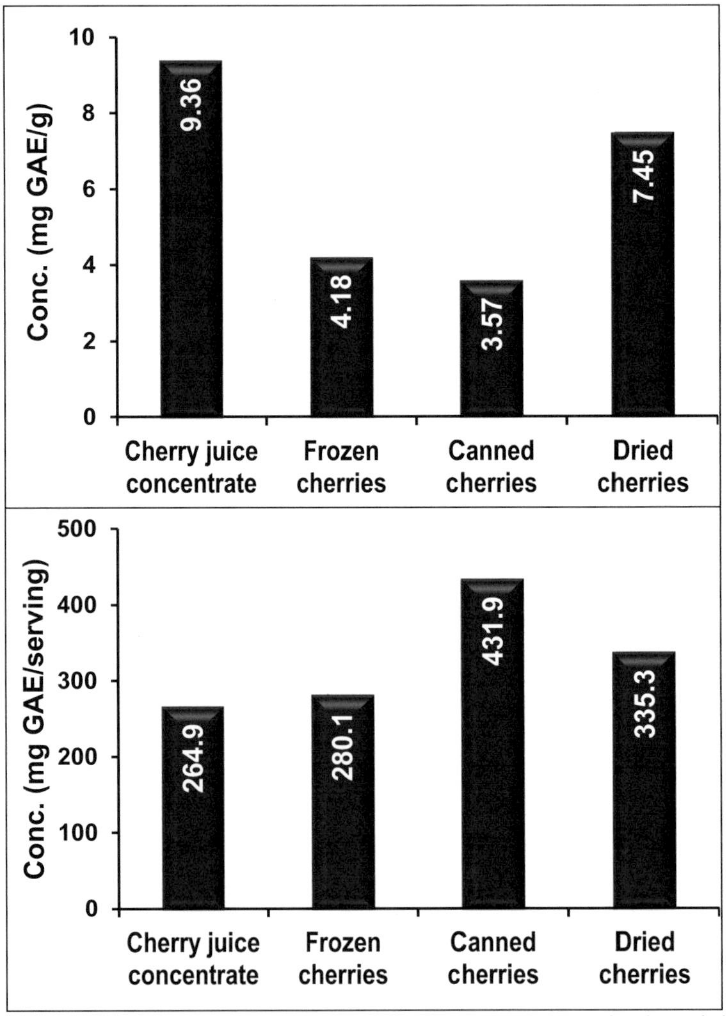

Figure 7: Total phenolic content of tart cherry products (top – per g fresh weight, bottom – per serving; one serving of cherry concentrate = 1 ounce or 28.3 g, frozen cherries = ½ cup or 67 g, canned cherries = ½ cup or 121 g, dried cherries = ¼ cup or 45 g). Adapted from Ou and colleagues [83].

The potential health benefits of phenolic compounds having antioxidant properties have generated a significant interest in their non-traditional uses or exploration of new sources: *e.g.*, yogurt with added cherry paste [89], use of phenolic extracts in fruit juices [90], and phenolics in white mulberry [91].

PEACHES AND NECTARINES

Peaches (*Prunus persica*) come under the genus *Prunus* of rose family (Rosaceae), and have attractive pink blossoms bearing juicy, sweet drupe fruit. The peach fruit originated in China, from where it was introduced to Persia and was later spread throughout Europe by the Romans. This fruit was introduced to North America by early Spanish settlers who brought several varieties. Peaches are generally distinguished as "clingstone" (pit adhering to the flesh) or "freestone" (pit relatively free and not adhering to the flesh); commercially significant 'Elberta' peaches are the latter type. Nectarines are smooth-skinned and fuzz-less peaches; the lack of fuzz is reported to be due to the presence of a single gene [57]. Both peaches and nectarines are relatively large fruits, typically 2–3.5 inches in diameter. Peaches are pubescent throughout their growing season, and are usually machine-brushed before marketing to get rid of most of the pubescence [4].

Table 8: Leading peach and nectarine producing countries, by quantity—million metric tons (1990–2010)

Country/Year	1990	1995	2000	2005	2010
China	1,279,140	2,769,877	3,851,919	7,649,678	10,828,348
Italy	1,719,740	1,328,680	1,655,250	1,693,150	1,590,660
U.S.A.	1,223,800	1,198,000	1,412,400	1,301,900	1,255,090
Spain	629,300	661,200	1,129,850	1,260,880	1,134,750
Greece	786,939	1,034,420	949,943	864,406	639,400
Turkey	350,000	340,000	430,000	510,000	534,903
Iran	70,664	160,000	350,000	439,771	500,000
Chile	196,000	275,000	260,000	311,000	357,000
France	491,932	529,127	480,657	402,646	324,401
Argentina	220,000	199,000	209,634	272,500	318,000
Egypt	37,442	267,000	240,193	360,000	273,256
India	70,000	89,000	120,000	173,989	244,000

From FAOSTAT [1]

In recent years, the world production of peaches and nectarines has remained fairly steady. China, Italy, the U.S.A., Spain, and Greece were the top-5 peach and nectarine producing countries in 2010 (Table **8**). China accounted for over 50% of total world peach and nectarine production, which is 9-fold its 1990 production. Leading peach and nectarine exporting countries in 2010 were Spain, Italy, the U.S.A., Greece, and Chile, while Germany, the Russian Federation, France, Poland, and the Netherlands were the leading importers [1].

In the United States, peach production is mostly centered in California, South Carolina, and Georgia, followed by Michigan, New Jersey and Pennsylvania. Clingstone peaches, grown exclusively in California, are used primarily for processing. Peaches contribute over $900 million to the Californian economy [93].

Production and Harvest

Cultivation

The peach tree grows better in pH 6-7 well-drained and loamy soils. Peach and nectarine cultivars are self-pollinating and can produce satisfactory crops without cross-pollination. Peaches bloom relatively early (*e.g.* around mid-March in Georgia), and exhibit lower cold tolerance than apple or pear. In most peach producing areas of the world, frost can be a problem since open flowers and young fruits are prone to serious damage on brief exposure to –2 °C or below. In the southeastern U.S. states, peach production has suffered a decline in the past few decades due to severe frosts in the late spring [57, 94].

Peach tree thinning is common for proper fruit size development; generally, thinning is done manually, by leaving one fruit per 6 inch shoot length. Occasionally, chemical blossom desiccants are used to thin flowers. Miranda-Jimenez and Royo-Diaz [95] reported that fruit productivity greatly improved, when trees were pruned early; that is, either at full bloom and/or soon after the fruit set. Although chemical thinning is more cost effective than manual pruning it results in an irregular distribution of fruits on shoots. Furthermore, thinning also helps to relieve water stress; however, it is shown to affect the quality of fruit [96]. Kumar and colleagues [97] reported that in the case of 'Flordaking' and 'Saharanpur Prabhat' peaches, lower fruit yields were obtained with higher levels of pruning, while medium pruning gave the highest 'Flordasun' yield. Levels of pruning had minimal impact on pulp and stone weight, pulp-to-stone ratio, moisture content, ascorbic acid, and sugar acid ratio. However, pruning can significantly increase fruit weight or size, soluble solids, and sugar and acid content.

In comparison to postharvest application, foliar sprays of calcium were reported to have a positive effect on peach and nectarine fruit quality [98]. Calcium sprays increase mesocarp and exocarp calcium thereby improving fruit firmness for better quality and postharvest storability of bagged fruit.

Fruit Classification, Maturity, and Harvest

The peach fruit can be classified based on the appearance and sensory attributes, such as: round, flat, beaked; skin surface − pubescent or smooth; "free" or "cling" stone type; flesh color, white, yellow or red; sweet, sour or astringent flavor; and "melting" or "non-melting" flesh [99]. Harvesting maturity for most peach and nectarine cultivars is determined according to skin ground color changes from green to yellow. A color chip guide is used in California to determine maturity of various cultivars employing a two-tier system: (1) Mature, corresponding to minimum maturity, and (2) Well-Mature and/or tree-ripe. In cases where skin ground color is masked owing to fully-developed red color before maturation, it is recommended to measure fruit firmness for an accurate assessment of maturity (maximum maturity index is applied in such cases). Maximum maturity is defined as the minimum flesh firmness (typically measured with hand-held 8-mm cylindrical probe penetrometer) at which fruit can be handled with no or minimal bruising damage [9].

Peaches and nectarines are harvested manually into bags or baskets and picked fruit is then transferred to bins on trailers and then transported to the packinghouse. At the packinghouse, fruit is cooled soon after harvest, then cleaned and sorted. Attention to detail on sorting line efficiency is especially important with peach fruits, where a variable range of colors, sizes/shapes is encountered.

Postharvest storage and physiology

Postharvest Storage

For both peaches and nectarines, optimum storage conditions consist of a temperature of −1 to 0 °C and relative humidity of 90-95% with about 50 ft^3 per minute airflow. Softening of fruit tissue is accelerated at higher temperatures, *e.g.*, the rate of respiration can be 10-times higher at 20 ºC than at 0 ºC. Depending on fruit maturity, the rate of ethylene production increases significantly from a low of <5.0 μL per kg per hour at 0 ºC to 160 μL per kg per hour at 20 ºC [9]. Polygalacturonase is responsible for fruit tissue softening resulting from depolymerization of pectic polysaccharide chains during storage [100].

Lysiak and colleagues [101] evaluated the storage stability of peaches by dipping fruit in a 2% $CaCl_2$ solution at 20 °C for 30 min followed by storage at 4 °C for 2 weeks in boxes with polyethylene film bags/wraps. This type of packaging minimized fruit weight loss and resulted in distinct improvements in storability from the $CaCl_2$ treatment and polyethylene barrier. The $CaCl_2$ improved firmness, generally maintained the soluble solids and increased the sugar-acid ratio.

Postharvest Disorders−Chilling Injury

Most mid-season and late-season peaches and nectarines are prone to chilling injury during postharvest storage. Development of chilling injury is faster and more intense in fruit stored at 2.2-7.6 °C (peaches) and 2.2 -7.8 °C (nectarines) compared to 0 °C or below [9]. Brovelli and colleagues [102] showed that, assessed on the development of flesh mealiness, the quality of non-melting flesh peaches was not affected as severely as a result of chilling exposure – a phenomenon common with melting flesh peaches. This very quality of non-melting flesh peaches is a major advantage in refrigerated storage and transportation to distant markets. Modified atmosphere (MA) storage of 12% CO_2 and 4% O_2 of peaches packaged in unperforated polypropylene bags could be helpful as it is related to lower weight loss, less senescence, lower incidence of chilling injury and decay, and delayed fruit ripening beyond typical shelf-life [103]. Controlled atmosphere (CA) storage under high CO_2 is an established technology to control chilling injury, while pre-storage heat treatment appears to be emerging as an alternative [104].

Processed Products

The two major processed products are canned and frozen peaches. Other products, juice, puree, dried peaches, jam and jelly are processed and marketed on a relatively smaller scale.

Canned Peaches

Clingstone peaches are the type most commonly used for commercial canning, as they are exceptional in their ability to retain flavor and texture consistency. Typically, peaches are canned within 24 hours of receipt at the processing plant to ensure that the peaches maintain better nutritional value and flavor [57, 93]. The optimum maturity of canning peaches is when the fruit is still firm and color is orange-yellow. Various processing steps involved in canning peaches are briefly described below [105]:

- ***Grading***: Since fruit received at the cannery consists of a wide size range, grading of fruit using a mechanical grader is necessary. For fruit stored before canning, pre-grading is carried out before the fruit has reached canning maturity to minimize fruit bruising.

- ***Halving and Pitting:*** Peaches are mostly canned as halves or thick slices. Pitting and cutting of peaches is done using automatic machines using sharp knives to minimize flesh damage.

- ***Peeling and Washing:*** Clingstone peaches are peeled using a lye solution. The peaches are sprayed or immersed in the lye solution (the lye solution concentration is 5-11% and it is applied for 45-60 seconds at 215-220 °F). The lye peeling conditions depend on fruit maturity. After hot lye peeling, the fruit pieces are sprayed thoroughly with cold water to remove the loose peel and residual lye. Freestone peaches on the other hand are peeled by steaming, by scalding in water, by hot lye, or by a combination of lye and steam.

- ***Quality Grading and Slicing:*** An inspection conveyor belt is used for the grading and sorting of poorly sized, partially-peeled, off-color, and otherwise imperfect halves or slices. Filling the graded/sorted halves directly from the inspection lines into cans to minimize additional injury of the fruit is highly recommended. Rejected fruit, not canned as halves, is directly conveyed to the slicing machine.

- ***Filling and Syruping*:** Peach halves from the above step should be filled into cans quickly, as extended exposure to the ambient air results in tissue discoloration. The canned peaches "Standards of Identity" specify the concentration of the cut-out syrup that match label names such as "heavy syrup" or "light syrup," *etc*. Syruping is accomplished by rotary or straight-line automatic syrupers or by pre-vacuumizing syrupers.

- ***Exhausting and Closing:*** The filled cans are given a steam exhaust for ~ 6 minutes at 190-195 °F, following which cans are closed immediately leaving a gross headspace of 5/16 inch.

- ***Processing and Cooling:*** Clingstone peaches are processed at 212 °F for 20-35, minutes depending on the can diameter and whether the fruit is in halves or slices, using continuous or reel type retorts. Some commercial processors also use a higher retorting temperature (240 °F) for 14-18 minutes. After thermal processing, cans are water-cooled immediately to a temperature of 95-105 °F.

Cut-out requirements for liquid media in canned freestone peaches are not incorporated in the grades of the finished product since syrup or any other liquid medium is not a factor of quality for the purpose of these grades [106]. The cut-out Brix measurements for the respective designations are shown in Table **9**.

Table 9: Cut-out strength (Brix levels) of syrups used in canned peaches

Designations	Brix measurements
-"Extra heavy syrup" or "Extra heavily sweetened fruit juice(s) and water" or "Extra heavily sweetened fruit juice(s)"	≥22° but < 35°
-"Heavy syrup;" or "Heavily sweetened fruit juice(s) and water" or "Heavily sweetened fruit juice(s)"	≥18° but < 22°
-"Light syrup;" or "Lightly sweetened fruit juice(s) and water" or "Lightly sweetened fruit juice(s)"	≥14° but < 18°
-"Slightly sweetened water;" or "Extra light syrup" or "Slightly sweetened fruit juice(s) and water" or "Slightly sweetened fruit juice(s)"	≥10° but < 14°

From USDA [106]

Frozen Peaches

Generally three types of peaches are frozen commercially: (1) varieties with a red tint around the pit cavity, *e.g.*, 'Rio Oso Gem', for retail packages of slices (2)

bulk-frozen for jam/preserve making; for this purpose fruits should not have a red colored pit cavity for uniform jam or marmalade color/appearance('Fay Elberta' peaches harvested on the slightly immature side are best for preserves), and (3) highly flavored varieties with firm-texture and resistant to browning or discoloration, which are best suited for the institutional market, mainly of pies. For quality freezing of peaches, an ideal variety should have the uniform shape and the bright color and textural firmness of 'Rio Oso Gem', it should be flavorful like 'Elberta', and be of the non-browning type like 'Sunbeam' [29]. The initial fruit preparation steps (washing, sorting, peeling, slicing and grading) are the same as discussed above for "Canned peaches."

The most commonly used system by most processors is air-blast tunnel freezing. The prepared peach product ready to be frozen is placed in thin layers on wire mesh trays, which are loaded on to racks. The tray racks are then moved into freezing tunnels where cold air is typically introduced from the opposite end of the side from where the product to be frozen enters. The process parameters of temperature and chilled air velocity are of critical importance to the freezing process. The chilled air temperature inside the freezing tunnel is usually between 0 °F and –30 °F (–18 °C and –34 °C, respectively). Air velocity, which also affects process efficiency, usually ranges from 100 ft per minute to 3,500 ft per minute. The length of freezing time is dependent on the size of product, *e.g.*, thickness of the slices [29].

Otero and colleagues [107] compared process effects of classical freezing and high-pressure-shift freezing (HPSF) method on the microstructure of peaches using a histochemical technique. In the HPSF method, peach slices were cooled to −20 °C under 200 MPa pressure, without ice formation, then the pressure was released to 0.1 MPa or atmospheric level. The rapid high level super-cooling led to uniform and quick ice nucleation. It was concluded that some quality problems linked to thermal gradients in traditional freezing (presence of large ice crystals or freeze-cracking) were minimized when using the HPSF method.

Dried Peaches

Both freestone and clingstone peaches can be used for the dried product. Dried peaches are the halved/pitted fruit product from which most of the water has been removed. Prior to packaging, the dried peaches are cleansed and may be treated with SO_2 to preserve their characteristic color. Only 1-2% of total peach production is processed as dried halves, bits, quarters or slices. Initial fruit preparation steps (washing, peeling, cutting and sorting) are the same discussed above under "Canned peaches." In order to minimize discoloration or browning, cut peaches are dipped in ascorbic acid or other food grade anti-browning solutions for 5-10 minutes (more details on anti-browning agents are given in the section under "Fresh-cut peaches and nectarines"). The prepared fruit is spread in single layers on trays (about 3-pounds/sq. ft.) and dried to a final moisture content of 25%.

Typically, forced-air tunnel dehydrators are employed for drying peaches. For peach drying in a countercurrent tunnel, temperatures should not exceed 155 °F. The total drying time of 24-30 hours is dependent on the size of the peach slices/pieces and the process temperature. Blanched peaches, having undergone mild heat treatment, are dried faster requiring only about 18 hours [108]. In developing countries, low-cost sun-drying is the most widely used method; however, the quality of the dried fruit is not comparable to that dried in high efficiency dehydrators running under controlled and sanitary conditions.

Souti and colleagues [109] investigated the feasibility of osmotic drying, as a pre-drying method, for peaches; their results showed that the use of this novel pre-treatment produced better quality dried peaches than those processed traditionally using solar dryers. Lerici and colleagues [110] reported that the osmotically dehydrated fruit products had a very good texture and retention of aroma and color. Although the a_w was lowered sufficiently enough to improve shelf-life, it was necessary to use further processing (freezing, drying, or pasteurization) to ensure that products were shelf-stable.

Peach Puree/Nectar

The major use of peach puree is in commercial baby food formulations. General processing steps for peach puree and nectar production are shown in Fig. **8.** Toralles and colleagues [111] studied the degradation of ascorbic acid in 12, 22, and 32 °Brix peach puree from the 'Jade' cultivar at 70-90 °C, under anaerobic conditions. Their results showed that the ascorbic acid degradation was significant, fitting zero and first-order, and that this degradation rate was highly dependent on temperature. Lavelli and colleagues [112] showed that carotenoid and phenolic contents were relatively lower in the peach nectar obtained from peeled fruit (cv. 'Elegant Lady' and 'Redhaven') than in that obtained from unpeeled peaches. It was further shown that the color of nectars was improved by using lye-peeled peaches at room temperature.

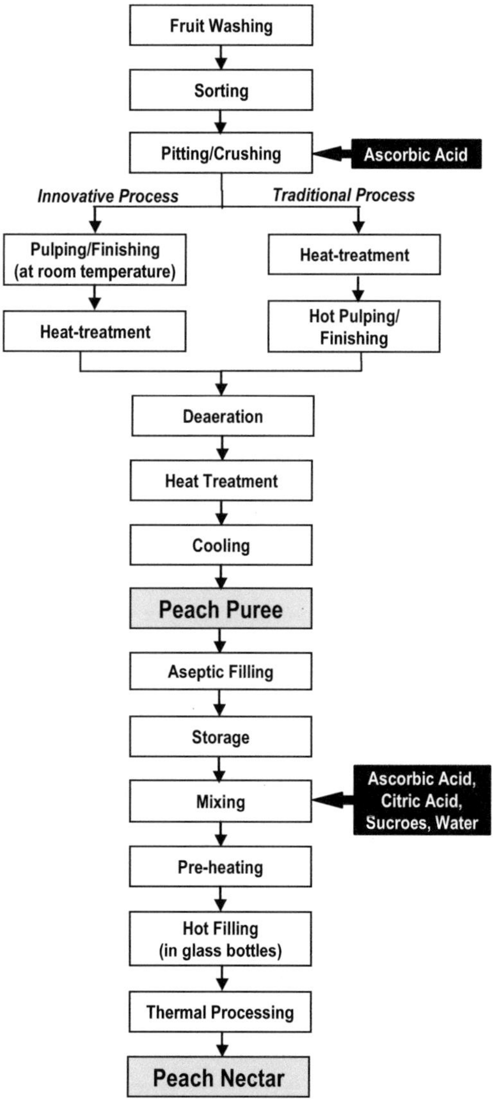

Figure 8: Flow diagram for the production of peach and nectarine purees and nectars. From Lavelli and colleagues [113].

Lavelli and colleagues [113] investigated the effects of a new and innovative method for processing peach and nectarine purees based on some quality attributes, namely, color, texture, carotenoids and phenol content, and antioxidant activity. This new process (pulping/finishing at ambient temperature) was compared to the traditional, or hot pulping/finishing process. The new process at a scaled-up level was shown to improve puree color (*i.e.*, higher Hunter color *L* values and lower Hunter color *a* values; Table **10**) and to maintain an almost similar content of carotenoids and other antioxidants.

Table 10: Hunter color values of peach and nectarine purees and nectars obtained from peeled and unpeeled fruits

	Puree		Nectar	
Hunter Color L	Unpeeled	Lye-peeled	Unpeeled	Lye-peeled
Red Haven Peach	46.96	53.62	41.41	42.38
Elegant Lady Peach	38.90	50.18	36.35	44.12
Stark Red Gold Nectarine	39.90	44.91	36.51	38.18
Hunter Color a				
Red Haven Peach	0.80	-5.97	-1.98	-7.35
Elegant Lady Peach	13.62	-2.90	5.65	-5.31
Stark Red Gold Nectarine	9.62	-2.56	2.61	-4.26
Hunter Color b				
Red Haven Peach	24.42	25.81	19.08	20.69
Elegant Lady Peach	16.69	24.35	15.08	19.67
Stark Red Gold Nectarine	19.52	23.77	17.58	19.27

From Lavelli and colleagues [112]

Peach Juice

Juice is produced on a fairly small basis compared to some other processed peach products. After preliminary cleaning and washing, fruit is heated in large kettles to about 45 °C with mechanical stirring. The addition of liquefying enzymes (pectinase and hemicellulase) facilitates juice extraction and results in higher juice yields and total soluble solids. Other steps used in peach juice production (filtration, clarification, and pasteurization) are similar to those used for other fruit juices. Pagan and colleagues [114] extracted juice from 'Caterino' peaches by treating pulp/mash with liquefying enzymes at a temperature range of 30 to 70 °C.

Both the juice yield and total soluble solids were higher using the enzymatic liquefaction process (to 55 ºC) than by non-enzyme extraction. However, at temperatures over 55 ºC, the use of enzymes was not effective, possibly due to their inactivation at higher temperatures.

The effect of two pectinolytic enzymes treatment on the volatile compounds composition of apricot, peach and pear juices were investigated by Riu-Aumatell and colleagues [115]. Over 80 compounds were detected, representing a wide range of chemical families (alcohols, aldehydes, ketones, terpenoids, esters, norisoprenoids, *etc.*). Enzyme treatment during juice processing enhanced the juice content of terpenes and norisoprenoids, which are aroma compounds.

Peach Jam and Jelly

Jam is made from crushed or chopped peaches; it holds its shape/consistency, but is less firm compared to jelly. Jelly is a mixture of clear fruit juice and sugar that is firm enough and can hold its shape. Granulated white sugar is often used for making peach jelly and jam and the use of pectin is optional [116]. Grigelmo-Miguel and Martin-Belloso [117] compared the quality attributes of conventional peach jam with that made by total or partial replacement of pectin with dietary fiber (DF), used as a thickener. Jam made with added DF was of a similar color but more viscous than conventional jam. From a sensory acceptance perspective, high DF peach jams were comparable to conventional jams. The type of sugar/sweetener and pectin used in jam making can have a significant effect on their texture, appearance, and flavor.

Fresh-Cut Peaches and Nectarines

Presently, a small amount of peach and nectarine production (<5%) is processed into fresh-cut form, however, these two fruits have greater potential to capture a significant share of the fresh-cut fruit market given the fact that over half of peaches and almost all nectarines are consumed fresh [118]. The optimal ripeness for processing fresh-cut peach slices is when the flesh has a firmness of 13-27 N (3-6 lb-force). Fresh-cut peach slices from different cultivars maintain good quality for 2-8 days when stored at 5 °C, 90-95% RH. The optimal flesh firmness for fresh-cut nectarines is the partially-ripe stage (27-49 N or 6-11 lb-force) or fully ripe stage (13-27 N or 3-6 lb-force); depending on the cultivar, fresh-cut nectarine slices maintain good eating quality for 2-12 days at 0 °C, 90-95% RH [9].

Gorny and colleagues [119] studied the impact of fruit ripeness stage and post-cutting storage conditions (temperature and RH) on the quality deterioration of fresh-cut slices from 'Flavorcrest' peaches and 'Zee Grand' nectarines. Their results showed ripe peaches and partially ripe nectarines were best suited for fresh-cut slice processing. When judged for sensory quality, fresh-cut peach and nectarine slices had a shelf life of 6-8 days, respectively, when stored at 0 °C, 90-95% RH. Koukounaras and colleagues [120] investigated the impact of short-time heat treatment on the quality of fresh-cut peach slices using different heat treatment parameters (heat intensity, application time). Their results showed an apparent beneficial effect of 4-hour pre-cutting heat treatment (50 °C, 10 minutes) on fresh-cut peach quality.

The passive modified atmosphere (MA) storage was shown to be useful in preserving the quality of fresh-cut peach slices that were pre-treated with mild heat, especially, with a significant improvement in firmness. In the heat-treated samples, lower levels of malic acid were observed, while succinic and citric acids levels were not affected [121].

By-Products

Peach processing produces significant amounts of by-products, mainly peels and kernels. Kernels are a rich source of specialty oil with some therapeutic properties and a good nutrient profile due to high amounts of oleic and linoleic acids [121, 122]. Oil extraction methods can have a significant impact on oil extractability and quality. Mezzomo and colleagues [122] assessed peach almond extraction yields using different methods: Soxhlet extractions by different solvents; hydrodistillation; ethanolic maceration followed by fractionation with different solvents, and supercritical fluid method at 30-50 °C and 100-300 bar, was performed with pure CO_2 and with a co-solvent. The yields of peach almond oil from all techniques was deemed adequate, with the supercritical fluid extraction method presenting advantages, mainly a high oleic acid content.

Innovative Processing

Blanching in hot water is a common method of inhibiting enzymes (especially, polyphenol oxidase and peroxidase) in fruits before drying, freezing, or canning. Kingsly and colleagues [123] investigated high-pressure processing (HPP) as an alternative to hot-water blanching; peach slices were pressure processed (50-700 MPa) at 25 °C, both with and without citric acid, for 2 holding times (5 and 10 min). Their results showed that HPP at >300 MPa, in combination with 1-1.2% citric acid, was an effective method for inactivation of polyphenol oxidase. Pressure-treated peach slices were then dried in a cabinet dryer at 70 °C; HPP pre-treatment enhanced the drying rate thereby reducing drying time. These results indicated that HPP has the potential to be used as an alternative to hot water blanching.

The color stability of fresh peach juice subjected to HPP was investigated during storage for 120 days by Zhou and colleagues [124]. Their results showed that browning by oxidation and polymerization of phenols, could be avoided by storing pressure treated peach juice at low temperatures.

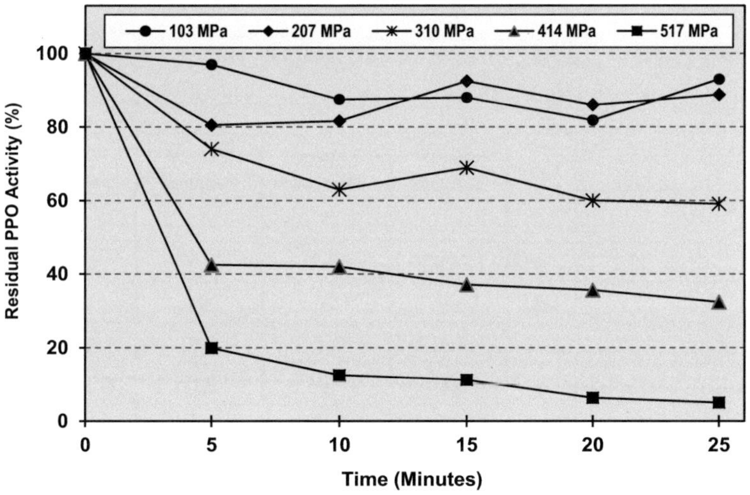

Figure 9: Residual polyphenoloxidase (PPO) in peach puree treated at high hydrostatic pressures and times. From Guerrero-Beltran and colleagues [125].

Guerrero-Beltran and colleagues [125] studied the shelf life of high hydrostatic pressure (HHP)-processed peach puree containing anti-browning agents; peach puree with or without 500 ppm of ascorbic acid (AA) or cysteine (250 ppm) were exposed to

HHP (517 MPa/5 min) followed by storage at 3, 21 and 35 °C for 30 days. Polyphenol oxidase activity decreased by 88-94.5% in different samples (Fig. **9**). HHP-processed purees retained their characteristic yellow (natural puree and puree with AA) and orange colors (puree containing cysteine) for 21 to 24 and 30 days, respectively. Only <10 cfu/g were counted in HHP-processed purees under 3 °C storage.

Nutritional Profile and Health Benefits

The nutritional profile of raw peaches and nectarines and processed peach products is shown in Table **11**. Peaches and nectarines are good sources of beta-carotene, which is thought to be a powerful anti-aging agent. Peaches and nectarines, especially when unpeeled, are considered a good source of dietary fiber, which is important to a healthy diet and can help control weight and lower cholesterol levels. Grigelmo-Miguel and colleagues [126] studied insoluble and soluble dietary fiber (DF) fractions in peach dietary fiber concentrates extracted from washed and dried peach pomace remaining after juice extraction process. Their results demonstrated that peach DF concentrates, in addition to having low energy/caloric values, could be an adequate source of DF, with approximately 2-to-1 ratio of insoluble:soluble DF.

Table 11: Composition of raw peaches, their processed products and raw Nectarines (per 100 g edible portion)

	Unit	Raw Peaches	Canned Peaches[a]	Frozen Peaches[b]	Dried Peaches[c]	Raw Nectarines
Proximate:						
Water	g	88.87	84.72	74.73	31.80	87.59
Energy	kcal	39	54	94	239	44
Protein	g	0.91	0.45	0.63	3.61	1.06
Total lipid (fat)	g	0.25	0.03	0.13	0.250.76	0.32
Fatty acids, total saturated	g	0.019	0	0.014	0.082	0.025
Carbohydrate, by difference	g	9.54	14.55	23.98	61.33	10.55
Fiber, total dietary	g	1.5	1.3	1.8	8.2	1.7
Sugars, total	g	8.39	13.25	22.18	41.74	7.89
Minerals:						
Calcium	mg	6	3	3	28	6
Iron	mg	0.25	0.36	0.37	4.06	0.28
Magnesium	mg	9	5	5	42	9

Table 11: contd...

Phosphorus	mg	20	11	11	119	26
Potassium	mg	190	97	130	996	201
Sodium	mg	0	5	6	7	0
Zinc	mg	0.17	0.09	0.05	0.57	0.17
Copper	mg	0.068	0.052	0.024	0.364	0.086
Manganese	mg	0.061	0.046	0.029	0.305	0.054
Vitamins:						
Vitamin A	IU	326	354	284	2163	332
Vitamin C, total ascorbic acid	mg	6.6	2.4	94.2	4.8	5.4
Thiamin	mg	0.024	0.009	0.013	0.002	0.034
Riboflavin	mg	0.031	0.025	0.035	0.212	0.027
Niacin	mg	0.806	0.593	0.653	4.375	1.125
Pantothenic acid	mg	0.153	0.05	0.132	0.564	0.185
Vitamin B-6	mg	0.025	0.019	0.018	0.067	0.25
Folate, total	µg	4	3	3	0	5
Vitamin E (α-tocopherol)	mg	0.73	0.49	0.62	0.19	0.77
Vitamin K (phylloquinone)	µg	2.6	0	2.2	15.7	2.2

[a] in light syrup (solids and liquids) ; [b] Sweetened; [c] sulfured, uncooked. From USDA [46]

Peach tree bark, as an herbal remedy, has been used for a wide variety of ailments. It is considered a blood moving herb, therefore, it has uses in encouraging menstruation in females experiencing delayed menses. Other health benefits include relieving bladder inflammation and other urinary tract problems; functioning as a mild laxative; as an expectorant for the lungs and sinus. It can also lessen chest pain and spasms [57].

Bioactive Compounds and Antioxidant Capacity

Remorini and colleagues [127] studied the influence of the 'Flavorcrest' peach rootstock (Ishtara, Mr. S 2/5, GF 677 and Barrier 1) and the harvesting time (early, middle, or late) on the physico-chemical characteristics and nutritional quality (*i.e.,* vitamin C, phenolics compounds, carotenoids, total antioxidant capacity). Their results demonstrated that phytochemical contents and antioxidant activity were best in late harvested fruit. Fruit on Mr. S 2/5 and on Barrier 1 rootstocks had higher antioxidant capacities and also higher phytochemical content. Moreover, peel removal from peaches resulted in a significant loss of total antioxidant activity. Cantin and colleagues [128] estimated antioxidant capacities and concentrations of total phenolics, anthocyanins, flavonoids, and vitamin C in 218 genotypes selected from 15 peach/nectarine breeding progenies. Their results revealed significant differences among progenies in terms of antioxidant profiles, which was shown to vary depending on fruit origin (peach or nectarine) and yellow or white fruit flesh color. The significance of genetic background in developing new genotypes that are rich in bioactive compounds should be considered in peach or nectarine breeding programs.

Gil and colleagues [129] investigated the levels of total phenolics, total ascorbic acid, beta-carotene, and AEAC (ascorbic acid equivalent antioxidant capacity) in selected cultivars of yellow-flesh and white-flesh peaches and nectarines. A strong correlation (0.93-0.96) was observed between total phenolic content and antioxidant activity. Yellow-flesh nectarines had significantly higher amounts of total phenolics, total ascorbic acid and AEAC than white-flesh fruit, however, a reverse trend was observed in the case of peaches; peels had higher concentrations in both fruits than the flesh irrespective of the flesh color (Table **12**). Asami and colleagues [130] reported that the peeling method can have a significant effect on the bioactive compounds, *e.g.*, lye peeling of peaches had 21% lower loss of total phenolics than that observed in manually peeled fruit. The role of phenolics in human health is discussed in further detail in the "Cherries" section.

Table 12: Total phenolics, total ascorbic acid, and antioxidant capacity in the peel and flesh tissue of peaches and nectarines

Fruit	**Total Phenolics (mg/kg)**		**Total Ascorbic Acid (mg/kg)**		**AEAC [a] (mg/kg)**	
	Peel	Flesh	Peel	Flesh	Peel	Flesh
Yellow-Flesh Peaches	485-1202	172-547	72-181	31-126	313-1107	93-432
White-Flesh Peaches	670-1836	228-1042	112-202	48-65	530-1789	146-1006
Yellow-Flesh Nectarine	427-1403	138-415	78-130	53-61	277-981	62-317
White-Flesh Nectarines	418-2020	91-901	93-200	42-122	230-1447	46-837

[a]Ascorbic Acid Equivalent Antioxidant Capacity
Adapted from Gil and colleagues [129]

PLUMS AND PRUNES

Plum is a common name for a tree of many species belonging to the genus *Prunus* of Rosaceae (rose) family and for its drupaceous (fleshy) fruit. Plums are generally cultivated in temperate zones with numerous varieties and hybrids that are suitable for many soils and sites. All prunes are botanically plums but not all plums are prunes [131]; however, in North America, *prunes* refer to a variety that can be and is normally dried without removing pits. *Plums* include any variety primarily used for fresh consumption or in canning, freezing, and production of jams and jellies. Somogyi [131] indicated that most plum varieties would ferment if dried as a whole fruit (with pits). However, if dried after the removal of pit, the product is called "dried plum" and not "prune."

According to Anon [132], plums can be divided into three groups, with the first two being the principal species of commercial plums. (1) *European-Asian (Prunus domestica)* – Familiar varieties of the European type are 'Stanley,' 'Reine Claude' (Green Gage) and the French and German prune (Fellenburg) types. The European-type plums, purple to black in color, are best for eating fresh and for canning. (2) *Japanese (Prunus salicina)* – Examples of Japanese plums are 'Methley,' 'Shiro,' 'Ozark Premier,' 'Burbank' and 'Elephant Heart.' Fruit color ranges from yellow to crimson. (3) *Damson (Prunus insititia)* – Plums in this group produce very tart fruit, which are used chiefly for cooking and preserving. Examples of Damson-type plums are 'French Damson' and 'Shropshire.' Most of the cultivated plums in the United States are derived from European and Japanese varieties, *e.g.*, *P. salicina,* introduced from Japan in 1870.

World production of plums was about 11 million metric tons in 2010 [1]. China was the leading plum producer with over 50% share of the total world production, followed by Romania, U.S.A., Serbia, and Chile (Table **13**). Spain, Chile, the U.S.A., Italy, and South Africa were the top-5 plum exporting countries in 2010, whereas, the Russian Federation, U.K., Germany, the Netherlands, and the U.S.A. were the leading plum importing countries [1]. The use of prunes and prune products as food ingredients have remained steady, most likely due to: 1) improved production and processing technologies, and 2) discoveries of health benefits of phytochemicals in fruits (like antioxidant benefits of phenolic compounds), which are present in significant quantities in plums and prunes.

Table 13: Leading plum producing countries, by quantity—million metric tons (1990–2010)

Country/Year	1990	1995	2000	2005	2010
China	945,455	2,229,339	3,941,952	5,229,202	5,664,826
Romania	449,500	252,495	549,627	622,357	624,884
U.S.A.	661,300	675,000	819,000	431,820	492,964
Serbia	--[1]	--	--	--	426,846
Chile	110,000	140,000	172,000	250,000	298,000
France	191,570	285,541	203,571	214,342	280,415
Iran	118,936	120,529	142,606	165,760	269,139
Turkey	188,000	187,000	195,000	220,000	240,806
Italy	139,200	104,273	179,833	185,404	207,497
India	50,000	58,000	78,000	148,662	200,000
Spain	126,200	124,200	168,045	251,812	192,000
Bosnia and Herzegovina	--	42,000	26,750	95,971	157,562

[1]not reported. From FAOSTAT [1]

Production and Harvesting

Trees of some plum cultivars are capable of self-pollination even if grown as a single isolated plant(s); other plum varieties require cross-pollination for fruit set and development. Plum trees usually begin to bear fruit 3-5 years from planting and have a useful life of 15-20 years. Depending on the type of cultivars, average yields can range from 3 to 5 bushels per tree. However, other factors like fertilizer use, irrigation, pruning and training, and plant protection practices play a big role in yield per tree [133, 134]. Southwick and colleagues [135] reported the use of gibberellin at pre-harvest stage to delay maturity and improve fruit firmness.

Fruit skin color is typically used as an indicator of maturity and color intensities are cultivar-specific. A color chip guide is used to determine harvest maturity. In California, a two-tier maturity system is currently used to determine maturity; 1) Mature (minimum maturity), and 2) Well Mature. Instrumental measurement of fruit firmness has been recommended for those plum cultivars in which skin ground color is masked by full red or dark color growth before maturation. Flesh firmness, measured with a penetrometer, can be used to determine a maximum maturity index, which is the stage at which fruit can be harvested without suffering bruising damage during postharvest handling. For objective measurements of maturity, soluble solids content and sugar-to-acid ratio provide more reliable markers [136]. Ripening typically results in increased fruit weight and soluble solids, decreased fruit firmness, darker color of fruits, increased concentration of total sugars, decreased concentration of total acids, and increased concentration of anthocyanins [137].

Plums and fresh prunes are generally machine-harvested; however, in some areas they are hand-picked into bags and then dumped into bins on trailers that move between tree rows. Each method of harvest has some advantages as well as some disadvantages over the other.

Postharvest Physiology and Storage

After washing, the fruits are sorted to eliminate those with visual defects and, sometimes, to divert fruit of high surface color to a high quality pack (sizing separate fruit by either weight or dimension). Ideal storage conditions for plums are maintained at a temperature of 0 °C and 90-95% relative humidity; under such conditions, storage life is 2-4 weeks [136]. Martinez-Romero and colleagues [138] reported that forced-air cooling after harvesting and before transportation to the packinghouse, and during storage helped maintain fruit quality and prolong shelf life. Their research indicated that forced-air cooling led to a reduction in the respiration rate of mechanically damaged plums.

During postharvest storage of fruits, ethylene triggers many, though not all, aspects of fruit ripening. Abdi and colleagues [139] reported that within a fruit species, such as plum, high ethylene producing fruits would soften and ripen at a faster rate than low ethylene producing cultivars. The ethylene action inhibitor 1-methylcyclopropene (1-MCP) has been shown to delay ripening and improve the postharvest quality of climacteric fruits [139, 140].

Application of 1-methylcyclopropene (1-MCP) before air storage of plums could be the best way to reduce the ripening process for short or medium storage periods (40 and 60 days), while controlled atmosphere (CA) storage plus 1-MCP treatment could be used for long-term (80 days) quality retention [141]. Ozkaya and Dundar [142] studied the effect of 1-MCP on different quality parameters and the respiration rate of 'Black Diamond' plums during storage. Treatment with 1-

MCP was effective in reducing weight loss and maintaining firmness during 30-day storage at 0 °C and 5 days of shelf life. The other quality parameters (titratable acidity, color, total soluble solids and individual sugars) were not affected significantly by 1-MCP. Edible coatings have been used for a variety of fruits to extend their shelf life. Eum and colleagues [143] treated 'Sapphire' plums with a coating based on carbohydrate (Versasheen) with sorbitol as plasticizer and stored at 20 °C and 85% RH. The coating treatment reduced the transmission rate of CO_2, O_2, and H_2O. The loss of firmness was delayed as a result of coating treatment, thereby improving the maintenance of quality of the plums. Luo and colleagues [144] explored the potential of hot-air treatment to delay the ripening of 'Qingnai' plums as a potential technology to expand the marketing of green plums. Heat treated fruits changed from green to yellow at a slower rate than control fruits. The hot air treatment extended the postharvest life of 'Qingnai' plums by up to 6 days, from 12 to 18 days.

Chilling injury and internal browning are two physiological disorders reported in plums. Most plum and fresh prune cultivars exhibit flesh translucency associated with flesh browning the same as for chilling injury, or "gel breakdown", a term used in South Africa for chilling injury [145, 146]. Internal browning is a physiological disorder that originates before harvesting and is associated with high temperatures during fruit maturation and delayed harvest. Polyphenol oxidase can initiate flesh discoloration in injured or bruised fruits [147, 148]. This enzyme not only affects sensory color but can also be detrimental to nutritional quality.

Plum and Prune Processed Products

About one-half of plums are consumed fresh while the rest are processed. Major processed plum products are dried prunes, prune juice, and whole canned plums. Other processed forms, such as paste, sauce, and plum juice have not been developed nor marketed on a scale similar to the same products from other fruits such as apples, cherries, pears, apricots, citrus fruits, *etc.* [149]. However, with numerous recent scientific claims of many health benefits of the phytochemicals (plums being rich in many of these), there is great potential for their use in a variety of new processed products.

Dried Plums/Prunes

The fruit is harvested at full maturity when soluble solids content reaches at least 22% (22 °Brix). Other maturity indices include fruit firmness, flesh and skin color. Modern dehydrators have replaced the old methods of sun drying prunes in the United States [131]. However, in many of the plum producing developing countries, sun drying still continues to be one of the cheapest sources of drying plums. Different processing operations involved in the production of prunes, as described by Somogyi [131], are:

- Harvested plums are stored temporarily in refrigerated storage or preferably delivered directly to the Processing Plant.
- Plums subjected to air classification to remove leaves, stems, and extraneous matter.
- Washing with ~ 20 ppm chlorinated water.
- Washed plums loaded onto drying trays and trays transferred to dryers.
- Drying for 24-36 hours at about 140-165 °F (to a final moisture of about 16 ± 3% from initial moisture of 80%).
- Dried prunes stored at ambient temperature for about 1-2 weeks for moisture equilibration.
- Packaged in high density polyethylene lined corrugated boxes.
- Stored in dry, cool (preferably refrigerated, 40-55 °F) conditions.

Fruits are dried to about 18% moisture, which has sufficiently low water activity to avoid problems of microbial spoilage allowing long-term storage [150]. Typically, forced-air dehydrators are used for drying plums, with a total drying process time of 24-36 hours, depending on the size and soluble solids content of the prunes. The operating temperature for the tunnel is 145-165 °F (dry bulb temperature) with the wet bulb 15 °F lower than the dry bulb at the cool end. The dried prunes yield is about 33%. For the prunes that are marketed as "Dried prunes with pits", the dried fruits from the above step are rehydrated to 24-30% moisture, pasteurized, inspected and bulk- or retail-packaged for food service or consumer use, respectively. Potassium sorbate is the most commonly used preservative when moisture contents of the finished prunes are higher than 25%. Some products made from dried prunes, on a smaller scale, are prune juices, juice concentrates, whole pitted prunes, canned prunes (pitted), and various forms of dry and low moisture products, such as diced prunes, prune bits, prune paste, low-moisture prune granules, low-moisture prune powder, prune fiber, and prune fillings.

In developing countries, plums can be dried under sun or using solar dryers. The main reasons for the use of these indigenous methods are the high energy requirements and equipment costs for commercial equipment used in developed countries. Dissa and colleagues [151] developed a prototype indirect solar dryer (Fig. **10**). When exposed to solar radiation, the solar energy collector panel converts a portion of radiant energy received on its surface into heat. The air crossing the porous matrix of its absorber then receives a portion of this energy by convective heat exchange; the resulting heated air circulates in the unit to dry mango slices arranged in a single layer in mesh trays [151].

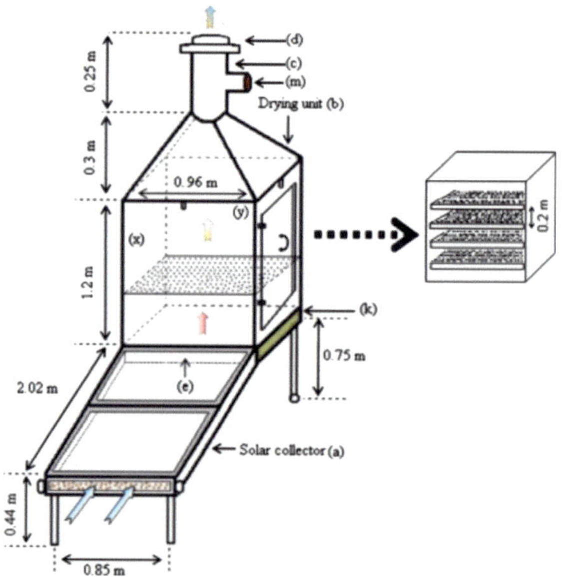

Figure 10: Prototype of indirect solar dryer. Labeled parts are: (a) solar collector; (b) drying unit; (c) chimney; (d) air extractor; (e) air entrance; (k) drying unit bottom side; (x) drying unit small side; (y) drying unit large side; (m) air recycling pipe. From Dissa and colleagues [151].

Most of the recent research related to prune processing has been focused on ways to improve efficiency of drying methods with due consideration given to the retention of best quality characteristics (both chemical and nutritional) in the finished product.

The effect of dipping treatment on air-drying of plums was studied by Doymaz [152]; results showed that dipping of plums in 5% potassium carbonate and 2% ethyl oleate for one minute was effective in removing the natural wax coating and hence reduced the drying time by about 30% compared to untreated samples. Gabas and colleagues [153] studied the rheological properties of prunes as a function of drying conditions. Their data showed that prunes exhibited a more pronounced elasticity at low moisture content and low drying temperatures. Higher moisture contents and higher temperatures resulted in more viscous and less rigid prunes. Sabarez and colleagues [154] used solid phase micro-extraction (SPME) in conjunction with GC-MS to monitor changes in some major volatile flavor compounds under simulated commercial drying conditions (80 °C air temp. 35% RH, 5 m/s air velocity) for 'd'Agen' plums. They observed that the aroma profile was significantly modified during drying, while substantial loss of the original volatile flavors occurred.

Sanders [155] noted a variety of functional attributes of dried plums that can replace traditional ingredients/additives (listed in parentheses): humectancy (mono- and di-glycerides), natural color (caramel color, molasses), flavor enhancement (salt, artificial flavors), natural sweetness (refined sugars), natural preservatives (calcium propionate), fat replacement (emulsifiers, modified starches, fats). Dried plum products are suitable for use in many bakery products [155], which, in addition to added nutritional benefits, can also improve desirable sensory attributes:

- Diced dried plums and extruded bits (breads, muffins, cookies, cakes, fillings)
- Dried plum paste (fillings, breads, bagels, cookies)
- Dried plum juice concentrate (breads, pastries, cakes, muffins, cookies, fillings)

- Dried plum powder/granules/flakes (dry mixes, bagels, reduced-fat/fat-free mixes)

- Dried plum puree (reduced-fat/fat-free bakery cakes, cookies, muffins, fillings)

Prune Juice and Concentrate

Prune juice is essentially a water extract of dried prunes with a Brix of 18.5°. It is a brownish to reddish brown liquid having the taste and flavor of prunes. The process involves the direct heating of dried prunes in appropriate volumes of water (which could be 4 to 5 times the weight of the fruit) to extract fruit solids without burning or affecting flavor and color. Traditionally, the process can take several hours (1 hr boil, 10 hrs simmering) under atmospheric cooking. A short duration (10-15 min) pressure-cooking may speed-up juice extraction. After extraction, other unit operations of filtration (to remove pits and un-dissolved solids), pasteurization (88 °C for 1 min) and packaging can be similar to the processing of other juices. The juice thus made can also be concentrated in a vacuum evaporator. For canning and bottling, the product should be hot filled (88 °C) and sealed before processing in boiling water for 20-35 min, based on the size of the can.

Prune juice, processed as above, can be concentrated to 60-72 °Brix depending on the intended end use. Concentrate with lower soluble solids is frozen and used for reconstitution into single strength juice. The high Brix (65% soluble solids and above) is shelf-stable/self-preserving without any need for freezing or added preservatives; the latter quality is of great benefit for shipping long distances, for export markets. For better process efficiency, prune juice is depectinized before concentration by the addition of commercial pectic enzymes [131]. The juice is concentrated in a high vacuum evaporator; many types of which are available commercially. The process is conducted at a temperature of 48 °C or lower.

Prune juice concentrate offers many benefits and applications in different food systems [156]; it 1) extends the shelf life of bread products, serving as an anti-staling agent, 2) sweetens and colors natural baked goods, 3) is a natural substitute for preservatives, 4) can be a good sugar substitute, and natural color/flavor enhancer, 5) can be used as filling for hard candies and chocolate, 6) maintains moisture in chewy cakes and cookies, and 7) can be used as a binding agent in cereal bars.

Plum Juice

Plum juice has not been processed on a scale similar to prune juice, probably due to its high acidity. Siddiq and colleagues [157] developed a method using pectinase enzymes to press juice from Stanley plums that significantly increased juice yields. Chantanawarangoon and colleagues [158] evaluated antioxidant capacity, total phenolics and total anthocaynins in juices prepared from 10 plum cultivars and suggested that plum juices with high antioxidant capacity, total phenolics and total anthocyanins could be new healthy beverages for consumers and also constitute potential sources of nutraceutical supplements. Will and Dietrich [159] reported that varietal/cultivar differences can have a significant effect on the polyphenol content of plum juices, as shown in Table **14**.

Table 14: Concentration of polyphenols in plum juice from four plum cultivars

Polyphenol	**Conc. (mg/L)**
Procyanidin B-1	63-81
(+)-Catechin	28-109
(−)-Epicatechin	22-54
Neochlorogenic acid	1,259-2,222
Coumaroylglucose	46-103
Chlorogenic acid	33-78
Quercetin-3-rutinoside	15-32

Adapted from Will and Dietrich [159]

Canned Prunes

Processing consists of washing dried prunes, blanching for 4 min in hot water (to start prune softening), filling in cans, and adding syrup. Cans are exhausted for 12-15 min at 170 °F before sealing. Thermal processing is carried out at 212 °F for 12 min. Canned prunes can be more prone to forming hydrogen springer than other canned fruits, which can be minimized with high vacuum as a result of extended exhaustion before can closure [160]. Bolin and colleagues [161] reported that when apple juice was used as the stewing medium for canned prunes, a desirable flavor was produced. Flavor was not improved by the addition of citric acid and ascorbic acid; instead both had rather an adverse effect.

Plum Puree/Paste

Plum paste is another product with limited commercial production, however, with known benefits of polyphenols and dietary fiber, it offers great potential for increasing production and uses in many food formulations.

Wang and colleagues [162] developed a procedure to produce pastes from 'Stanley' plums and studied the effects of processing conditions on chemical, physical and sensory properties of the pastes. 'Stanley' plums were processed by heat concentration into 2 pastes of 25 and 30 °Brix. Soluble solids of these 2 pastes were increased to 40 and 45 °Brix, respectively, by the addition of sugar. Heat concentration resulted in a significant decrease in titratable acidity, total anthocyanins and total pectin. Pastes showed pseudoplastic behavior within the shear rate range of 20-100 rpm. Sugar addition had a darkening effect on color, but no noticeable effect on the rheological properties of the pastes. Sensory evaluation indicated that preference could be adequately predicted by flavor and color under suitable °Brix/acid ratio. Raina and colleagues [163] also reported that concentration of plum paste had significant effects on the rheological properties and acidity of the resultant paste. It was not feasible to concentrate the paste beyond 35°Brix.

Jam and Jelly

Processing plums into jam and jelly is not very common and these products have not gained sufficient consumer acceptance to result in commercial production. However, there is potential for the production of jams and jelly in combination with other fruits, which are lower in phenolic compounds and antioxidant activity. Kim and Padilla-Zakour [164] investigated the changes in total phenolics, antioxidant capacity and anthocyanins from fresh fruits (cherries, plums, raspberries) to process shelf-stable jams, with the objective of finding if high sugar and acid levels may provide protection to phenolic compounds. Their results showed that jam processing had no consistent effect on phenolics, although antioxidant capacities generally decreased. Nonetheless, plums can be used as a source of phenolics in jams prepared from other fruits that may be low in these phytochemicals.

Minimally Processed or Fresh-Cut Plums

In recent years, a market for fresh-cut plums has developed; however, only less than 5% of plums are marketed through this channel. Fresh-cut plums can maintain good quality for 2-5 days, depending on the cultivar and ripeness stage (firmness) when stored at 0 °C in packages that minimize loss of water [136]. Given consumer preference for convenient and healthy food choices, fresh-cut plums have great marketing potential.

Nutrient Profile and Health Benefits

The nutritional profile of plums and plum products is shown in Table **15** [46]. Plum and prune products, in addition to being low in fat content like other fruits, contain fairly significant concentrations of major nutrients.

Table 15: Composition of plums, prunes and their processed products (per 100 g edible portion)

Nutrient	Unit	Plums	Canned[1]	Dried (Prunes)	Prune Juice[2]
Proximate:					
Water	g	87.23	83	69.73	81.24
Energy	kcal	46	63	107	71
Protein	g	0.7	0.37	0.96	0.61
Total lipid (fat)	g	0.28	0.1	0.16	0.03
Carbohydrate, by diff.	g	11.42	16.28	28.08	17.45
Fiber, total dietary	g	1.4	0.9	3.1	1
Sugars, total	g	9.92	15.35	24.98	16.45
Minerals:					
Calcium	mg	6	9	19	12
Iron	mg	0.17	0.86	0.41	1.18
Magnesium	mg	7	5	18	14
Phosphorus	mg	16	13	30	25
Potassium	mg	157	93	321	276
Sodium	mg	0	20	1	4
Zinc	mg	0.1	0.08	0.19	0.21
Copper	mg	0.057	0.038	0.123	0.068
Manganese	mg	0.052	0.032	0.131	0.151
Vitamins:					

Table 15: contd...

Vitamin A, IU	IU	345	231	342	3
Vit. C, total ascorbic acid	mg	9.5	0.4	2.9	4.1
Thiamin	mg	0.028	0.016	0.024	0.016
Riboflavin	mg	0.026	0.039	0.1	0.07
Niacin	mg	0.417	0.297	0.723	0.785
Pantothenic acid	mg	0.135	0.072	0.107	0.107
Vitamin B-6	mg	0.029	0.027	0.218	0.218
Folate, total	µg	5	3	0	0
Choline, total	µg	1.9	1.3	4.4	2.7
Vitamin E (α-tocopherol)	mg	0.26	0.18	0.19	0.12
Vitamin K (phylloquinone)	µg	6.4	4.3	26.1	3.4
ß-Carotene	µg	190	127	173	2
Lutein + zeaxanthin	µg	73	49	65	40

[1]in light syrup (solids and liquids); [2] canned
From USDA [46]

The values shown here are for fruits grown and processed in the United States, therefore, some differences can be anticipated in the composition of plum and prune products in other parts of the world due to different climatic and soil conditions, agricultural practices, varieties, postharvest handling and processing techniques, *etc.*

Ki and Jong [165] reported that drying was the mildest processing method for maintaining original levels of dietary fiber in vegetable and fruit products as demonstrated by a 7.35% and 3.45% increase in total dietary fiber in prunes and raisins, respectively. Somogyi [131] gave examples of variations in fiber content (from 2.04 to 16.1 g/100 g prunes with 23.3-32.4% moisture) reported by various sources, and indicated that these variations could not be attributed solely to the natural variations in the composition of fruits. He concluded that the lack of standardized tests, especially for the extraction of soluble fibers, might be responsible for such discrepancies.

Fresh plums and prunes are an excellent healthy food due to their low fat and good source of dietary fiber. Prunes are a moist and convenient snack, and are also easy way to get more natural fiber into the diet. Health experts have been advising people of all ages to consume more dietary fiber, based on research findings suggesting that fiber may prevent cancer, diabetes, heart disease and obesity. Prune fiber mostly consists of a soluble fraction (about 80%), mainly pectin, hemicellulose, cellulose, and some lignin. Prunes could be classified as a unique health food that is not only high in dietary fiber but also exhibits the highest antioxidant activity.

Halloran and colleagues [166] reported that dried plums contain proanabolic factors that can dramatically increase bone volume and restore bone that has already been lost due to aging. Dried plums can provide effective prophylactic and therapeutic agents for the treatment of osteoporosis. Gallaher and Gallaher [167] suggested that consuming dried plums or prunes, which are high in pectin with substantial antioxidative activity, might help slow the development of atherosclerosis.

Bioactive Compounds and Antioxidant Capacity

Plums are reported to be high in natural phenolic phytochemicals, such as flavonoids and phenolic acids, which function as effective natural antioxidants in the human diet and which have been reported to reduce the risk of cancer and other chronic diseases [168]. Cultivar differences are known to affect the phenolic content, as shown in Fig. **11** for some varieties of plums grown in Michigan [157]. Slimestad and colleagues [169] analyzed six plum cultivars, grown in Norway, for phenolic composition. Neochlorogenic acid was found to be the most predominant phenolic acid in all cultivars, whereas, cyanidin 3-rutinoside accounted for >60% of the total anthocyanin content. Minor amounts of flavonols (rutin and quercetin 3-glucoside) were also detected. The total antioxidant capacity among different plum cultivars ranged from 290 to 814 μmol of Trolox equivalent per 100 g on fresh-weight basis.

According to Shahidi and Naczk [47], the presence of phenolic compounds in foods has an important effect on the oxidative stability and microbial safety of these products. In addition, many phenolics in foods possess important biological activity related to their inhibitory effects on mutagenesis and carcinogenesis. Therefore, in recent years, rapid progress has been made regarding different aspects of polyphenols in food. Prunes are shown to have the highest antioxidant capacity, expressed as oxygen radical absorbance capacity (ORAC, Fig. **12**) that measures the ability of a food to subdue oxygen free radicals by comparing its absorption of peroxyl or hydroxyl radicals to that of trolox, a water-soluble vitamin E analog [170]. Wang and colleagues [171] reported total antioxidant capacities that were 4.4-times higher in plums than apples, the latter being the most commonly consumed fruits.

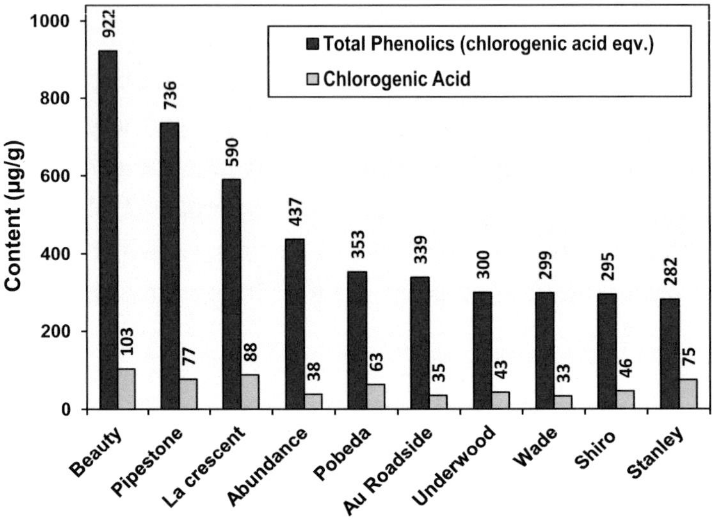

Figure 11: Total phenolics and chlorogenic acid contents in different plum cultivars. Adapted from Siddiq and colleagues [157].

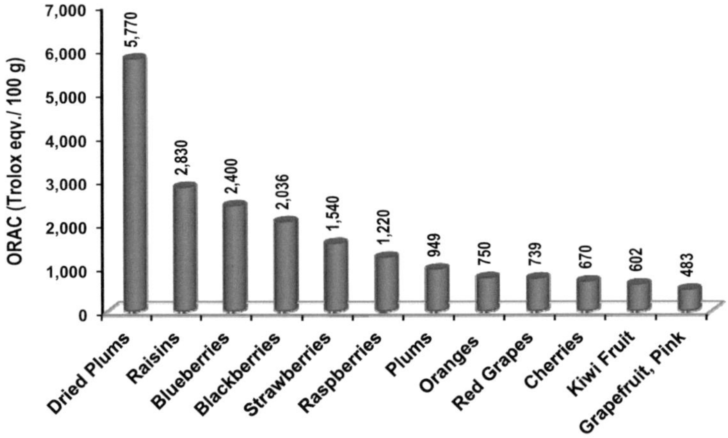

Figure 12: Oxygen Radical Absorbance Capacity (ORAC) values of fruits with antioxidant potential. Adapted from Keeton and colleagues [170].

ACKNOWLEDGEMENTS

Declared None.

CONFLICT OF INTEREST

The author confirms that this chapter contents have no conflict of interest.

REFERENCES

[1] FAOSTAT Food and Agriculture Organization of the United Nations (2012), World Primary Crops Data, viewed 18 September (2012), <http://www.faostat.org>.

[2] Anon. (2007), Apricot. *The Columbia Electronic Encyclopedia*, 6th. Columbia University Press, viewed 10 September (2012) <http://www.columbia.edu>.

[3] Boriss, H, Brunke, H & Kreith, M (2006), Commodity Profile: Apricots. University of California (Davis) Agricultural Marketing Resource Center Bulletin, viewed 23 August (2012), <http://aic.ucdavis.edu/profiles/Apricot-2006.pdf>.

[4] Magness, JR, Markle, GM & Compton, CC (1971), Food and Feed Crops of the United States. New Jersey Agricultural Experiment Station Bulletin 828.

[5] APC Apricot Producers of California (2010), California Apricots. Available at http://www.apricotproducers.com/ (accessed August 14, 2012).

[6] Rieger, M (2012a), Apricot – *Prunus armeniaca*, viewed 20 November (2011), <http://www.fruit-crops.com/apricot-prunus-armeniaca/>.

[7] Perez-Gonzales, S (1992), Associations among morphological and phenological characters representing apricot germplasm in central Mexico. *Journal of the American Society Horticultural Sciences*, 117, 486-490.

[8] Tagliavini, M & Marangoni, B (2002), Major nutritional issues in deciduous fruit orchards of northern Italy. *Horticultural Technology*, 12, 26-31.

[9] Crisosto, CH & Kader, AA (2004), Apricots, Peach, Nectarine. In: Gross, KC, Wang, CY & Saltveit, M (Eds),*The Commercial Storage of Fruits, Vegetables, and Florist and Nursery Stocks*-- Agriculture Handbook Number 66. United States Dept. of Agriculture, Agriculture Research Service, Washington, DC.

[10] Femenia, A, Sanches, ES, Simal, S & Rosello, C (1998), Developmental and ripening related effects on the cell wall of apricot (*Prunus armeniaca*) fruit. *Journal of the Science of Food and Agriculture*, 77, 483-493.

[11] Gomez, E & Ledbetter, CA (1997), Development of volatile compounds during fruit maturation: characterization of apricot and plum x apricot hybrids. *Journal of the Science of Food and Agriculture*, 74, 541-546.

[12] Fan, X, Argenta, L & Mattheis, JP (2000), Inhibition of ethylene action by 1-methylcyclopropene prolongs storage life of apricots. *Postharvest Biology and Technology*, 20,135-142.

[13] Botondi, R, DeSantis, D, Bellincontro, A, Vizovitis, K & Mencarelli, F (2003), Influence of ethylene inhibition by 1-methylcyclopropene on apricot quality, volatile production, and

glycosidase activity of low- and high-aroma varieties of apricots. *Journal of Agricultural and Food Chemistry*, 51, 1189-1200.

[14] Zhang, H, Wang, J, Pang, H-M & Li, XW (2009), Influence of 1-MCP processing on postharvest physiological catabolism of Liguang apricot. *Xinjiang Agricultural Sciences*, 46, 281-284.

[15] Martinez-Romero, D, Serrano, M, Carbonell, A, Burgos, L, Riquelme, F & Valero, D (2002), Effects of post-harvest putrescine treatment on extending shelf life and reducing mechanical damage in apricot. *Journal of Food Science*, 67, 1706-1712.

[16] Bussi, C, Besset, J & Girard, T (2003), Effects of fertilizer rates and dates of application on apricot (cv. Bergeron) cropping and pit burn. *Scientia Horticulturae*, 98, 139-147.

[17] Haciseferogullari, H, Gezer, I, Ozcan, MM & MuratAsma, B (2007), Post-harvest chemical and physical-mechanical properties of some apricot varieties cultivated in Turkey. *Journal of Food Engineering*, 79, 364-73.

[18] Ozakan, M, Cemeroglu, B (2002a), Desulfiting dried apricots by exposure to hot air flow. *Journal of the Science of Food and Agriculture,* 82, 1823-1828.

[19] Ozakan, M & Cemeroglu, B (2002b), Desulfiting dried apricots by hydrogen peroxide. *Journal of Food Science,* 82, 1631-1635.

[20] Piga A, Poiana M, Pinna I, Agabbio M & Mincione A (2004), Drying performance of five Italian apricot cultivars. *Science des Aliments.* 24, 247-259.

[21] Nisar-Alizai, M & Ahmad, Z (1997), Comparative investigation of sun-drying of apricots and their products produced in N.W.F.P. and northern areas of Pakistan. *Sarhad Journal of Agriculture*, 13, 501-509.

[22] [CLC 2000] Cayuga Landscape Co (2000), Fruits—Apricots, viewed 18 September (2012), <http://www.cayugalandscape.com/gardencenter/fruits.html>.

[23] Lopez A (1981a), Canning of Fruits—Apricots. In: A Complete Course in Canning, Book II: Processing Procedures for Canned Food Products, 11th Ed. Baltimore: The Canning Trade. p 137-139.

[24] Brecht, JK, Kader, AA, Heintz, CM & Norona, RC (1982), Controlled atmosphere and ethylene effects on quality of California canning apricots and clingstone peaches. *Journal of Food Science*, 47, 432-436.

[25] Sharma, TR, Sekhon, KS & Saini, SPS (1992), Studies on canning of apricot. *Indian Journal of Food Science and Technology*, 29, 22-25.

[26] Mallidis, CG & Katsaboxakis, C (2002), Effect of thermal processing on the texture of canned apricots. *International Journal of Food Science and Technology,* 37, 569-572.

[27] Chitarra, AB, Labavitch, JM & Kader, AA (1989), Canning-induced fruit softening and cell wall pectin solubilization in the 'Patterson' apricot. *Journal of Food Science*, 54, 990-992, 1046.

[28] Toker, I & Bayindirli, A (2003), Enzymatic peeling of apricots, nectarines and peaches. *LWT–Food Science and Technology*, 36, 215-221.

[29] Boyle, FP, Feinberg, B, Ponting, JD & Wolford, ER (1977), Freezing Fruits. In: Desrosier ND, Tressler DK (Eds), *Fundamentals of Food Freezing*. The AVI Publ Co., Westport, CT. pp. 151-164.

[30] Scorza, R, Hui, YH (1996), Apricots and Peaches. In: Somogyi LP, Barret DM, Hui YH (Eds), *Processing Fruits: Science and Technology: Vol. 2—Major Processed Products*. Technomic Publ. Co., Inc., Lancaster, PA. pp. 37-76.

[31] Forni, E, Sormani, A, Scalise, S & Torreggiani, D (1997), The influence of sugar composition on the color stability of osmodehydrofrozen intermediate moisture apricots. *Food Research International*, 30, 87-94.

[32] Chauhan, SK, Tyagi, SM & Singh, D (2001), Pectinolytic liquefaction of apricot, plum and mango pulps for juice extraction. *International Journal of Food Properties*, 4, 103-109.

[33] Zhang, RA & Xing, J (2004), Development of composite juice drink of Prunus mume, date, apricot, grape and Russian olive. *Food Science and Technology*, 3, 76-79.

[34] Vaillant, F, Millan, A, Dornier, M, Decloux, M & Reynes, M (2001), Strategy for economical optimisation of the clarification of pulpy fruit juices using crossflow microfiltration, *Journal of Food Engineering*, 48, 83-90.

[35] McHugh, TH, Huxsoll, CC & Krochta, JM (1996), Permeability properties of fruit puree edible films. *Journal of Food Science*, 61, 88-91.

[36] Lopez A (1981b), Jams, Jellies, and Related Products. In: A Complete Course in Canning, Book II: Processing Procedures for Canned Food Products, 11th Ed. Baltimore: The Canning Trade. pp. 359.

[37] Benamara, S, Messoudi, Z, Bouanane, A & Chibane, H (1999), Formulation and analysis of color of a light apricot jam. *Industries Alimentaires et Agricoles,* 116, 27-33.

[38] Anon. (2012), Apricots, viewed 25 September (2012), <http://en.wikipedia.org/wiki/Apricot>.

[39] Seker, I, Ozboy-Ozbas, O, Gokbulut, I, Ozturk, S & Koksel, H (2010), Utilization of apricot kernel flour as fat replacer in cookies. *Journal of Food Processing and Preservation*, 34, 15–26.

[40] Rahma, EH, El-Adawy, TA, Lasztity, R, Gomaa, MA, El-Badawey, AA & Gaugecz, J (1994), Biochemical studies of some non-conventional sources of proteins. VI. Physicochemical properties of apricot kernel proteins and their changes during detoxification. *Nahrung,* 38, 3-11.

[41] Sharma, A & Gupta, MN (2004), Oil extraction from almond, apricot and rice bran by three-phase partitioning after ultrasonication. *European Journal of Lipid Science and Technology*, 106: 183-186.

[42] Fathollahzadeh, H, Tabatabaie, H & Mobli, H (2010), Effective conditions for extracting higher quality kernels from the *Sonnati salams* Apricot. *International Journal of Food Engineering*, 6, 1-12.

[43] Patrignani, F, Vannini, L, Kamdem, LS, Lanciotti, R & Guerzoni, ME (2009), Effect of High Pressure Homogenization on *Saccharomyces cerevisiae* inactivation and physico-chemical features in apricot and carrot juices. *International Journal of Food Microbiology,* 136, 26–31.

[44] Evrendilek, GA, Tok, FM, Soylu, EM & Soylu, S (2008), Inactivation of *Penicillium expansum* in sour cherry juice, peach and apricot nectars by pulsed electric fields. *Food Microbiology*, 25, 662–667.

[45] Evrendilek, GA, Tok, FM, Soylu, EM & Soylu, S (2009), Effect of pulsed electric fields on germination tube elongation and spore germination of *Botrytis cinerea* inoculated into sour cherry juice, apricot and peach nectars. *Italian Journal Food Science*, 21, 171-182.

[46] [USDA] United States Dept. of Agri. (2012), USDA Nutrient Database, viewed 6 November (2012), <http://www.nal.usda.gov>.

[47] Shahidi, F & Naczk, M (Eds) (1995), *Food Phenolics—Sources, Chemistry, Effects and Applications*. Technomic Publ. Co., Inc., Lancaster, PA. pp. 1-5, 95.

[48] Jimenez, AM, Martinez-Tome, M, Egea, I, Romojaro, F & Murcia, MA (2008), Effect of industrial processing and storage on antioxidant activity of apricot (*Prunus armeniaca* v. bulida). *European Food Research and Technology*, 227, 125-134.

[49] [AHA] American Heart Association (2003), Efficiency and safety of low-carbohydrate diets, viewed 6 July (2012), <http://www.americanheart.org>.

[50] Rieger, M (2012b), Cherry – Prunus avium, Prunus cerasus, viewed 14 August (2012), <http://www.fruit-crops.com/cherry-prunus-avium-cerasus/>.

[51] Anon. (2013a), Cherry, viewed 26 January (2013), <http://en.wikipedia.org/wiki/Cherry>.

[52] McLellan, MR & Padilla-Zakour OI (2005), Sweet Cherry and Sour Cherry Processing. In: Barrett, DM, Somogyi, L, Ramaswamy, H (Eds), *Processing Fruits Science and Technology*. CRC Press, Boca Raton, FL. pp. 497-508.

[53] Seymour, EM, Singer, AAM, Kirakosyan, A, Kaufman, PB, Warber, S & Bolling, SF (2008), Tart cherry-enriched diets reduce hepatic lipid content, hepatic PPAR expression, metabolic syndrome and oxidative stress in Dahl-SS rats. *Journal of Medicinal Foods,* 11, 252–259.

[54] Wang, H, Nair, MG, Strasburg, GM, Booren, AM & Gray, JI (1999), Novel antioxidant compounds from tart cherries (*Prunus cerasus*). *Journal of Natural Products,* 1999, 62, 86-88.

[55] Steger-Mate, M (2012), Sweet and Tart Cherries. In: Handbook of Fruits and Fruit Processing (Eds: Sinha, NK, Sidhu, JS, Barta, J, Wu, JSB & Cano, MP. Ames, IA, USA: John Wiley & Sons, Ltd. pp. 433-446.

[56] Siddiq, M, Iezzoni, A, Khan, A, Breen, P, Sebolt, A, Dolan, KD & Ravi, R (2011), Characterization of new tart cherry (*Prunus cerasus* L.) selections based on fruit quality, total anthocyanins, and antioxidant capacity. *International Journal of Food Properties*, 14, 471-480.

[57] Rieger, M (2012c), Peach – *Prunus persica*, viewed 11 December (2012), <http://www.fruit-crops.com/peach-prunus-persica/>.

[58] Matheis, J & Fellman, J (2004), Cherries (Sweet). In: Gross, KC, Wang, CY & Saltveit, M (Eds),*The Commercial Storage of Fruits, Vegetables, and Florist and Nursery Stocks*--Agriculture Handbook Number 66. United States Dept. of Agriculture, Agriculture Research Service, Washington, DC.

[59] Wei, J, Qi, X, Guan, J & Zhu, X (2011), Effect of cold storage and 1-MCP treatment on postharvest changes of fruit quality and cell wall metabolism in sweet cherry. *Journal of Food, Agriculture and Environment,* 9, 118-122.

[60] Diaz-Mula, HM, Serrano M & Valero, D (2012), Alginate coatings preserve fruit quality and bioactive compounds during storage of sweet cherry fruit. *Food and Bioprocess Technology,* 5: 2990-2997.

[61] Linke, M, Herppich, WB & Geyer, M (2010), Green peduncles may indicate postharvest freshness of sweet cherries. *Postharvest Biology and Technology*, 58, 135-141.

[62] [NASS] National Agricultural Statistical Service. (2011). Noncitrus fruits and nuts, viewed 12 December (2011), <http://usda01.library.cornell.edu.proxy2.cl.msu.edu/usda/current/NoncFruiNu/NoncFruiNu-07--07-2011.pdf>.

[63] Woodroof, JG (Ed) (1986), Commercial Fruit Processing. The AVI Pub Co., Westport, CT. 678p.

[64] Carle, R, Borzych, P, Dubb, P, Siliha, H & Maier, O (2001), A new process for firmer canned cherries and strawberries. *Food Australia*, 53, 343-348.

[65] [USDA] United States Dept. of Agri. (1976) United States Standards for Grades of Canned Red Tart Pitted Cherries, viewed on May 21, 2013, http://www.ams.usda.gov/AMSv1.0/getfile?dDocName=STELDEV3007196.

[66] Šumić, Z, Tepić, A, Vidović, S, Jokić, S & Malbaša, R (2013), Optimization of frozen sour cherries vacuum drying process. *Food Chemistry*, 136, 55-63.

[67] Alibas, I (2007), Energy consumption and colour characteristics of nettle leaves during microwave, vacuum and convective drying. *Journal of Agricultural Engineering Research,* 96, 495–502.

[68] Doymaz, I & Ismail, O, (2011), Drying characteristics of sweet cherry. Food and Bioproducts Processing, 89, 31-38.

[69] Konopacka, D, Jesionkowska, K, Klewicki, R & Bonazzi, C, (2009), The effect of different osmotic agents on the sensory perception of osmo-treated dried fruit. *Journal of Horticultural Science and Biotechnology*, Special Issue, 80–84.

[70] Torreggiani, D, & Bertolo, G (2001), Osmotic pre-treatments in fruit processing: chemical, physical and structural effects. *Journal of Food Engineering*, 49, 247–253.

[71] Chaovanalikit, A & Wrolstad, RE (2004), Total anthocyanins and total phenolics of fresh and processed cherries and their antioxidant properties. *Journal of Food Science,* 69, FCT67–FCT72.

[72] Altuntas, J, Evrendilek, GA, Sangun, MK & Zhang, HQ (2010), Effects of pulsed electric field processing on the quality and microbial inactivation of sour cherry juice. *International Journal of Food Science & Technology* 45, 899-905.

[73] Place C, Harte JB, Dolan KD & Siddiq, M (2010), Long-term shelf-life evaluation of sensory and antioxidant quality of Michigan montmorency cherry juice concentrate. Presented at the 72nd annual meeting of the Institute of Food Technologists, Chicago, Illinois.

[74] FDA Food and Drug Administration, USA (1980), Maraschino Cherries, viewed 26 January (2013), <http://www.fda.gov/ICECI/ComplianceManuals/CompliancePolicyGuidanceManual/ucm074535.htm>.

[75] Anon. (2013b). Maraschino Cherries. Available at http://en.wikipedia.org/wiki/Maraschino_cherry (viewed January 26, 2013).

[76] Siddiq, M, Ravi, R, Sulaiman, R, Dolan KD & Harte, JH (2008), Value-added processing of fruit-based extruded porridge and snacks. East Lansing, Mich, USA: Bean Improvement Cooperative, Annual Report, 51, 150-151.

[77] Adil, IH, Yener ME & Bayindirli, A (2008), Extraction of total phenolics of sour cherry pomace by high pressure solvent and subcritical fluid and determination of the antioxidant activities of the extracts. *Separation Science and Technology,* 43, 1091-1110.

[78] Popa, VM, Misca, C, Bordean, D, Raba, DN, Stef, D & Dumbrava, D (2011), Characterization of sour cherries (*Prunus cerasus*) kernel oil cultivars from Banat. *Journal of Agroalimentary Processes and Technologies,* 17, 398-401.

[79] Beattie, J, Crozier, A & Duthie, GG (2005), Potential health benefits of berries. *Current Nutrition and Food Science*, 1, 71–86.

[80] Hertog, MGL, Feskens, EJM, Hollman, PCH, Katan, MB & Kromhout, D (1993), Dietary antioxidant flavonoids and risk of coronary heart disease: the Zutphen Elderly Study. *Lancet,* 342, 1007–1011.

[81] [CMI] Cherry Marketing Institute (2010), The Cherry Nutrition Report, viewed 9 November (2012), <http://www.choosecherries.com/>.

[82] Howatson, G, Bell, PG, Tallent, J, Middleton, B,; McHugh, MP& Ellis, J (2012), Effect of tart cherry juice (*Prunus cerasus*) on melatonin levels and enhanced sleep quality. *European Journal of Nutrition,* 51, 909-916.

[83] Ou, B, Bosak, KN, Brickner, PR, Iezzoni, DG & Seymour, EM (2012), Processed tart cherry products comparative phytochemical content, *in vitro* antioxidant capacity and *in vitro* anti-inflammatory activity. *Journal of Food Science*, 77, H105-H112.

[84] Blando, F, Gerardi, C & Nicoletti I (2004), Sour cherry (*Prunus cerasus* L.) anthocyanins as ingredients for functional foods. *Journal of Biomedical and Biotechnology*, 5, 253-258.

[85] McGhie, TK & Walton, MC (2007), The bioavailability and absorption of anthocyanins: Towards a better understanding. *Molecular Nutrition and Food Research,* 51, 702–713.

[86] Zafra-Stone, S, Yasmin, T, Bagchi, M, Chatterjee, A, Vinson, JA & Bagchi, D (2007), Berry anthocyanins as novel antioxidants in human health and disease prevention. *Molecular Nutrition and Food Research,* 51, 675-683.

[87] Kirakosyan, A, Seymour, EM, Llanes, DEU, Kaufman, PB & Bolling, SF (2009), Chemical profile and antioxidant capacities of tart cherry products. *Food Chemistry,* 115, 20–25.

[88] Bonerz, D, Wurth, K, Dietrich, H, & Will, F (2007), Analytical characterization and the impact of ageing on anthocyanin composition and degradation in juices from five sour cherry cultivars. *European Food Research and Technology,* 224, 355–364.

[89] Celik, S, Bakrici, I, & Scedilat, IG (2008), Physicochemical and organoleptic properties of yogurt with Cornelian cherry paste. *International Journal of Food Properties,* 9, 401-408.

[90] Arabshahi-Delouee, S & Urooj, A (2007), Application of phenolic extracts from selected plants in fruit juice. *International Journal of Food Properties,* 10, 479-488.

[91] Gungor, N & Sengul, M (2008), Antioxidant activity, total phenolic content and selected physicochemical properties of white mulberry (*Morus alba* L.) fruits. *International Journal of Food Properties,* 11, 44-52.

[92] Anon (2008), Peach. In: *The Columbia Electronic Encyclopedia,* 6th ed. Columbia University Press, viewed 11 November (2012), <http://www.columbia.edu>.

[93] [CCPB] California Cling Peach Board (2010), California Cling Peach Industry Facts, viewed 21 November (20120), <http://www.calclingpeach.com/html/nav/industry.html>.

[94] Logan, J, Mueller, MA & Searcy, MJ (2000), Microclimates, peach bud phenology, and freeze risks in a topographically diverse orchard. *HortTechnology*, 10, 337-340.

[95] Miranda-Jimenez, C & Royo-Diaz, JB (2002), Fruit distribution and early thinning intensity influence fruit quality and productivity of peach and nectarine trees. *Journal of the American Society Horticultural Sciences,* 127, 892-900.

[96] Lopez, G, Behboudian, MH, Vallverdu, X, Mata, M, Girona, J & Marsal, J (2010), Mitigation of severe water stress by fruit thinning in 'Henry' peach: Implications for fruit quality. *Scientia Horticulturae*, 125, 294-300.

[97] Kumar, M, Rawat V, Rawat, JMS & Tomar, YK (2010), Effect of pruning intensity on peach yield and fruit quality. *Scientia Horticulturae*, 25, 218-221.

[98] Serrano, M, Martínez-Romero, D, Castillo, S, Guillén, F & Valero, D (2004), Effect of preharvest sprays containing calcium, magnesium and titanium on the quality of peaches and nectarines at harvest and during postharvest storage. *Journal of the Science of Food and Agriculture*, 84, 1270-1276.

[99] Brovelli, EA, Brecht, JK, Sherman, WB, Sims, CA & Harrison, JM (1999), Sensory and compositional attributes of melting- and nonmelting-flesh peaches for the fresh market. *Journal of the Science of Food and Agriculture*, 79, 707-712.

[100] Tijskens, LMM, Rodis, PS, Hertog, MLATM, Kalantzi, U & Dijk, C (1998), Kinetics of polygalacturonase activity and firmness of peaches during storage. *Journal of Food Engineering*, 35, 111-126.

[101] Lysiak, G, Florkowski, WJ & Prussia, SE (2008), Postharvest calcium chloride application and moisture barrier influence on peach fruit quality. *HortTechnology,* 18, 100-105.

[102] Brovelli, EA, Brecht, JK, Sherman, WB & Sims, CA (1998), Quality of fresh market melting- and nonmelting-flesh peach genotypes as affected by postharvest chilling. *Journal of Food Science*, 63, 730-733.

[103] Fernandez-Trujilio, JP, Martinez, JA & Artes, F (1998), Modified atmosphere packaging affects the incidence of cold storage disorders and keeps 'flat' peach quality. *Food Research International*, 31, 571-579.

[104] Murray, R, Lucangeli, C, Polenta, G & Budde, C (2007), Combined pre-storage heat treatment and controlled atmosphere storage reduced internal breakdown of 'Flavorcrest' peach. *Postharvest Biology and Technology*, 44, 116-121.

[105] Downing, DL (Ed) (1996a), Canning of Fruits–Peaches. In: *A Complete Course in Canning, Book III: Processing Procedures for Canned Food Products*, 13th ed. CTI Pub, Inc., Timonium, Maryland. pp. 172-178.

[106] [USDA] United States Dept. of Agri. (1985), United States Standards for Grades of Canned Clingstone Peaches, viewed 21 October (2012), <http://www.ams.usda.gov/standards/cnpeachc.pdf>.

[107] Otero, L, Martino, M, Zaritzky, N, Solas, M & Sanz, PD (2000), Preservation of microstructure in peach and mango during high-pressure-shift freezing. *Journal of Food Science*, 65, 466-4670.

[108] Brekke, JE & Nury, FS (1964), Fruits. In: van Arsdel WB, Cople MJ (Eds), *Food Dehydration: Volume II—Products and Technology*. The AVI Publ Co., Westport, CT. 487p.

[109] Souti, M, Sahari, MA & Emam-Jomeh, Z (2003), Improving the dehydration of dried peach by applying osmotic method. *Iranian Journal of Agricultural Sciences,* 34, 283-291.

[110] Lerici, CR, Mastrocola, D & Nicoli, MC (1988), Use of direct osmosis as fruit and vegetables dehydration. *Acta Alimentaria Polonica* 14, 35-40.

[111] Toralles,RP, Vendruscolo, JL, Vendruscolo, CT, Del Pino, FAB & Antunes, PL (2008), Determinação das constantes cinéticas de degradação do ácido ascórbico em purê de pêssego: efeito da temperatura e concentração. *Ciência e Tecnologie de Alimentatia* 28, 18-23.

[112] Lavelli V, Pompei C & Casadei MA (2009), Quality of nectarine and peach nectars as affected by lye-peeling and storage. *Food Chemistry,* 115, 1291-8129.

[113] Lavelli, V, Pompei, C & Casadei, MA (2008), Optimization of color and antioxidant activity of peach and nectarine puree, scale-up study from pilot to industrial plant. *Journal of Agricultural and Food Chemistry*, 56, 7091-7099.

[114] Pagan, J, Soliva, R, Plque, MT & Ibarz, A (1997), Extraction of peach juice using enzymic liquefaction. *Alimentaria Equipos Tecnologia,* 16(8), 65-70.

[115] Riu-Aumatell, M, Lopez-Tamames, E & Buxaderas, S (2005), Assessment of the volatile composition of juices of apricot, peach, and pear according to two pectolytic treatments. *Journal of Agricultural and Food Chemistry*, 53, 7837-7843.

[116] Willenberg, BJ & Hughes, KV (2004), Jam and Jelly Basics: Tempt your taste buds with natural sweets, viewed 2 November (2012), <http://muextension.missouri.edu>.

[117] Grigelmo-Miguel, N & Martin-Belloso, O (2000), The quality of peach jams stabilized with peach dietary fiber. *European Food Research and Technology*, 211, 336-341.

[118] Siddiq, M., Liavoga, A. and Greiby, I. (2012) Peaches and Nectarines, in Handbook of Fruits and Fruit Processing, Second Edition (eds N. K. Sinha, J. S. Sidhu, J. Barta, J. S. B. Wu and M. P. Cano), Wiley-Blackwell, Oxford, UK. p 535–49.

[119] Gorny, JR, Hess-Pierce, B & Kader, AA (1998), Effects of fruit ripeness and storage temperature on the deterioration rate of fresh-cut peach and nectarine slices. *HortScience,* 33, 110-113.

[120] Koukounaras, A, Diamantidis, G & Sfakiotakis, E (2008), The effect of heat treatment on quality retention of fresh-cut peach. *Postharvest Biology and Technology*, 48, 30-36.

[121] Steiner, A, Abreu, M, Correia, L, Beirão-da-Costa, S, Leitão, E, Beirão-da-Costa, M, Empis, J & Moldão-Martins, M (2006), Metabolic response to combined mild heat pre-treatments and modified atmosphere packaging on fresh-cut peach. *European Food Research and Technology*, 222, 217-222.

[122] Mezzomo, N, Mileo, BR, Friedrich, MT, Martínez, J & Ferreia, SRS (2010), Supercritical fluid extraction of peach (*Prunus persica*) almond oil: process yield and extract composition. *Bioresource Technology*, 101, 5622-5632.

[123] Kingsly, ARP, Balasubramaniam, VM & Rastogi, NK (2009), Influence of high-pressure blanching on polyphenoloxidase activity of peach fruits and its drying behavior. *International Journal of Food Properties,* 12, 671-680.

[124] Zhou, J-Q, Zhao, G-Y, Zhang, P-Q & Bai, Y-H (2008), Color stability of fresh peach juice prepared by heating at high pressure during storage. *Modern Food Science and Technology,* 24, 548-551.

[125] Guerrero-Beltran, JA, Barbosa-Canovas, GV & Swanson, BG (2004), High hydrostatic pressure processing of peach puree with and without antibrowning agents. *Journal of Food Processing and Preservation*, 28, 69-85.

[126] Grigelmo-Miguel, N, Gorinstein, S & Martin-Belloso, O (1999), Characterization of peach dietary fiber concentrate as a food ingredient. *Food Chemistry*, 65, 175-181.

[127] Remorini, D, Tavarini, S, Degl'Innocenti, E, Loreti, F, Massai, R & Guidi, L (2008), Effect of rootstocks and harvesting time on the nutritional quality of peel and flesh of peach fruits. *Food Chemistry*, 110, 361-367.

[128] Cantín, CM, Moreno, MA & Gogorcena, Y (2009), Evaluation of the antioxidant capacity, phenolic compounds, and vitamin c content of different peach and nectarine [*Prunus persica* (L.) Batsch] breeding progenies. *Journal of Agricultural and Food Chemistry*, 57, 4586-4592.

[129] Gil, MI, Tomas-Barberan, A, Hess-Pierce, B & Kader, AA (2002), Antioxidant capacities, phenolic compounds, carotenoids, and vitamin C contents of nectarine, peach, and plum cultivars from California. *Journal of Agricultural and Food Chemistry*, 50, 4976-4982.

[130] Asami, DK, Hong, Y-J, Barrett, DM & Mitchell, AE (2003), Processing-induced changes in total phenolics and procyanidins in clingstone peaches. *Journal of the Science of Food and Agriculture*, 83, 56-63.

[131] Somogyi, LP (1996), Prunes and Plums. In: Somogyi, LP, Barrett, DM, Hui, YH (Eds), *Processing Fruits: Science and Technology, Vol. 2*. Technomic Publ. Co., Inc., Lancaster, PA. pp 95-116.

[132] Anon. (2004), Fact Sheet: Plum Culture. University of Rhode Island, Horticulture Program, viewed 7 December (2012), <http://www.uri.edu/ce/factsheets/sheets/plums.html>

[133] DeJong, TM, Day, KR & Doyl, JF (1992), Evaluation of training/pruning systems for peach, plum and nectarine trees in California. *Acta Horticulturae,* 322, 99-105.

[134] Keulemans, J (1990), Cropping behavior, flowerbud formation, pollination and fruit set of different plum cultivars in Belgium. *Acta Horticulturae,* 283, 117-129.

[135] Southwick, SM, Moran, RE, Yeager, JT & Glozer, K (2000), Use of gibberellin to delay maturity and improve fruit quality of 'French' prune. *Journal of Horticultural Science and Biotechnology,* 75, 591-597.

[136] Crisosto, CH & Kader, AA (2000), Plum and fresh prune postharvest quality maintenance guideline. Extension Bulletin, viewed January 5, 2012), <http://kare.ucanr.edu/files/123829.pdf>.

[137] Usenik, V, Kastelec, D, Veberic, R & Stampar, F (2008), Quality changes during ripening of plums (*Prunus domestica* L.). *Food Chemistry,* 111, 830-836.

[138] Martinez,-Romero, D, Castillo, S & Valero, D (2003), Forced-air cooling applied before fruit handling to prevent mechanical damage of plums (*Prunus salicina* Lindl.). *Postharvest Biology and Technology,* 28, 135-142.

[139] Abdi, N, McGlasson, WB, Holfor, P, Williams, M & Mizrahi, Y (1998), Responses of climacteric and suppressed-climacteric plums to treatment with propylene and 1-methylcyclopropene. *Postharvest Biology and Technology*, 14: 29-39.

[140] Dong, L, Lurie, S & Zhou, HW (2002), Effect of 1-methylcyclopropene on ripening of 'Canino' apricots and 'Royal Zee' plums. *Postharvest Biology and Technology*, 24, 135-145.

[141] Menniti, AM, Donati, I & Gregori, R (2006), Responses of 1-MCP application in plums stored under air and controlled atmospheres. *Postharvest Biology and Technology,* 39, 243-246.

[142] Ozkaya, O & Dundar, O (2009), Response of 1-methylcyclopropene (1-MCP) treatments on some quality parameters of plum during storage. *Journal of Food Agriculture and Environment,* 7, 233-236.

[143] Eum H, Hwang D, Linke M, Lee S & Zude M (2009), Influence of edible coating on quality of plum (*Prunus salicina* Lindl. cv. 'Sapphire'). *European Food Research and Technology*, 229, 427-434.

[144] Luo, Z, Zhang, L, Xu, T, Xie, J & Xi, Y (2010), Effect of hot-air treatment on the ripening of 'Qingnai' plum (*Prunus salicina* Lindl.). *Journal of Horticultural Science and Biotechnology,* 85, 12-16.

[145] Crisosto, CH, Mitchell, FG & Ju, Z (1999), Susceptibility to chilling injury of peach, nectarine, and plum cultivars grown in California. *HortScience*, 34, 1116-1118.

[146] Taylor, MA, Rabe, E, Jacobs, G & Dodd, MC (1995), Effect of harvest maturity on pectic substances, internal conductivity, soluble solids and gel breakdown in cold stored 'Songold' plums. *Postharvest Biology and Technology*, 5, 285-294.

[147] Siddiq, M, Sinha, NK & Cash, JN (1993), Characterization of polyphenol oxidase from Stanley plums. *Journal of Food Science,* 57, 1177-1179.

[148] Siddiq, M, Sinha, NK, Cash, JN & Hanum, T (1996), Partial purification of polyphenol oxidase from plums (*Prunus domestica* L. cv. Stanley). *Journal of Food Biochemistry,* 20, 111-123.

[149] Chang, T-S, Siddiq, M, Sinha, NK & Cash, JN (1994), Plum juice quality affected by enzyme treatment and fining. *Journal of Food Science*, 59, 1065-1069.

[150] Newman, GM, Price, WE & Woolf, LA (1996), Factors influencing the drying of prunes. I. Effects of temperature upon the kinetics of moisture loss during drying. *Food Chemistry,* 57, 241-244.

[151] Dissa, AO, Bathiebo, J, Kama, S, Savadogo, PW, Desmorieux H, Koulidiati J (2009), Modelling and experimental validation of thin layer indirect solar drying of mango slices. *Renewable Energy*, 34, 1000-1008.

[152] Doymaz, I (2004), Effect of dipping treatment on air-drying of plums. *Journal of Food Engineering,* 64:465-470.

[153] Gabas, AL, Menegalli, FC, Ferrari, F & Telis-Romero, J (2002), Influence of drying conditions on the rheological properties of prunes. *Drying Technology,* 20, 1485-1502.

[154] Sabarez, HT, Price, WE & Korth, J (2000), Volatile changes during dehydration of d'Agen prunes. *Journal of Agricultural and Food Chemistry*, 48, 1838-1842.

[155] Sanders, SW (1993), Dried Plums: A multi-functional bakery ingredient. American Society of Bakery Engineers; Bulletin 228. pp. 1-23.

[156] Anon. (2010), Prune Juice Concentrate: A Versatile Ingredient-- A wide range of functionality with all the health benefits of dried plums. Stapleton-Spence Packing Company, San Jose, California, viewed 6 December (2012), <www.stapleton-spence.com>.

[157] Siddiq, M, Arnold, JF, Sinha, NK & Cash, JN (1994), Effect of polyphenol oxidase and its inhibitors on anthocyanin changes in plum juice. *Journal of Food Processing and Preservation*, 18, 75-84.

[158] Chantanawarangoon, S, Kim, D-O & Padilla-Zakour, OI (2004), Antioxidant capacity and polyphenolic compounds of plum juices. Presented at the annual meeting of the Institute of Food Technologists, Las Vegas, NV.

[159] Will, F & Dietrich, H (2006), Optimised processing technique for color- and cloud-stable plum juices and stability of bioactive substances. *European Food Research and Technology*, 223, 419-425.

[160] Downing, DL (Ed) (1996b), Canning of Fruits–Plums, Dried Prunes. In: *A Complete Course in Canning, Book III: Processing Procedures for Canned Food Products*, 13th ed. CTI Pub, Inc., Timonium, Maryland. pp. 187-188.

[161] Bolin, HR, Stafford, AE & Guadagni, DG (1971), Innovations with canned stewed dried prunes. *Canner/Packer*, 140(7), 18-19.

[162] Wang, W-M, Siddiq, M, Sinha, NK & Cash, JN (1995), Effect of processing conditions on the physicochemical and sensory characteristics of Stanley plum paste. *Journal of Food Processing and Preservation*, 19, 65-81.

[163] Raina, CS, Bawa, AS & Ahmed, J (1999), Rheological study and sensory characteristics of plum paste. *Indian Food Packer,* 53(2), 15-19.

[164] Kim, D-O & Padilla-Zakour, OI (2004), Jam processing effects on phenolics and antioxidant capacity in anthocyanins-rich fruits: Cherries, plums and raspberry. Presented at the annual meeting of the Institute of Food Technologists, Las Vegas, NV.

[165] Ki, HS, Jong, BE 2000, Total dietary fiber contents in processed fruit, vegetable, and cereal products. *Food Science and Biotechnology,* 9, 1-4

[166] Halloran, BP, Wronski, TJ, Von Herzen, DC, Chu, V, Xia, X, Pingel, JE, Williams, AA & Smith, BJ (2010), Dietary dried plum increases bone mass in adult and aged male mice. *Journal of Nutrition,* 140, 1781-1787.

[167] Gallaher, CM & Gallaher, DD (2009), Dried plums (prunes) reduce atherosclerosis lesion area in apolipoprotein E-deficient mice. *British Journal of Nutrition,* 101, 233-239.

[168] Machlin, LJ (1995), Critical assessment of the epidemiological data concerning the impact of antioxidant nutrients on cancer and cardiovascular disease. *Critical Reviews in Food Science and Nutrition* 35, 41-50.

[169] Slimestad, R, Vangdal, E & Brede, C (2009), Analysis of phenolic compounds in six Norwegian plum cultivars (*Prunus domestica* L.). *Journal of Agricultural and Food Chemistry*, 57, 11370-11375.

[170] Keeton, JT, Rhee, KS, Hafley, BS, Nunez, MT, Boleman, RM & Movileanu, I (2002), Evaluation of ham and roast beef products containing fresh plum juice concentrate, dried plum juice concentrate, or spray dried plum powder. Final Report to the California Prune Board (USA), pp. 1-23.

[171] Wang, H, Cow, G & Prior, RL (1996), Total antioxidant capacity of fruits. *Journal of Agricultural and Food Chemistry*, 44, 701-705.

Index

Index

A

B

C

Innocenzo Muzzalupo & Sabrina Micali (Eds.)

D

E

Q

R

S

T

U

V

W

X

Y

Z

Made in the USA
Las Vegas, NV
04 November 2026

57438640R00240